JEFF BARBANELL - MYERS 344 - Galovich

# MATHEMATICS
## THE MAN-MADE UNIVERSE

# A SERIES OF BOOKS IN MATHEMATICS

Editors:

R. A. Rosenbaum
G. Philip Johnson

# MATHEMATICS

## THE MAN-MADE UNIVERSE

*An Introduction to the Spirit of Mathematics*

## SECOND EDITION

### SHERMAN K. STEIN
*University of California, Davis*

W. H. FREEMAN AND COMPANY
*San Francisco*

Printed in the United States of America
Library of Congress Catalog Card Number: 68-8634
International Standard Book Number: 0-7167-0436-6

9 8 7 6 5

*To my wife, Hadassah*

# PREFACE
## to the Second Edition

In the six years since the first edition appeared, many teachers have sent me suggestions based on their experience with classes for which *Mathematics: The Man-made Universe* was the text. On the basis of their suggestions and our experience at Davis I have made many changes. I will describe only a few of the major ones.

First, the exercises are now separated into three groups. The first group usually consists of routine exercises that offer the reader a chance to check his understanding of the definitions and basic ideas. Exercises in the second group, separated from the first by one solid circle (●), generally require the reader to apply ideas from the chapter. Two solid circles precede the third group, in which the exercises offer the greatest challenge, present alternative approaches, or develop ideas tangential to the central theme of the chapter.

Second, I have introduced two new chapters, Chapters 3 and 17. Chapter 3, Questions on Weighing, motivates by a concrete example the question: Is the greatest common divisor of two natural numbers a linear combination of them? In this chapter, which contains no theorems and no proofs, the student has occasion to review his arithmetic, meet the negative integers and a little algebra, and develop confidence in his mathematical ability. I think that most classes should begin with this chapter, and treat the two earlier chapters later, if there is time. Chapter 3 provides a strong intuitive basis for the basic Lemma 4 of Chapter 5, which previously seemed to come out of the blue.

The new Chapter 17, Construction by Straightedge and Compass, treats geometric ideas with the aid of the algebra of complex numbers. I have introduced it for two reasons: to give additional attention to geometry; and to exhibit an interesting application of the complex numbers. After all, one of the underlying themes of the book is "each number system has its uses." The first edition had ample illustrations of the uses of the integers, rational numbers, real numbers, and finite fields. But the only application of the complex numbers had been to provide a root for the equation $x^2 + 1 = 0$. For many students this is not enough of a vindication. Hence the appearance of Chapter 17 (and the description in Chapter 16 of how Steinmetz introduced the complex numbers into the theory of alternating currents).

I have also made many minor changes. Definitions have been clarified, proofs have been simplified, repeating decimals have been moved into Chapter 6, and new historical material has been incorporated. Hardly a page of the first edition has survived unscathed.

My thanks to all who have contributed to this revised edition: to my colleagues Henry Alder and Don Chakerian for reading the manuscript, and Roland Hoermann for his translation of a letter of Gauss, to Jack Robertson of Washington State University, Donna Seaman, Olympic College, Marguerite Dunton and Huguette Bach of Sacramento State College, Ted Tracewell of Hayward State College, and to Martin Davis of the Courant Institute, and Guy Hirsch, Université Libre de Bruxelles, for their many suggestions.

*December 1968*                                                    SHERMAN K. STEIN

# PREFACE
## to the First Edition

We all find ourselves in a world we never made. Though we get used to the kitchen sink, we do not understand the atoms that compose it. The kitchen sink, like all the objects surrounding us, is a convenient abstraction.

Mathematics, on the other hand, is completely the work of man. Each theorem, each proof, is the product of the human mind. In mathematics all the cards can be put on the table. In this sense, mathematics is concrete, whereas the world is abstract.

This book exploits that concreteness to introduce the general reader to mathematics. The "general reader" might be either the college student or the high school student, whatever his special interest might be, or the curious adult. This book grew out of a college course designed primarily to give students in many fields an appreciation of the beauty, extent, and vitality of mathematics. I had searched several years for a suitable text, but those I found were either too advanced or too specialized.

The subjects, chosen from number theory, topology, set theory, geometry, algebra, and analysis, can be presented to the reader having little mathematical background (some chapters use only grammar school arithmetic). Each topic illustrates some significant idea and lends itself easily to experiments and problems.

The reader is advised to take advantage of the concrete nature of mathematics as he reads each theorem and proof; to take nothing on faith; to be suspicious and vigilant; to examine each step of the reasoning; and to take seriously such suggestions as "the reader may provide an example of his own" or "the reader should check this theorem for some special cases before going on to the proof." It would be wise to read this book with pencil and paper always at hand.

The exercises at the end of each chapter vary in difficulty; some are just routine checks, whereas others raise questions that no one has answered. They give the reader a chance to test and apply his understanding of the material. Many of the exercises offer either alternate proofs of theorems proved in the text or further results. Some point out relations to other chapters.

Using the Map and Guide (page xi), the reader or teacher may choose his own route through this book. The recommended route takes the chapters in order.

I would like to thank my students at the University of California at Davis for their comments on mimeographed versions of most of the chapters; my colleagues, Henry Alder and Curtis Fulton, for their encouragement and advice; and the artist William Brown, who read several chapters and agreed that a proof can be beautiful.

It is a special pleasure to acknowledge the invaluable assistance of Robert Blair of Purdue University and George Raney of Wesleyan University, both of whom read the complete manuscript and made countless suggestions.

*August 1962*                                                    SHERMAN K. STEIN

# MAP AND GUIDE

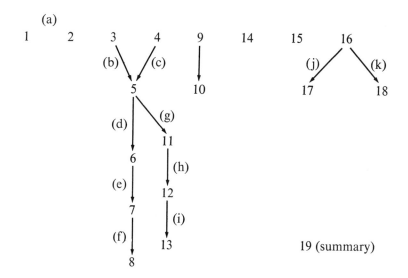

Many chapters refer for comparisons, contrasts, or exercises to earlier chapters. The map does not record these relations.

Even if there is time to cover the whole book, a reader on his own or an instructor of a class may wish to begin with this sequence: 3, 4, 5, 6, 7, 15, 11, 12.

(a) Though these two chapters use only the addition of natural numbers and the contrast between "odd" and "even," the arguments offer a challenge to many students. They may be delayed or omitted.

(b) Chapter 3 is a psychological prelude to Chapter 5, but is not a logical prerequisite.

(c) Chapter 5 needs only the definition of "prime" from Chapter 4.

(d) Chapter 6 applies "unique factorization" to questions of rationality.

(e) Chapter 7 applies the distinction between "rational" and "irrational" to a geometric question.

(f) One could go directly from Chapter 6 to Chapter 8; however Chapter 7 is, psychologically, a prelude to Chapter 8.

(g) Most of Chapter 11 does not need Chapter 5.

(h) Chapter 11 is used in constructing some examples in Chapter 12.

(i) One could go from Chapter 11 directly to Chapter 13, including from Chapter 12 simply the definition of "table."

(j) Chapter 17 needs only part of Chapter 16, the algebra of complex numbers.

(k) Chapter 18 needs only part of Chapter 16, the distinction between "algebraic" and "transcendental."

Throughout the text, E stands for exercise and R for reference. A starred reference presupposes more mathematical training than the general reader is expected to have; it is intended primarily as an aid to a teacher seeking background information. The contents of some references overlap.

I call the reader's attention to the appendices, with which he may wish to familiarize himself in advance and refer to when convenient.

# CONTENTS

# MATHEMATICS
## THE MAN-MADE UNIVERSE

*Chapter* **1**

# THE FIFTEEN
# PUZZLE

In this chapter and the next we introduce the reader to the mathematical way of thinking and, in particular, to the concepts of proof and theorem. All that we will need in these two chapters is the distinction between the *odd numbers*, beginning with one (1, 3, 5, . . .) and the *even numbers*, beginning with zero (0, 2, 4, . . .). From such a simple notion we will deduce important and surprising consequences. Indeed, the ancient duality of odd and even, which separates the *natural numbers* 0, 1, 2, 3, 4, 5, . . . into two types, will be of use several times in the course of this book; for example, in paths over highway systems, algebras, coloring of maps, and roots of polynomials—topics appearing in Chapters 9, 12, 14, and 16, respectively.

The only properties of the odd and even numbers that we will need now are these: The sum of two even numbers is even; the sum of two odd numbers is even; the sum of an even number and an odd number is odd. Exploiting these three properties, we will analyze the well-known "Fifteen Puzzle."

The Fifteen puzzle, sold in almost every dime store, drugstore, and toy store, consists of a square frame holding 15 small movable squares, numbered 1 through 15, and one empty square. We are allowed to switch the blank square with a square next to it as often as we wish. The challenge is to move by a sequence of switches from a given position, such as

| 1 | 2 | 3 | 4 |
|----|----|----|----|
| 5 | 6 | 7 | 8 |
| 9 | 10 | 11 | 12 |
| 13 | 14 | 15 |  |

to another specified position, such as

| 15 | 14 | 13 | 12 |
|----|----|----|----|
| 11 | 10 | 9  | 8  |
| 7  | 6  | 5  | 4  |
| 3  | 2  | 1  | ≦  |

The reader who would like to test his luck, but who has no Fifteen puzzle handy, can use paper, pencil, and eraser or fifteen numbered slips of paper placed, for convenience, far apart.

An editorial footnote in a research paper published in 1879 remarks about this puzzle:

> The "15" puzzle for the last few weeks has been prominently before the American public, and may safely be said to have engaged the attention of nine out of ten persons of both sexes and of all ages and conditions of the community. But this would not have weighed with the editors to induce them to insert articles upon such a subject in the *American Journal of Mathematics*, but for the fact that the principle of the game has its root in what all mathematicians of the present day are aware constitutes the most subtle and characteristic conception of modern algebra, viz: the law of dichotomy applicable to the separation of the terms of every complete system of permutations into two natural and indefeasible groups, a law of the inner world of thought, which may be said to prefigure the polar relation of left- and right-handed screws or of objects in space and their reflexions in a mirror.

Before we analyze the Fifteen puzzle, let us consider a simpler problem—one that also involves switches. This is presented by a weaver of hatbands. After we solve the weaver's problem, we will apply the tools we have developed to the Fifteen puzzle.

Let us turn to the weaver's problem. To make his hatbands, this weaver braids several threads together, interchanging two at a time. Moreover, no two of his threads are of the same color. For instance, when he has only two threads, his pattern looks like

"After one switch," the weaver tells us, "neither thread is directly below its starting position. After the second switch, each is directly below its starting

position. After three switches, each is again out of place; after four, they are back in place. Since I want these bands to go around a hat and not show the seam, my designs must have an even number of switches—at least if I have only two threads.

"Now, it seems to me that with more than two threads I should be able to make a seamless band with an odd number of switches. But, hard as I have tried, I have found no such design. Can you help me?"

Let us try to help the weaver find his design. Here is an experiment with three threads:

This diagram records the effect of four switches. Each switch interchanges two threads, as the weaver demands. The remaining thread is brought straight down (this is recorded by the vertical lines). Each of the three threads has been returned to its starting place, and so no seam will show. Regrettably, however, the number of switches is even; we did not find the weaver a design having an odd number of switches.

Let us try again. Here is a seamless design made with four threads:

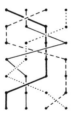

The vertical lines record threads that are not switched; since there are four threads, two will not be switched at each stage.

The reader will notice that each of the four threads is back in place after six switches, again an even number. The reader is invited, at this point, to do some experimenting of his own with three, four, five, or more threads and to try designing a seamless hatband having an odd number of switches.

After several attempts we find ourselves in a disturbing position. We have found no design with an odd number of switches; we feel none can be found, but we are not sure of this. Perhaps there does exist a design for a seamless hatband made with very many threads and an odd number of switches.

What are we going to tell the weaver? He showed us why no two-thread seamless design having an odd number of switches could be found. It seems likely that if no designs exist at all—designs having any number of threads and an odd number of switches—then we ought to be able to explain why there are none. Experiments in mathematics prove nothing; they only point to a possible truth. Since no one can ever write down the endless list of all possible designs, we must somehow see into all designs at once and find a general principle that will tell us why the number of switches can never be odd.

We might find a clue to the more general question in the weaver's analysis of designs using only two threads. In order to simplify our diagrams, let us number his threads "1" and "2." The starting position is 1, 2. After one switch he has the backward position 2, 1; 2 is now to the left of 1. After another switch he has the original order 1, 2. The weaver's design of four switches we record simply as

$$
\begin{array}{cc}
1 & 2 \\
2 & 1 \\
1 & 2 \\
2 & 1 \\
1 & 2
\end{array}
$$

We could translate the weaver's analysis into: after an odd number of switches, 1 and 2 are backward; after an even number of switches they are forward, in their usual order, 2 to the right of 1.

With this as a clue, let us look at our four-switch experiment with three threads, which we now record as

$$
\begin{array}{ccc}
1 & 2 & 3 \\
2 & 1 & 3 \\
3 & 1 & 2 \\
1 & 3 & 2 \\
1 & 2 & 3
\end{array}
$$

Looking only at threads 1 and 2, we notice that, unfortunately, they do not alternate "forward" and "backward" with each of the four switches. In fact, after each of the last three they are forward. Clearly, if there were more threads, then threads 1 and 2 might not be displaced at all. If we are to extend the weaver's analysis, we will have to pay attention to threads other than just 1 and 2.

In order to treat all the threads without favoritism, perhaps we should examine each pair of threads and see whether it is forward or backward. This might help; but then again it might not.

At the beginning position (position 1 2 3), each pair [(1, 2), (1, 3), (2, 3)] is forward; none is backward. At the next position, 2 1 3, the pair (2, 1) is backward, and (2, 3) and (1, 3) are forward. At 3 1 2 the pairs (3, 1) and (3, 2) are backward, and (1, 2) is forward. At 1 3 2 we have the one backward pair (3, 2), whereas (1, 3), and (1, 2) are forward. Finally, all are forward again.

The number of backward pairs for each arrangement runs successively through 0, 1, 2, 1 and returns to 0. With just two threads these numbers alternate 0, 1, 0, 1, and so on, as the weaver has told us. Clearly we do not have quite so simple a situation in our experiment with three threads.

Now let us look at our design with four threads and six switches, which we record as

$$
\begin{array}{cccc}
1 & 2 & 3 & 4 \\
3 & 2 & 1 & 4 \\
3 & 4 & 1 & 2 \\
2 & 4 & 1 & 3 \\
4 & 2 & 1 & 3 \\
1 & 2 & 4 & 3 \\
1 & 2 & 3 & 4
\end{array}
$$

For simplicity, let us call the number of backward pairs of an arrangement $B$; for example, $B$ of 1, 2, 3, 4 is 0. As the reader may check, $B$ for each of the seven arrangements is successively 0, 3, 4, 3, 4, 1, 0. For example, the third arrange-, ment, 3, 4, 1, 2, has the backward pairs (3, 1), (3, 2), (4, 1), (4, 2). Thus $B$ is 4, or we might write $B(3, 4, 1, 2) = 4$. Similarly, $B(1, 2, 3, 4) = 0$.

Now look at the two sequences we found; namely, 0, 1, 2, 1, 0 and 0, 3, 4, 3, 4, 1, 0. Naturally each begins and ends with 0. But, chaotic as they are, they do have something else in common: *each alternates even, odd, even, odd, and so on.* For two-thread designs the alternation is restricted simply to 0, 1, 0, 1, . . . , which is again even, odd, even, odd, . . . . So we have a very promising clue for solving the weaver's problem. *All would be answered* if we could prove this statement:

*If one switch is made in an arrangement of natural numbers, then the number of backward pairs always changes by an odd number.*

We have reduced the question from one concerning weavers' designs that may involve billions of switches to one concerning the effect of a single typical switch. The thorough study of what happens when just one switch is made can be carried out with enough generality (if we make the bookkeeping sufficiently flexible) to cover completely the effect of any switch whatsoever.

We must scrutinize what happens to $B$ when one switch is made in an arrangement. To be sure that our thinking will be valid for any switch in any arrangement, we will put hoods over all the natural numbers in question. This we can do quite easily by calling the left one of the two switched natural numbers "*c*" and the other "*d*." All we know is that *c* and *d* are two natural numbers; *c* may be less than *d*, or *d* may be less than *c*. The arrangement will look like

where the lines indicate that there may be more natural numbers in the arrangement left of *c*, between *c* and *d*, or right of *d*. After switching *c* and *d*, we obtain the new arrangement

_____*d*_____*c*_____ .

Let us look at those pairs that might have influenced the *change* in *B* as we went from the first arrangement to the second. First of all, if neither number of a pair is *c* or *d*, then such a pair contributes nothing to the change in *B*; it is unaffected by switching *c* and *d*. Second, what about pairs that include either *c* or *d* but not both, such as $(b, c)$ or $(b, d)$? If *b* lies to the left of both *c* and *d*,

_____*b*_____*c*_____*d*_____ ,

then in the new arrangement,

_____*b*_____*d*_____*c*_____ ,

*b* remains left of *c* and *d*. Thus if $(b, c)$, for example, is forward in the first arrangement, it remains so in the second. Such pairs do not, of course, influence the *change* in *B*. Similar reasoning holds if *b* is right of both *c* and *d*.

But let us look closely at the pairs $(c, b)$ and $(b, d)$, where *b* is between *c* and *d*. The first arrangement is

_____*c*_____*b*_____*d*_____ ,

and the second is

_____*d*_____*b*_____*c*_____ .

If we assume that *c* is smaller than *d*, there are then three possibilities:

1. *b* is smaller than *c* (hence smaller than *d*);
2. *b* is larger than *c* but smaller than *d*;
3. *b* is larger than *d* (hence larger than *c*).

In case 1 our bookkeeping runs like this. At first $(c, b)$ is backward, and $(b, d)$ is forward. In the new arrangement $(d, b)$ is backward, and $(b, c)$ is forward. The total effect on the *change* in *B* is *zero*.

Let us look at the bookkeeping for case 2. At the start, $(c, b)$ and $(b, d)$ are forward. In the new arrangement $(d, b)$ and $(b, c)$ are backward. The total effect on the *change* in *B* is *two*.

The bookkeeping in case 3, similar to that for case 1, is left to the reader.

Now we consider the pair $(c, d)$. If $c$ is less than $d$, then $(c, d)$ is forward at the beginning and backward after the switch. The effect of this pair on the *change* in $B$ is *one*.

Combining all the possibilities, we see that the *change* in $B$ is an even number plus one, hence *odd*. The reader may dispose of the case in which $c$ is larger than $d$ in a similar manner.

We have kept our bookkeeping flexible enough to cover any switch in an arrangement. Having proved that the change is odd in all possibilities, we can summarize our conclusion formally in this

THEOREM. *When one switch is made in an arrangement, the number of backward pairs changes by an odd number.*

For simplicity we will call an arrangement with an even number of backward pairs "*even*," and one with an odd number of backward pairs "*odd*." From our theorem follow several corollaries.

COROLLARY 1. *One switch turns an even arrangement into an odd arrangement, and an odd arrangement into an even arrangement.*

From Corollary 1 it follows that if we start with an even arrangement and apply an odd number of switches, we end up with an odd arrangement. In particular, we have this reply to the weaver's question, in the form of

COROLLARY 2. *A seamless hatband woven of any number of threads must have an even number of switches.*

The reader may have noticed something peculiar about our reasoning. Though the weaver's question involves only the notion of odd and even, our solution involves, in addition, the notion that one natural number is smaller or larger than another natural number. He might complain, "Though the weaver is now happy, I am not. It seems to me that we should be able to find a proof that stays within the given notions—a proof that doesn't make use of those backward pairs." Actually there is such a proof, but reading it requires more mathematical background than the typical reader is assumed to have (see R 3).

With Corollary 1 in our possession we are ready to explain the Fifteen puzzle. The reader who tries to reach the second position from the first, described at the beginning of the chapter, will never succeed. We will show *why* it is impossible to go from the first to the second by moves of the Fifteen puzzle.

Place the puzzle on top of a red and white checkerboard having squares the size of the movable squares of the puzzle, with a red square in the lower right corner. The blank square begins on red. On the first move, it changes to white. Clearly, with every switch it changes color. In the final position, the blank square, having been returned to the lower right corner, is again red. So the *number of moves* in any solution of this particular problem must be *even*.

Next, we shall relate the positions and the moves of a Fifteen puzzle to arrangements and switches of sixteen numbers in a row. Calling the blank square 16, we can read off the opening position as we read a printed page, from left to right and top to bottom

$$1, 2, 3, 4, 5, 6, 7, 8, 9, 10, 11, 12, 13, 14, 15, 16.$$

We wish to reach the arrangement that would read

$$15, 14, 13, 12, 11, 10, 9, 8, 7, 6, 5, 4, 3, 2, 1, 16.$$

Every time we move the blank square we obtain a new arrangement of the sixteen numbers, differing from a preceding arrangement by one switch (of 16 with one of the numbers $1, 2, \ldots, 15$). Since $B$ for the opening position is zero, it is even. The final arrangement has 105 backward pairs (as the reader may check) and hence is odd. If this particular Fifteen puzzle had a solution, then the number of moves would have to be *odd*, since we are traveling from an even arrangement to an odd arrangement.

Since no number is both *odd* and *even*, we can conclude that there is no way of going from the first to the second position through a sequence of switches of the blank square with its neighbors.

Similar reasoning establishes the following

COROLLARY 3. *It is impossible to go from one position of the Fifteen puzzle to another if either*

    (*a*) *the blank square has the same color in both positions but the B of one position is even and that of the other is odd*;

*or*

    (*b*) *the blank square is red in one position and white in the other but the B's of the two positions are both odd or both even*.

The corollary, pessimistic in character, does not say when the Fifteen puzzle is solvable. It turns out that if the $B$-reasoning and the red-white reasoning do not lead to a contradiction, then one position can be reached from the other. The technique for doing this is sketched in E 38.

The contrast of the odd and even arrangements is an important concept in mathematics, as the footnote we quoted at the beginning of the chapter points out. We cannot go thoroughly into the "prefigured relations" of left and right on the line, clockwise and counterclockwise in the plane, left-handed and right-handed screws (or gloves) in space, all of which rest on the part of algebra called "the theory of determinants," which in turn depends on the distinction between even and odd arrangements. An adequate treatment of determinants would require a small book all to itself.

We can, however, give an inkling of the relation of clockwise and counter-clockwise to our notion of even and odd arrangements. Draw a triangle on a piece of paper, and label its corners 1, 2, and 3, such that a bug traveling over the three corners of the triangle in the order 1, 2, 3 would move clockwise, as in this figure:

There are six routes a bug might choose: 1 2 3 (just mentioned), 1 3 2, 2 1 3, 2 3 1, 3 1 2, and 3 2 1. Some will be clockwise, some counterclockwise; some will be even arrangements, some odd. The reader who checks the six cases will see that the even arrangements are also the clockwise, whereas the odd arrangements are the counterclockwise. A similar relation exists for a pyramid (see E 42, E 43, and E 44).

The Fifteen puzzle has taken us far: to threads and seamless hatbands, through the relative size of natural numbers into backward pairs and the odd and even arrangements, and finally to clocks and counterclocks. In the next chapter we set out from the weaver's two threads and soon find ourselves on quite a different path.

## Exercises

1. Is 0 a natural number? Is 0 an even number?

2. Draw the series of switches in three threads that would be recorded by 1 2 3, 1 3 2, 3 1 2, 2 1 3, 1 2 3.

3. Draw the series of switches in four threads recorded by 1 2 3 4, 1 3 2 4, 4 3 2 1, 4 3 1 2, 1 3 4 2, 1 2 4 3, 1 2 3 4.

4. Compute $B(2, 1)$, $B(2, 4, 1, 3)$, $B(4, 5, 2, 1, 3)$.

5. Compute $B(3, 2, 5, 1, 4)$, $B(2, 3, 4, 6, 7, 5, 1)$.

6. Compute $B(3, 2, 1)$, $B(4, 3, 2, 1)$, $B(5, 4, 3, 2, 1)$.

7. (a) Draw an "interesting" seamless hatband with five threads and at least six switches.
   (b) Check that $B$ changes by an odd number at each stage.

8. What can we say about the number of switches whenever we go from the arrangement 2, 7, 5, 1, 4, 3, 6 to the arrangement 6, 4, 1, 5, 3, 7, 2?

9. What can we say about the number of switches whenever we go from the arrangement 1, 2, 3, 4, 5, 6 to the arrangement 6, 5, 4, 3, 2, 1?

10. (a) We proved the theorem only with the assumption that $c$ is less than $d$. Take care of the case in which $c$ is larger than $d$.
    (b) Prove Corollary 3.

11. (a) Draw the position of the Fifteen puzzle recorded by the arrangement

    $$1, 2, 3, 4, 5, 6, 7, 8, 9, 10, 11, 12, 13, 14, 15, 16.$$

    (b) Draw the position of the Fifteen puzzle recorded by the arrangement

    $$1, 2, 3, 4, 13, 14, 15, 16, 9, 10, 11, 12, 5, 6, 7, 8.$$

    (c) Can one go from (a) to (b) by moves of the Fifteen puzzle?

12. (a) Show how we can go from the arrangement 5, 4, 3, 2, 1, 10, 9, 8, 7, 6 to the arrangement 10, 9, 8, 7, 6, 5, 4, 3, 2, 1 in five switches.
    (b) Is it possible to go from the first arrangement to the second in 142 switches? In 143 switches?

13. Explain this statement about the Fifteen puzzle: If we get thirteen of the numbers and the blank square in place but the two remaining numbers not in place, then the puzzle is not solvable.

14. True or false: If the given position of the Fifteen puzzle has odd $B$ and a red blank square, then we cannot reach a position which has an even $B$ and a red blank square. Explain.

15. Draw the position of the Fifteen puzzle recorded by the arrangement 8, 7, 4, 2, 3, 6, 5, 10, 1, 9, 16, 15, 12, 13, 14, 11.

For each of Exercises 16 through 20 decide whether the problem can be solved. If your answer is "no" give your argument; if "yes" solve it. (The instructor may not expect you to record all the moves of your solution.)

16.

| 1 | 2 | 3 | 4 |
|---|---|---|---|
| 5 | 6 | 7 | 8 |
| 9 | 10 | 11 | 12 |
| 13 | 14 | 15 |  |

to

|  | 1 | 2 | 3 |
|---|---|---|---|
| 4 | 5 | 6 | 7 |
| 8 | 9 | 10 | 11 |
| 12 | 13 | 14 | 15 |

17.

| 12 | 11 | 10 | 9 |
|---|---|---|---|
| 13 | 2 | 1 | 8 |
| 14 | 3 |  | 7 |
| 15 | 4 | 5 | 6 |

to

| 9 | 10 | 11 | 12 |
|---|---|---|---|
| 8 | 1 | 2 | 13 |
| 7 |  | 3 | 14 |
| 6 | 5 | 4 | 15 |

18.

| 1 | 2 | 3 | 4 |
|---|---|---|---|
| 5 | 6 | 7 | 8 |
| 9 | 10 | 11 | 12 |
| 13 | 14 | 15 |  |

to

| 13 | 14 | 15 |  |
|---|---|---|---|
| 1 | 2 | 3 | 4 |
| 5 | 6 | 7 | 8 |
| 9 | 10 | 11 | 12 |

19.

| 1 | 5 | 2 | 4 |
|---|---|---|---|
| 6 |  | 8 | 7 |
| 3 | 10 | 12 | 14 |
| 9 | 11 | 13 | 15 |

to

| 1 | 6 | 2 | 4 |
|---|---|---|---|
| 5 |  | 8 | 7 |
| 3 | 10 | 12 | 14 |
| 9 | 11 | 13 | 15 |

20.

| 1 | 2 | 3 | 4 |
|---|---|---|---|
| 5 | 6 | 7 | 8 |
| 9 | 10 | 11 | 12 |
| 13 | 14 | 15 |  |

to

| 1 | 2 | 3 | 4 |
|---|---|---|---|
| 13 | 14 | 15 |  |
| 9 | 10 | 11 | 12 |
| 5 | 6 | 7 | 8 |

●

21. Scattered in a room are ten people sitting on ten fixed chairs. Every so often two people switch places while the remaining eight stay fixed. After many switches everyone is in his original chair. What can be said about the number of switches made?

22. (a) Without computing any *B*, show that the number of switches in going from

| 1 | 2 | 3 | 4 |
|---|---|---|---|
| 5 | 6 | 7 | 8 |
| 9 | 10 | 11 | 12 |
| 13 | 14 | 15 |  |

to

| 2 | 1 | 4 | 3 |
|---|---|---|---|
| 6 | 5 | 8 | 7 |
| 10 | 9 | 12 | 11 |
| 14 | 13 |  | 15 |

must be even. (Show that, if we are allowed to lift numbers out of the box and switch any pair, we could get from the first to the second in 8 switches.)

(b) Prove that the puzzle in (a) is not solvable.

23. Without computing any $B$, determine whether one can go from

| 1 | 2 | 3 |
|---|---|---|
| 4 | 5 | 6 |
| 7 | 8 | ⧹⧹ |

to

| 1 | 4 | 7 |
|---|---|---|
| 2 | 5 | 8 |
| 3 | 6 | ⧹⧹ |

in what might be called "The Eight Puzzle." (See E 22.)

24. Fill in the blank with as small a number as possible and explain: We can always go from one arrangement of the natural numbers 1 through 20 to any other arrangement of them in _____ or fewer switches.

25. (a) How many arrangements of 1, 2, 3, 4 have their $B$ equal to 0?
    (b) How many arrangements of 1, 2, 3, 4 have their $B$ equal to 1? To 2?
    (c) The same for $B$ equal to 3, 4, 5, or 6.

26. (a) List all possible arrangements of 1, 2, 3, 4 (there are 24).
    (b) How many have their $B$ even?
    (c) How many have their $B$ odd?

27. The same as E 26 for the six arrangements of 1, 2, 3 (instead of the arrangements of 1, 2, 3, 4).

28. (a) Inspecting your list of E 26, show that, when you switch the 1 with the 2 in any arrangement with even $B$, you obtain an arrangement with odd $B$ (what result in the chapter promises this?).
    (b) Why does (a) tell us that there are at least as many arrangements of 1, 2, 3, 4 with odd $B$ as with even $B$?
    (c) Similarly, show why there must be at least as many arrangements with even $B$ as with odd $B$.
    (d) From (b) and (c) conclude that there must be the same number of arrangements with even $B$ as with odd $B$.

29. Prove that the number of arrangements of 1, 2, 3, 4, 5, 6, 7, 8, 9, 10 with even $B$ is the same as with odd $B$. (*Hint:* Use the ideas of E 28.)

30. What is the smallest number of switches required to go from the arrangement 1, 2, 3, 4, 5, 6, 7, 8 to the arrangement 8, 7, 6, 5, 4, 3, 2, 1? How do you know that it is the smallest?

31. What is the smallest number of switches required to go from the arrangement 1, 2, 3, 4, 5, 6, 7 to the arrangement 7, 6, 5, 4, 3, 2, 1? How do you know that it is the smallest?

32. How would you define the even natural numbers using (a) only addition, (b) only multiplication, (c) the decimal system?

33. What type of number do you obtain whenever you add (a) two even numbers, (b) two odd numbers, (c) an odd number and an even number? (d) Why?

34. What type of number do you obtain whenever you multiply (a) two even numbers, (b) two odd numbers, (c) an odd number and an even number? (d) Why?

THEOREM 1. *If we divide a given line segment into little segments by means of dots, and then label these dots (including the two ends of the given line segment) either A or B, then there is this relation between the number of complete segments and the two end dots of the given line segment:*

*1. If the two end dots are labeled A and B, then there is an odd number of complete segments.*

*2. If the two end dots are both labeled A (or are both labeled B), then there is an even number of complete segments.*

(In this theorem and in our discussion a *little* segment means a segment that contains no other segment drawn.)

The reader may be wondering, "Must we use a cat to prove Theorem 1? Surely there is a proof that can be expressed within the structure of the original line segment that we are dividing, since Theorem 1 concerns just that line." It is easy to satisfy such a scrupulous reader. We may imagine dividing the line into little segments by drawing and labeling dots one after the other. Though these divisions can be made in many ways, we can separate them into three types and see what *change* results in the number of *complete* segments each time we draw a dot.

Each time we draw a dot we introduce it into a segment whose ends are *A* and *A*, *B* and *B*, or *A* and *B*.

In the first case we have

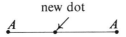

If we label the new dot *A*, we create no complete segments; if we label it *B*, we create two new complete segments. In both cases the change is even. The same holds if we draw a dot in a segment having ends *B* and *B*, as the reader may check.

In the last case we have

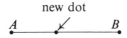

If we label the new dot *A*, then we have replaced a complete segment with one complete segment and with one segment having ends *A* and *A*. Thus we have no change in the total number of complete segments. A similar conclusion holds if we label the new dot *B*.

Consequently every time we draw and label a dot, the number of complete segments changes by an even number. If we start out with the line labeled *AB* (one complete segment), we continue to have an odd number of complete segments. If we start with the line labeled *AA*, or the one labeled *BB* (zero complete segments), we still have an even number of complete segments. This reasoning, pulling no cat out of the hat, should please that demanding reader.

Before we go on, let us rephrase Theorem 1 in order to emphasize the role of even and odd:

THEOREM 1 (rephrased). *If we divide a given line segment into little segments by means of dots, and then label these dots either A or B (including the two at the ends of the line), then there is this relation between the number of complete segments and the two end dots of the line:*

1. *If an odd number of the two end dots is labeled A, then there is an odd number of complete segments.*

2. *If an even number of the two end dots is labeled A, then there is an even number of complete segments.*

(*Of course we could have used the letter B instead of A in statements 1 and 2.*)

A mathematician who proves something about the line is not at peace until he learns whether his result is related to (usually more interesting) truths in the higher dimensions of the plane and space. If he finds no analogue in higher dimensions, he is disappointed. If he does find analogues, then he feels that he has penetrated a mystery that the line half concealed, half revealed. Perhaps he will find to his dismay that the proof that worked so easily on the line gives him no clue for dealing with the plane or with space. If so, he will seek a proof for the line that does help him treat the higher dimensions.

Now where do we go from Theorem 1, which is so close to that case of the weaver's problem that involves only two threads? The weaver himself suggested the deeper problem that lay beyond: designs with three or more threads. But Mackie gives us no clue as to what might be the theorem corresponding to the three letters *A*, *B*, and *C*, instead of just *A* and *B*. If there is anything to be discovered, we shall have to discover it for ourselves.

Let us try. The basic idea was the *complete segment*. With three letters at our disposal we might think of a **complete triangle,** *a triangle furnished with all three of the letters A, B, and C at its corners*. Instead of cutting up a line segment into little segments, perhaps we should experiment with cutting up a triangle into little triangles by means of dots. Then we will label these dots either *A*, *B*, or *C*, and see what, if anything, we can find out.

This time we begin with a big triangle and draw dots inside it and on its border. We keep joining these dots with straight lines until we have cut up the big triangle into little triangles that, together, cover the big triangle. The dots become corners of the little triangles. (A *little* triangle means a triangle that contains no other triangle drawn.)

Next we label each dot including the three corners of the original triangle *A*, *B*, or *C* until each little triangle has a letter at each of its three corners. Is there any relation between the number of complete little triangles and some property of the border of the big triangle? If there is, then we will have an analogue in the plane of Theorem 1. Let us make some experiments and, keeping odd and even in mind, try to discover a relation between complete triangles and the border—if there is any such relation.

Not knowing what might be the relation between the number of complete triangles and the border, we record various data in the hope of finding a clue. Here is one experiment:

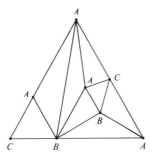

The data from this experiment appear in the first line of the table below. Four other experiments (not drawn here) lead to additional facts, which we record in the appropriate columns.

| Complete Triangles | Dots A | Dots B | Dots C | On the Border Edges: AA | BB | CC | AB | AC | BC |
|---|---|---|---|---|---|---|---|---|---|
| 3 | 3 | 1 | 2 | 1 | 0 | 0 | 1 | 3 | 1 |
| 2 | 2 | 3 | 3 | 0 | 1 | 1 | 2 | 2 | 2 |
| 5 | 2 | 6 | 1 | 0 | 4 | 0 | 3 | 1 | 1 |
| 3 | 3 | 3 | 4 | 0 | 0 | 3 | 5 | 1 | 1 |
| 3 | 4 | 4 | 1 | 3 | 3 | 0 | 1 | 1 | 1 |

The reader is invited to make his own experiments.

Now let us see what we can find out. Our first instinct is to see whether the number of dots *A* on the border influences the number of complete triangles, as it did in Theorem 1 (rephrased). The answer is no; with two dots *A* we have both two and five complete triangles, as our list of experiments shows. Similarly, we can dispose of dots *B* and *C* and edges *AA*, *BB*, and *CC*.

But there seems to be a simple relation between complete triangles and edges

*AB, AC,* or *BC.* Directing our attention to edges *AB,* for example, we are tempted to conjecture this analogue of rephrased Theorem 1:

THEOREM 2. *If we divide a triangle into little triangles by means of dots, and then label these dots either A, B, or C (including the three at the corners of the big triangle), then there is this relation between the number of complete little triangles and the edges on the border:*

*1. If an odd number of the edges on the border have their two ends labeled A and B, then there is an odd number of complete triangles.*

*2. If an even number of the edges on the border have their two ends labeled A and B, then there is an even number of complete triangles.*

Of course we could replace *AB* by *BC* or *AC* and still obtain a reasonable conjecture.

It certainly is not obvious that our conjectured Theorem 2 is always true. After all, if it were true, it would say in particular that if we have an odd number of edges of the type *AB* on the border, then there must be at least one complete triangle. It is hard to imagine that the border can influence the inside of the triangle so much. Therefore the reader may be surprised at the results of his own experiments if he starts at the border with an odd number of *AB* edges and tries to avoid forming a complete triangle as he labels the dots inside the big triangle.

After a few more experiments the reader may feel certain that Theorem 2 is indeed true. But how shall we go about proving it? If we look back at our catless proof of Theorem 1, we might think of watching what happens if a dot is drawn in a little triangle: does the number of complete triangles *change* by an even number?

For example, what happens when we draw a dot in a complete triangle? No matter how we label it, we end up with one complete triangle, as the reader may see by glancing at this picture:

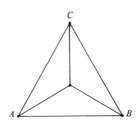

In this case there is no change in the number of complete triangles.

As another example, if we were to draw a dot inside

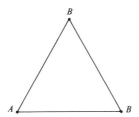

we would end up with either zero or two complete triangles, again an even *change*.

The reader may easily check the other cases and see that, each time, the change in the number of complete triangles is even. Unfortunately this gets us nowhere. Though such reasoning worked on the line, it is inadequate for triangles. Some divisions of the triangle can, of course, be obtained by the process described above; for example this division can be obtained by using the process twice:

But a division such as

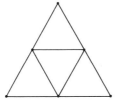

cannot be obtained by such a method.

If we are to prove Theorem 2, we will have to use new ideas. Perhaps the best thing to do now is to go back and try to find a new proof of Theorem 1, a proof which neither uses a cat nor draws a dot at a time. Perhaps the idea behind such a new proof might help us with Theorem 2. So let us backtrack for a moment and study the line once again.

Let us look at the various kinds of little segments that can appear in a division of a line segment. There are three types:

$\overset{A}{\bullet}\rule{2cm}{0.4pt}\overset{A}{\bullet}$     $\overset{B}{\bullet}\rule{2cm}{0.4pt}\overset{B}{\bullet}$     and the complete     $\overset{A}{\bullet}\rule{2cm}{0.4pt}\overset{B}{\bullet}$
$\underset{B}{\bullet}\rule{2cm}{0.4pt}\underset{A}{\bullet}$

A complete little segment contributes one $A$, whereas the other types contribute either zero $A$'s or two $A$'s. More briefly, a complete segment contributes one $A$, whereas all the other little segments contribute an even number of $A$'s. From this fresh observation we might be able to get a new proof of Theorem 1. Let us try.

Imagine that we have a division of a line segment into little segments. Let us say that $T$ of these little segments are complete. Of course, $T$ might be zero, but in any case, we want to find a relation between $T$ and the number of $A$'s at the ends of the original line. We want to use our fresh observation.

Consider a typical little segment. Place pebbles on it near any of the ends labeled $A$. Do this for each little segment. For example, if we have this division,

$\overset{A}{\bullet}\rule{3cm}{0.4pt}\overset{B}{\bullet}\rule{1cm}{0.4pt}\overset{A}{\bullet}\rule{0.5cm}{0.4pt}\overset{A}{\bullet}\rule{0.5cm}{0.4pt}\overset{B}{\bullet}\rule{2cm}{0.4pt}\overset{B}{\bullet}$

we end up with the pebbles arranged as follows:

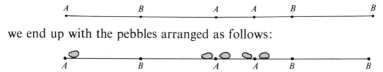

Any dot labeled $A$ inside the original line segment has two pebbles near it, because it is the end of two little segments. Any dot labeled $A$ at an end of the original line segment collects one pebble. These observations hold not only in this example but in any division of a line segment into little segments.

Let us now do some bookkeeping for the general case. We count the total number of pebbles in two different ways and equate these two counts.

1. We first determine the total number of pebbles by counting the number of pebbles on all the little line segments: Each complete little segment has one $A$ and hence has one pebble on it; all the other little segments have an even number (either zero or two). Thus, if $T$ denotes the number of complete little segments, the total number of pebbles is

$$T + \text{an even natural number}$$

2. We next determine the total number of pebbles by counting the pebbles near each dot labeled $A$: Any dot labeled $A$ at the end of the original segment has one pebble near it. Any inside dot labeled $A$ has two pebbles near it. Thus, the total number of pebbles is

number of $A$'s at ends of original segment $+$ some even natural number

Since the counts for the total number of pebbles obtained in these two ways must give the same answer, we obtain

$$T + \begin{array}{c} \text{an even} \\ \text{natural number} \end{array} = \begin{array}{c} \text{number of } A\text{'s at ends} \\ \text{of original segment} \end{array} + \begin{array}{c} \text{some even} \\ \text{natural number} \end{array}$$

Since the sum of two even numbers is even, and the sum of an even number and an odd number is odd, we have arrived at a new proof of Theorem 1.

Does this third proof of Theorem 1 give us a method for attacking Theorem 2? The key is that a complete little segment contributes one $A$, whereas all other little segments contribute an even number of $A$'s. We will now examine closely the way in which little triangles contribute edges of the type $AB$, for Theorem 2 concerns such edges. There are ten types of little triangles:

Of the ten triangles, seven have no $AB$ edges, two have two $AB$ edges, and the complete triangle has one $AB$ edge. In short: a complete triangle has one $AB$; all the other little triangles have an even number of $AB$'s. Can we base a proof of Theorem 2 on this fact?

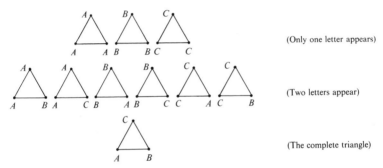

(Only one letter appears)

(Two letters appear)

(The complete triangle)

Yes, we can, in fact, merely by replacing in the proof of Theorem 1 just given "segment" with "triangle" and "dot labeled $A$" with "edge labeled $AB$." So doing, we obtain the following

PROOF OF THEOREM 2. Consider a typical little triangle. Place a pebble on it near any of its edges labeled $AB$. Do this for each little triangle, as indicated in this example:

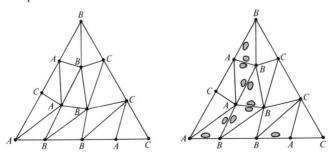

Let us now do some bookkeeping for the general case. We count the total number of pebbles in two different ways and equate these two counts.

1. We first determine the total number of pebbles by counting the number of pebbles on all the little triangles: Each complete little triangle has one $AB$ edge and hence has one pebble on it; all the other little triangles have an even number (either zero or two). Thus, if $T$ denotes the number of complete little triangles, the total number of pebbles is

$$T + \text{an even natural number}$$

2. We next determine the total number of pebbles by counting the pebbles near each edge labeled $AB$: Any $AB$ edge on the border of the original triangle has one pebble near it. Any inside edge labeled $AB$ has two pebbles near it. Thus the total number of pebbles is

number of $AB$'s on border of original triangle + some even natural number

Since the counts for the total number of pebbles obtained in these two ways must be identical, we obtain

$$T + \begin{matrix}\text{an even} \\ \text{natural number}\end{matrix} = \begin{matrix}\text{number of } AB\text{'s on border} \\ \text{of original triangle}\end{matrix} + \begin{matrix}\text{some even} \\ \text{natural number}\end{matrix}$$

Thus, if the number of $AB$'s on the border is even, then so is the number of complete triangles. And if the number of $AB$'s on the border is odd, so is the number of complete triangles. We have succeeded in proving Theorem 2; and only our third proof of Theorem 1 was of use. For this reason we can claim that the third proof of Theorem 1 is preferable, since it extended to the plane, whereas the other two did not.

The third proof has another distinction. It treats as a whole the division of the line into segments, whereas the others involve the manner in which the dots are inserted. The first two proofs are adequate for the line, since the divisions into little segments are relatively simple, but they fail to cope with the more complex divisions of the triangle.

From Theorem 2 we can deduce immediately

THEOREM 3. *If there is an odd number of AB's on the border of the big triangle, then there must be at least one complete little triangle.*

Since Theorem 3 is just a consequence of Theorem 2, we might think that it would be easier to prove. But I know of no elementary proof of Theorem 3 that bypasses the more general result—Theorem 2. It frequently happens in mathematics that it is simpler to prove more than to prove less; the general case exposes more and might even suggest the ideas on which to base a proof, whereas the special case might simply mislead. Suppose, for example, that the weaver of Chapter 1 had asked us: "Can I make a seamless hatband with 315 switches and 315 threads?" We would not have known where to turn. But he asked the seemingly tougher question, "Can I design a seamless hatband with an odd number of switches?" This helped us to concentrate on the number of switches, not the number of threads.

Historically, one special form of Theorem 2 is of importance here. It is the basis of several results in topology, a branch of mathematics that we shall discuss in later chapters. In honor of the mathematician who discovered it in 1928, it is named

SPERNER'S LEMMA. *Label the three corners of the big triangle A, B, and C. Label dots inside the big triangle either A, B, or C. Label dots on the edge AB of the big triangle either A or B. Label dots on the edge BC of the big triangle either B or C. Label dots on the edge AC of the big triangle either A or C. Then the number of complete little triangles is odd.*

The reader can easily prove Sperner's lemma with the aid of Theorems 1 and 2.

Among the consequences of Sperner's lemma is the following. Let us say that in the morning you pass by a circular puddle of water completely covered with oil. A gentle breeze blows all day, moving the oil about the surface of the puddle, but not breaking the oil film. Then (it can be proved) when you pass by the puddle in the evening, at least one molecule of oil will be back exactly where you saw it in the morning. (We mean an "ideal" oil, composed of an infinite number of "molecules," each of which coincides with a point.) Stating this a little more formally, we have

BROUWER'S FIXED POINT THEOREM. *Every smooth distortion of a circular disk into itself leaves at least one point fixed.*

The proof of Brouwer's theorem from Sperner's lemma is rather involved; it can be found in R 1. Brouwer's theorem, in turn, has many applications inside and outside of mathematics. Recently it has even assumed importance in economic theory. In the 1959 *SIAM Review*, A. T. C. Koopmans and A. F. Bausch report that

> The extent to which equilibria exist in actual markets is a matter for observation. The concern here is whether the [mathematical] model—extensively used to draw various conclusions about equilibrium—is capable of insuring the possibility of an equilibrium under the conditions it specifies. . . . There is little point in knowing that a competitive equilibrium, if possible, achieves efficient allocations of resources, if in fact it is a logical impossibility. The problem of the existence of a

competitive equilibrium . . . is a topological one, and all recent proofs have made use of some topological result equivalent to Brouwer's fixed point theorem, or a generalization thereof.

Sperner's lemma also plays a key role in defining the dimension of a space. (This is touched on in Appendix F.)

In 1945 A. W. Tucker explored another type of labeling problem, which also has topological applications. This problem refers to squares and four labels rather than triangles and three labels, and is more fundamental than Sperner's in the sense that it concerns only regularly spaced dots in the plane and thus can be translated into a statement about the natural numbers. (See E 28–33 and R 2.)

We have come quite far from the problem that opened the chapter. Though the riddles of the cat and the weaver are really the same, they lead to theorems whose only similarity is that they both involve the fundamental distinction between the even and the odd natural numbers. One led us into algebra, the other into topology.

### Exercises

1. Prove that Theorem 1 (rephrased) remains true if we consider dots labeled $B$ instead of $A$.

2. Does Theorem 2 remain true if we look at border edges labeled $BC$ instead of $AB$? Justify your answer.

3. Prove that if the number of edges $AA$ on the border of the big triangle is odd, then there is at least one little triangle with the letter $A$ appearing on at least two of its three corners.

4. Prove that if the number of edges $AA$ on the border of the big triangle is odd, then there is an odd number of little triangles with the letter $A$ appearing on at least two of its three corners. (*Hint:* Do bookkeeping with contributions of $AA$ edges.)

5. Contrast E 3 and E 4 with Theorems 3 and 2.

6. If only the letter $C$ appears on the border of the big triangle, what can we say about the number of complete little triangles?

7. Prove Sperner's lemma.

8. Prove that if dots are drawn on the border of a triangle and labeled either $A$ or $B$, then there is an even number of edges having both an $A$ and a $B$ by
   (a) making use of Theorem 3;
   (b) employing an argument that uses only the border.

9. Prove that if a triangle is cut into an odd number of little triangles by dots, then there must be an odd number of edges on the border of the big triangle.

10. Does Theorem 1 remain true if the line is drawn curved, or have we used straightness in our proof?

11. Does Theorem 2 remain true if all the lines are drawn curved, or have we used straightness in our proof?

The same type of argument used in the proof of Theorem 2 may be used in other situations, such as cutting a circular disk, a rectangle, or the surface of a sphere (ball) into small pieces, perhaps triangles or rectangles whose sides may be curved. Exercises 12–20 provide some illustrations.

12. Translate Theorem 2 into a theorem about cutting a circular disk into "triangles" (allowing some triangles to have one curved side). It should begin, "If we divide a circular disk into little triangles. . . ."

13. On the border of a circle, dots are drawn and labeled either $A$, $B$ or $C$. Find a relation between the number of arcs $AB$, the number of arcs $BC$, and the number of arcs $AC$.
    (a) Experiment three times.
    (b) Make a conjecture.

14. (a) Prove E 13(b) using Theorem 2 and E 2.
    (b) Prove E 13(b) by studying what happens each time a dot is drawn and labeled.
    (c) Prove E 13(b) using pebbles.

15. On the border of a circle, dots are drawn and labeled either $A$, $B$, $C$, or $D$. Is there any relation between the number of little segments $AB$ and the number of little segments $AC$? Explain.

16. The surface of a sphere is divided into little triangles (with curved sides) by dots, which are then labeled $A$, $B$, $C$, or $D$. Prove that if the number of $ABC$ triangles is odd, then so is the number of $DBC$ triangles. (*Hint:* Count contributions of $BC$ edges carefully.)

17. The surface of a sphere is divided into little triangles (with curved sides) by dots, which are labeled $A$, $B$, $C$, or $D$. Prove that the number of little triangles with more than one $A$ at its corners is even.

18. Is it possible to cut the surface of a sphere into little triangles by dots in such a way that there is an odd number of little triangles?
    (a) Experiment (on a ball, if one is available).
    (b) Make a conjecture.
    (c) Prove it.
    (*Hint:* Do bookkeeping by placing three pebbles in each triangle, one near each edge.)

19. The surface of a sphere is divided into triangles (with curved sides) by means of dots, which are then labeled $A$, $B$, or $C$. How many complete triangles are there?
    (a) Experiment three times (perhaps on a ping-pong ball); record the number of complete triangles each time.
    (b) Make a conjecture.
    (c) Prove your conjecture.

20. The surface of a sphere is divided into quadrilaterals by dots labeled $A$ or $B$. Prove that the number of quadrilaterals labeled

with two adjacent $A$'s and two adjacent $B$'s, is even. (*Suggestion:* Use the pebble technique with $AA$ edges.)

●

21. (a)  Divide a triangle into little triangles by means of dots.
    (b)  Count the number of little triangles in (a).
    (c)  With the *same* dots, divide the big triangle into little triangles in another way. (If you wish, you might make a copy of the big triangle with its dots.)
    (d)  Count the number of little triangles in (c).
    (e)  Using the same dots, carry out the experiment again, and count the number of little triangles.
    (f)  Make a conjecture. (A proof can be based on material in Chapter 14, in particular E 18 of Chapter 14.)

Exercises 22–25 refine Theorems 1 and 2 and relate them to "left and right" and "clockwise and counterclockwise."

22. If Mackie was at the foot of the stairs in the morning and at the top in the evening, we could deduce something stronger than Theorem 1. We could be sure that she went up the stairs once more than she went down them. In terms of the line segment picturing Mackie's diary, we could say that *the number of complete little segments with B right of A is one more than the number of complete little segments with B left of A.* Prove, without using Mackie, this refined version of Theorem 1. (Observe that the left end of the big line segment is $A$ and that the right end is $B$; then examine what happens each time a dot is introduced and labeled.)

23. Left and right are to the line segment as clockwise and counterclockwise are to the triangle. Let us call a complete triangle clockwise if a bug making a journey $A$ to $B$ to $C$ along its vertices moves clockwise; and similarly for counterclockwise. Assume that the dots by which the big triangle is cut up are labeled as in Sperner's lemma.
    (a)  Make five experiments in which you count the number of clockwise, and the number of counterclockwise, complete little triangles.
    (b)  What do you notice?

24. Let us strengthen Sperner's lemma as we strengthened Theorem 1 in E 22.

   THEOREM. *Assume that the dots are labeled in accordance with the restriction in Sperner's lemma and that the three corners of the big triangle are labeled A, B, and C in clockwise order. Then there is one more clockwise complete little triangle than the number of counterclockwise complete little triangles.*

    (a)  Show that the theorem agrees with your experiments.
    (b)  Show that this theorem implies Sperner's lemma.

25. We now outline the steps for proving the theorem of E 24. Consider a bug who always travels around triangles in a clockwise direction. Whenever he moves along an edge that has both an $A$ at one end and a $B$ at the other, he does so in one of two possible ways: either he meets the $A$ first and then the $B$ (which we record as $\overrightarrow{AB}$) or he meets the $B$ first and then the $A$ (which we record then as $\overrightarrow{BA}$). Do bookkeeping with contributions of $\overrightarrow{AB}$ and $\overrightarrow{BA}$. Using suitable drawings, show that

   (a) In a clockwise complete triangle the bug sweeps out one $\overrightarrow{AB}$.

   (b) In a counterclockwise complete triangle the bug sweeps out one $\overrightarrow{BA}$.

   (c) In a triangle whose corners are labeled $A$, $A$, and $B$, the bug sweeps out one $\overrightarrow{AB}$ and one $\overrightarrow{BA}$.

   (d) In a triangle whose corners are labeled $B$, $B$, and $A$, the bug sweeps out one $\overrightarrow{AB}$ and one $\overrightarrow{BA}$.

   (e) In all other triangles the bug sweeps out no $\overrightarrow{AB}$'s and no $\overrightarrow{BA}$'s.

   (f) Now prove the theorem of E 24 by counting contributions of $\overrightarrow{AB}$ and $\overrightarrow{BA}$ (recall E 22).

• •

26. What do you think must be true about the dots of a division of the big triangle into little triangles if they can be labeled in such a way that each little triangle is complete? Consider (a) dots inside the big triangle and (b) dots on the border of the big triangle. Express your answer in terms of the number of edges that meet at the dot.

27. (a) What would be the analogue of Theorem 2 for a solid triangular pyramid?
    (b) Prove it.

   Exercises 28–33, devoted to Tucker's theorem and some of its interpretations, are derived from R 2 and R 3.

28. How close can you come to arranging some playing cards in a square (say twenty-five cards in a five-by-five square) in such a way that both of these conditions hold:
    (a) No two cards of the same color, but of different suits, are adjacent (horizontally, vertically, or diagonally). That is, hearts are not adjacent to diamonds; spades are not adjacent to clubs.
    (b) On the border opposite cards are of the same color but different suits.
    (Two cards on the border are "opposite" if the line through their centers passes through the center of the square. In geography the term is "antipodal.")

29. Twenty-five weather stations are arranged in a square, five in each row (like the cards of E 28). The sixteen stations on the border we will call "border stations." Let us say that at each station the wind is blowing in one of the four directions

N, E, S, W. Determine whether it is possible for the wind to be blowing at the twenty-five stations in such a way that

(a) At adjacent stations the wind does not blow in opposite directions.

(b) At opposite border stations the wind blows in opposite directions. ("Opposite" and "adjacent" are defined as in E 28.)

Experiment. For convenience tilt North 45° from the lines of the stations.

30. Experiment with E 28 and E 29 and six-by-six squares.

31. How closely are E 28 and E 29 related?

32. Sperner's lemma concerns three symbols, $A$, $B$, and $C$. Tucker's theorem concerns four symbols. For convenience we will denote them 1, $-1$, 2, and $-2$ rather than $A$, $B$, $C$, $D$. Tucker's theorem, presented in E 33, answers this question: Can you place several 1's, $-1$'s, 2's, and $-2$'s in a square array in such a way that

(a) the sum of two adjacent numbers is never 0,

(b) opposite numbers on the border have a sum 0? ("Opposite" and "adjacent" are defined in E 28.)

Experiment. How is this exercise related to E 28 and E 29?

33. We now outline a proof for Tucker's theorem, which asserts that an arrangement of the type sought in E 32 does not exist.

Assume that such an arrangement *does* exist. Call the big square $ABA^*B^*$, and cut it into triangles as shown in this figure (which corresponds to the six-by-six case):

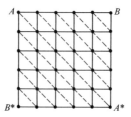

The numbers 1, 2, $-1$, $-2$ are located at the dots, not in the individual squares. The portion of the border $ABA^*$ we will call the "semiperimeter $ABA^*$." Imagine, now, that we have a labeling of the dots in accord with E 32(a), (b).

The symbols $j$, $k$, $l$ will be replaceable by 1, 2, $-1$, $-2$ in the following definitions.

(a) We now define numbers $A(1)$, $A(2)$, $A(-1)$, $A(-2)$ as follows: $A(j) = 1$ if vertex $A$ is labeled $j$ (hence $A^*$ has the label $-j$); if $A$ is not labeled $j$, we set $A(j) = 0$. Show that

$$A(1) + A(-1) + A(2) + A(-2) = 1.$$

(b) Let $S(j, k)$ equal the number of little segments in the semiperimeter $ABA^*$ that have one end labeled $j$ and the other end $k$ (this, in turn, equals the number of segments in the semiperimeter $A^*B^*A$ that have one end labeled $-j$ and the other end $-k$). Show that $S(1, -1) = 0$, $S(2, -2) = 0$.

(c) Let $S(j)$ equal the number of vertices in the semiperimeter $ABA^*$, excluding $A$ and $A^*$, that have the label $j$ (hence the number of vertices along $A^*B^*A$, excluding $A^*$ and $A$, that have the label $-j$). Using pebbles near vertices labeled 1 on the semiperimeter $ABA^*$, show that

$$A(1) + 2S(1) + A(-1) = S(1, 2) + S(1, -2) + 2S(1, 1).$$

(d) Counting pebbles placed at vertices labeled $-2$ on the semiperimeter $ABA^*$, show that

$$A(-2) + 2S(-2) + A(2) = S(1, -2) + S(-1, -2) + 2S(-2, -2).$$

(e) From (a), (c), and (d) deduce that

$$S(1, 2) + S(-1, -2)$$
$$= 1 + 2[S(1) + S(-2) - S(1, 1) - S(1, -2) - S(-2, -2)].$$

(f) Let $E(1, 2)$ equal the number of segments, excluding those on the border, that have one end labeled 1 and the other 2. Let $T(j, k, l)$ equal the number of triangles whose corners are labeled $j$, $k$, $l$. Why, for instance, does $T(1, -1, 2) = 0$? Placing pebbles in each triangle along an edge whose ends are labeled 1 and 2, show that

$$S(1, 2) + 2E(1, 2) + S(-1, -2) = 2T(1, 1, 2) + 2T(1, 2, 2).$$

(g) From (f) deduce that $S(1, 2) + S(-1, -2)$ is even; from (e) deduce that $S(1, 2) + S(-1, -2)$ is odd.

34. Does the proof given in E 33 extend to rectangular arrays?

## References

1. *P. S. Aleksandrov, *Combinatorial Topology*, vol. 1, Graylock Press, Rochester, 1956. (Chapter 5 is devoted to Sperner's lemma and its applications.)

2. A. W. Tucker and H. S. Bailey, Topology, *Scientific American*, 1950, pp. 8–24. (This is a readable brief introduction to several topics in topology.)

3. *A. W. Tucker, Some topological properties of disk and sphere, *Proc. First Canadian Mathematical Congress*, Univ. Toronto, 1946, pp. 285–309. (This is a technical discussion of the relation between Tucker's theorem, described in E 33, and various topological theorems.)

*Chapter* **3**

# QUESTIONS ON WEIGHING

We will raise some important questions in this chapter but not answer them until we have acquired the mathematical tools to be developed in Chapter 5. Our goal is to provide a down-to-earth setting for the theory in Chapter 5, to refresh our arithmetic, and to develop some ease in working in the language of algebra.

The questions concern weighing; we introduce them by a few examples. Say that we have a two-pan scale of the type seen in chemistry labs and statues of "Justice":

Furthermore, we have an unlimited supply of 5- and 7-ounce measures. Now, supposing that potatoes weigh only whole numbers of ounces, rather than any amount as they actually do, let us ask which potatoes we would be able to weigh with our balance and our two types of measures.

For instance, using only the 5-ounce measures, we can weigh 5, 10, 15, 20, 25, 30, 35, . . . ounces. Or using only the 7-ounce measures, we can weigh 7, 14, 21, 28, 35, . . . ounces. Moreover, we could place one of each type together on a pan:

Thus we can weigh 12 ounces, $12 = 5 + 7$. Or we could put one of each type of weight alone on a pan:

In this way we can weigh 2 ounces, since a potato of this weight, together with the 5-ounce measure, balances the 7-ounce measure.

Can we weigh a 3-ounce potato? Yes, by placing two 5-ounce measures on one pan and a 7-ounce measure with the potato:

The balancing records the equation $3 + 7 = 2 \cdot 5$.

Can we weigh a 4-ounce potato? Yes. For instance, by placing two 5-ounce measures with the potato and two 7-ounce measures on the other pan:

The corresponding equation is

$$4 + 2 \cdot 5 = 2 \cdot 7.$$

Or we could place three 7-ounce measures with the potato, which then balance five 5-ounce measures. The equation in this case is $4 + 3 \cdot 7 = 5 \cdot 5$.

Can we weigh a 1-ounce potato? Even this can be done, as the reader may prefer to work out for himself *before* reading the next sentence. Two 7-ounce measures and the potato on one pan balance three 5-ounce measures on the other pan. The reader may wish to pause and devise still other ways of measuring this 1-ounce potato with 5- and 7-ounce measures.

Once we know that we can weigh a 1-ounce potato, then we know that we can weigh any number of ounces. For instance, we can weigh a 6-ounce potato as follows. First recall that two 7-ounce measures and a 1-ounce potato balance three 5-ounce measures:

Repeating this arrangement of measures six-fold

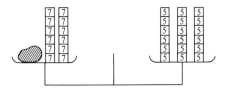

we weigh a six-ounce potato. That is, from the relation $1 + 2 \cdot 7 = 3 \cdot 5$ we conclude that $6(1 + 2 \cdot 7) = 6(3 \cdot 5)$ or

$$6 + 12 \cdot 7 = 18 \cdot 5.$$

Of course, this may not be the simplest method for weighing a 6-ounce potato. Indeed, $6 + 3 \cdot 5 = 3 \cdot 7$, so we could have managed by placing three 5-ounce measures with the potato and three 7-ounce measures on the other pan. But the reasoning at least assures us that if we can weigh a 1-ounce potato with a supply of two types of measures, then we can weigh any whole number of ounces with those measures.

So much for the combination 5 and 7. Suppose we turn to another combination, 8 and 21. Even if we have only 8-ounce and 21-ounce measures available, we can measure a 1-ounce potato, since

$$1 + 3 \cdot 21 = 8 \cdot 8.$$

Eight 8-ounce measures on one pan will balance the potato and three 21-ounce measures on the other pan.

The reader may now suspect that perhaps any pair of measures can weigh a 1-ounce potato. But this is not so. If, for example, we have only 6-ounce and 8-ounce measures, then we could never hope to measure a 1-ounce potato, or, for that matter, any odd number of ounces. (The reader should pause to think about why this is so.)

We are now in a position to ask some basic questions. Suppose we have at our disposal an unlimited supply of measures of two types. How can we decide whether we can weigh a 1-ounce potato with them? For instance, can we use 539-ounce and 1619-ounce measures to weigh a 1-ounce potato? More generally we can ask: *What potatoes can we weigh with an unlimited supply of two given types of measures?* Recall that we assume all weights in question are natural numbers and that the potato weighs a whole number of ounces.

The question really concern numbers, not potatoes. Let us gradually translate the second question into the language of numbers: Denote the weights of the two measures by $A$ and $B$ ounces respectively. In our first combination we had $A = 5$ and $B = 7$. The weight of the potato will be denoted by $W$ ounces.

There are several methods of weighing the potato. One consists in putting several $A$-ounce measures on the pan with the potato and several $B$-ounce measures on the other pan. How many of each will depend on $W$, $A$, $B$, and our arithmetic. Say that we use $X$ of the $A$-ounce measures and $Y$ of the $B$-ounce measures:

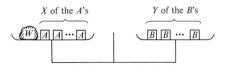

The corresponding equation is

$$W + XA = YB,$$

an equation that asserts merely that the scale in the figure balances. (We omit the multiplication sign between letters.)

For $A = 5$ and $B = 7$, let us see what $X$ and $Y$ are for various $W$. When $W = 1$, for instance, we have, $1 + 4 \cdot 5 = 3 \cdot 7$. Here $X = 4$ and $Y = 3$. Also $1 + 11 \cdot 5 = 8 \cdot 7$, so that $X = 11$ and $Y = 8$ also perform the weighing when $W = 1$.

Another manner of weighing consists in placing only *B*-ounce measures with the potato and only *A*-ounce measures on the other pan. For $A = 5, B = 7$, the equation $1 + 2 \cdot 7 = 3 \cdot 5$ illustrates this. We have $W + XB = YA$, where $W = 1, X = 2, Y = 3$.

A third method consists in placing the potato on one pan and the measuring weights on the other pan. For example, if $W = 12$ we have $12 = 5 + 7$; if $W = 27$ we have $27 = 4 \cdot 5 + 1 \cdot 7$. This method corresponds to an equation of the type $W = XA + YB$.

No other practical method exists, for there would be no point in placing measures of equal weight on *both pans*, since they could be removed without affecting the balance. Thus there are only three types of equations we need to consider:

$$W + XA = YB, \qquad W + XB = YA, \qquad W = XA + YB.$$

In all of these, $X$ and $Y$ are to be natural numbers, possibly including zero. Our question about potatoes now becomes one about natural numbers: Let $A$ and $B$ be natural numbers. For which natural numbers $W$ can we find natural numbers $X$ and $Y$ such that at least one of these equations holds:

$$W + XA = YB, \qquad W + XB = YA, \qquad W = XA + YB?$$

The potatoes are gone, but we are left with three equations to deal with. We can simplify matters further (reducing the three equations to one) by making use of the negative numbers, $-1, -2, -3, -4, -5, \ldots$, which lie to the left of $0$ on the number line. (See Appendix A.)

With the aid of the negative numbers, we will reduce the first two equations, $W + XA = YB$ and $W + XB = YA$, to the third type, $W = XA + YB$. Consider first, $W + XA = YB$. This can be rewritten as $W = (-X)A + YB$, which is of the third type. (For instance, we may rewrite $1 + 4 \cdot 5 = 3 \cdot 7$ as $1 = (-4)5 + 3 \cdot 7$.) Similarly, $W + XB = YA$, where $X$ and $Y$ are natural numbers, can be reduced to the third form by writing it as $W = YA + (-X)B$.

Our question about potatoes now reduces to this: *Let $A$ and $B$ be natural numbers. Which natural numbers $W$ can be expressed in the form $W = MA + NB$ for certain integers $M$ and $N$?*

Let us see which values of $M$ and $N$ describe our earlier work for the combination $A = 5$ and $B = 7$. The following table records the cases we considered, beginning with 5, 10, 15, . . . ; 7, 14, 21, . . . .

SOME WEIGHINGS WITH 5's AND 7's

| $W$ | Weighing | Algebraic Representation as $MA + NB$ | $M$ and $N$ |
|---|---|---|---|
| 5 | $5 = 1 \cdot 5$ | $5 = 1 \cdot 5 + 0 \cdot 7$ | $M = 1, \quad N = 0$ |
| 10 | $10 = 2 \cdot 5$ | $10 = 2 \cdot 5 + 0 \cdot 7$ | $M = 2, \quad N = 0$ |
| 7 | $7 = 1 \cdot 7$ | $7 = 0 \cdot 5 + 1 \cdot 7$ | $M = 0, \quad N = 1$ |
| 14 | $14 = 2 \cdot 7$ | $14 = 0 \cdot 5 + 2 \cdot 7$ | $M = 0, \quad N = 2$ |
| 12 | $12 = 5 + 7$ | $12 = 1 \cdot 5 + 1 \cdot 7$ | $M = 1, \quad N = 1$ |
| 2 | $2 + 5 = 7$ | $2 = (-1)5 + 1 \cdot 7$ | $M = -1, N = 1$ |
| 3 | $3 + 7 = 2 \cdot 5$ | $3 = 2 \cdot 5 + (-1)7$ | $M = 2, \quad N = -1$ |
| 4 | $4 + 2 \cdot 5 = 2 \cdot 7$ | $4 = (-2)5 + 2 \cdot 7$ | $M = -2, N = 2$ |
| 1 | $1 + 2 \cdot 7 = 3 \cdot 5$ | $1 = 3 \cdot 5 + (-2)7$ | $M = 3, \quad N = -2$ |

The row for "3", for instance, tells us

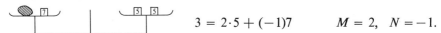

$$3 = 2 \cdot 5 + (-1)7 \qquad M = 2, \quad N = -1.$$

Just as a potato may be weighed in more than one way, so are $M$ and $N$ not necessarily unique. For the case $W = 6$ we had the two weighings

$$6 + 12 \cdot 7 = 18 \cdot 5 \qquad (6 = 18 \cdot 5 + (-12)7)$$

and

$$3 \cdot 5 + 6 = 3 \cdot 7 \qquad (6 = (-3)5 + 3 \cdot 7).$$

A little arithmetic produces still more representations of 6; for instance,

$$6 = 4 \cdot 5 + (-2)7; \qquad 6 = 11 \cdot 5 + (-7)7; \qquad 6 = (-10)5 + 8 \cdot 7.$$

Our first question, concerning the possibility of weighing a 1-ounce potato, now reads: *For which pairs of natural numbers $A$ and $B$ can we find integers $M$ and $N$ such that $1 = MA + NB$?*

Take the combination $A = 24$, $B = 73$. Since $3 \cdot 24 = 72$, we have $1 = (-3)24 + 1 \cdot 73$. Thus for the combination 24 and 73 $M$ and $N$ exist. But what about the combination 24 and 75? Can they measure 1? What about 21 and 34? What about 89 and 233? And, we may wonder: If a pair of measures can't measure a 1-ounce potato, what is the smallest positive weight they can measure?

There is another way of looking at these questions. Recall the representation $1 = 8 \cdot 8 + (-3)21$. The basis of this is that $8 \cdot 8$ differs from $3 \cdot 21$ by just 1. We may have found this representation by listing $1 \cdot 8 = 8$, $2 \cdot 8 = 16$, $3 \cdot 8 = 24$, $4 \cdot 8 = 32$, $5 \cdot 8 = 40$, $6 \cdot 8 = 48$, $7 \cdot 8 = 56$, $8 \cdot 8 = 64$, . . . and $1 \cdot 21 = 21$, $2 \cdot 21 = 42$, $3 \cdot 21 = 63$, . . . until we found an entry in one list that differed by 1 from an entry in the other list.

The numbers of the form "integer times 8" we will call the *multiples* of 8. The *multiples* of 21, or of any integer, are defined similarly. This illustration shows some of the multiples of 8 and some of the multiples of 21:

Thus the question "Can $A$ and $B$ measure 1?" can be phrased in terms of multiples: "Are there multiples of $A$ and of $B$ that differ by exactly 1?" Since the multiples of a number are regularly spaced on the number line, the question can also be interpreted geometrically: If we have an unmarked ruler $A$ inches long, and another $B$ inches long, can we measure a distance of one inch?

We have not obtained an answer to any of these general questions. The answers will come in Chapter 5, where the equation $W = MA + NB$ will turn out to be our basic tool for establishing fundamental properties of the natural numbers.

### Exercises

1. Is every natural number also an integer? Is 0 a positive integer? Is 0 a negative integer? Is every positive integer also a natural number?

2. Drawing pictures of the two-pan scale, show three different methods of weighing a 1-ounce potato with 4- and 7-ounce measures.

3. Drawing pictures of the two-pan scale, show three different methods of weighing a 1-ounce potato with 8- and 13-ounce measures.

4. What do you think is the smallest weight of a potato that can be measured with (a) 8- and 12-ounce measures, (b) 8- and 11-ounce measures, (c) 9- and 12-ounce measures? Why do you think so?

5. A 4-ounce potato and five 7-ounce measures balance three 13-ounce measures. Express this in the form (a) $W + XA = YB$, where all letters stand for natural numbers, (b) $W = MA + NB$, where $M$ and $N$ are integers.

6. Are there integers $M$ and $N$ such that (a) $1 = M \cdot 7 + N \cdot 10$, (b) $1 = M \cdot 8 + N \cdot 10$, (c) $1 = M \cdot 13 + N \cdot 22$, (d) $1 = M \cdot 6 + N \cdot 21$? If your answer is "yes," give values of $M$ and $N$; if "no," explain why.

7. (a) Show how to weigh a 1-ounce potato with 11-ounce and 18-ounce measures.
   (b) Use (a) to find a method of weighing a 3-ounce potato with 11-ounce and 18-ounce measures.

8. Draw the balanced loading of the two-pan scale that corresponds to each of these equations: (a) $3 + 6 \cdot 7 = 5 \cdot 9$, (b) $5 = 3 \cdot 11 + (-4)7$, (c) $23 = 2 \cdot 4 + 3 \cdot 5$.

9. Can a 1-ounce potato be weighed with 5- and 8-ounce measures in such a way that (a) the potato is with the 5's, and the 8's are in the other pan; (b) the potato is with the 8's, and the 5's are in the other pan?

10. Can a collection of 5-ounce measures on one pan ever balance a collection of 8-ounce measures on the other pan?

11. The products $5 \cdot 10$ and $3 \cdot 17$ differ by 1. Draw four different balanced loadings of the scales for weighing a 1-ounce potato that reflect this arithmetic fact. (The measures may be 5, 10, 3, or 17.)

12. Find integers $M$ and $N$, if you think they exist, such that
    (a) $1 = M \cdot 4 + N \cdot 9$,
    (b) $1 = M \cdot 9 + N \cdot 11$,
    (c) $1 = M \cdot 10 + N \cdot 15$,
    (d) $1 = M \cdot 23 + N \cdot 25$.

13. (a) Draw on a number line several multiples of 7 and several multiples of 12.
    (b) Do the multiples of 7 meet the multiples of 12 anywhere other than at 0?
    (c) What is the closest that a multiple of 7 can be to a multiple of 12 without actually coinciding with it?

14. Consider 5- and 12-ounce measures. Show that it is possible without using more than four 12-ounce measures to weigh (a) a 1-ounce potato, (b) a 2-ounce potato, (c) a 3-ounce potato, (d) a 4-ounce potato, (e) a 5-ounce potato, (f) a 6-ounce potato, (g) a 7-ounce potato. Will this be true for heavier potatoes?

15. (a) Find a multiple of 9 that differs from a multiple of 11 by 1.
    (b) Show them on the number line.

16. List (a) four multiples of 3, (b) five multiples of 2, (c) six multiples of 7.

17. (a) Show that 1 can be weighed with 8's and 11's.
    (b) Give some small values of $W$ that can be weighed with 16's and 22's.
    (c) Give some small values of $W$ that can be weighed with 24's and 33's.

18. Find four pairs of integers $M$ and $N$ such that (a) $12 = M \cdot 2 + N \cdot 5$, (b) $5 = M \cdot 2 + N \cdot 5$, (c) $13 = M \cdot 2 + N \cdot 5$.

19. What potatoes can be weighed with 4- and 7-ounce measures, if we never use more than three 7-ounce measures?

●

20. Consider weighings by 3- and 7-ounce measures. Draw a picture of balanced scales recording the equality $3 \cdot 7 = 7 \cdot 3$. (a) Show that if a potato can be weighed by placing it on the pan with 3-ounce measures it can also be weighed by placing it on the pan with 7-ounce measures. (b) Is this true more generally than for just 3's and 7's?

21. Find integers $M$ and $N$ such that $1 = M \cdot 5 + N \cdot 9$ and (a) $M$ is positive, $N$ negative; (b) $M$ is negative, $N$ positive. (c) Draw the corresponding loadings of the scales.

22. Find integers $M$ and $N$ such that $13 = M \cdot 4 + N \cdot 5$ and (a) $M$ and $N$ are positive; (b) $M$ is positive, $N$ negative; (c) $M$ is negative, $N$ positive. (d) Draw the corresponding loadings of the scales.

23. (a) What amounts of postage can you make with 5¢ stamps and 8¢ stamps?
    (b) What type of weighing problem would be equivalent to the postage problem raised in (a)?

(c) What requirement does the postage problem impose on $M$ and $N$ if the postage, $W$, equals $M \cdot 5 + N \cdot 8$?

24. What amounts of postage can be made with 4¢ and 7¢ stamps?

25. What amounts of postage can be made with 4¢ and 9¢ stamps?

26. What amounts of postage can be made if you are allowed to use at most one 1¢ stamp, at most one 2¢ stamp, at most one 4¢ stamp, at most one 8¢ stamp, at most one 16¢ stamp, and so on (each denomination being a product of 2's)?

27. Imagine that you have a two-pan scale and a supply of measured weights: one 1-ounce weight, one 3-ounce weight, one 9-ounce weight, one 27-ounce weight, and so on (each measure being a product of 3's). What weights can you measure?

28. You have nine 1¢ stamps, nine 10¢ stamps, and nine 100¢ stamps. What postages can you make?

29. (a) You have four 1¢ stamps, four 5¢ stamps, and four 25¢ stamps. What postages can you make?
    (b) You have four pennies, four nickels, four quarters. What amounts can you make from them?

● ●

30. You have two unmarked jugs, one holding 5 quarts, and the other 7 quarts. You walk down to the river and hope to come back with precisely one quart of water. How could you manage to do it?

31. See E 30. (a) If you had, instead, a 4-quart and a 6-quart jug, could you bring back exactly 1 quart? (b) Could you do it if you had a 7-quart and an 11-quart jug?

32. Devise more problems like E 30, and develop a conjecture as to when an $A$-quart and a $B$-quart jug can measure exactly one quart of water.

33. See E 30. Can you measure 1 quart if you (a) never pour water from the 7-quart jug into the 5-quart jug, (b) never pour water from the 5-quart jug into the 7-quart jug?

34. See E 30. What possible amounts of water can you bring back from the river in a single trip with the two jugs mentioned? (Surely not more than 12 quarts, since that is all the jugs hold.)

35. Our coinage consists of pennies, nickels, dimes, quarters, and half-dollars. Consider a country that has only two types of coins called "nickels" and "luckies," the nickels being worth 5 cents and the luckies being worth 7 cents.
    (a) If a customer has only "luckies" and the clerk has only nickels, can change be made for a one-cent purchase?
    (b) If a customer has only nickels and the clerk has only "luckies," can change be made for a one-cent purchase?
    (c) For what purchases can change be made?

36. Like E 35 except that the two types are "luckies" and "unluckies," one "unlucky" being worth 13 cents.

37. Like E 35 except that the two coins are dimes and "unluckies."

38. Like E 35 except that the two coins have the values 15¢ and 36¢.

39. See E 35. Examine the problem of making change in a country that has coins worth 6¢, 7¢, and 8¢.

40. (a) A banana costs 17 cents and an orange 13 cents. John trades some bananas for Bill's oranges. What is the smallest number of bananas involved?
    (b) John trades some bananas and one cent for some of Bill's oranges. What is the smallest number of bananas involved?

## Chapter 4

# THE PRIMES

The even numbers are those natural numbers that can be expressed as the sum of 2's; natural numbers that are not even are odd. The duality of even and odd rests, then, on the number 2. But 2 is important also as being the first of the primes. To these natural numbers, the object of study for well over two thousand years, we now turn.

In order to define the primes, we shall need the notion of the divisors of a natural number. *A natural number A* **divides** *a natural number B if there is a natural number Q such that B is the product of Q and A,*

$$B = QA.$$

*If A divides B, we call A a* **divisor** *of B and B a* **multiple** *of A.* Thus 6 has four divisors, 1, 2, 3, and 6; and the multiples of 2 are precisely the even numbers. Any natural number $A$ is a divisor of 0, since $0 = 0 \times A$. The only natural number with precisely one divisor is the number 1. Any other natural number has at least two divisors, namely itself and 1. *The natural numbers with only two divisors are called the* **primes.** The reader may check that the first twenty primes are

2, 3, 5, 7, 11, 13, 17, 19, 23, 29, 31, 37, 41, 43, 47, 53, 59, 61, 67, 71,

and may easily extend this brief list. A prime can also be described as a natural number (larger than 1) that is not the product of two smaller natural numbers.

As we compute more and more primes, various questions come to mind. How many primes are there? Calling two primes *adjacent* if there is no prime between them, we may ask, "How far apart can adjacent primes be?" (In the above list the largest gap between adjacent primes is 6.) How many pairs of primes differ by 1 (for example, 2 and 3)? How many pairs of primes differ by 2 (for example, 3 and 5, 5 and 7, 11 and 13)? How many pairs of primes differ by 3 (for example, 2 and 5)? How many pairs of primes differ by 4 (for example, 3 and 7, 7 and 11)? Some of these questions can be answered easily, but others are so hard that nobody knows their answer. Let us take them up one by one.

Our first question—"How many primes are there"?—was answered by the

Greeks some 2300 years ago. Their solution, which appears as Proposition 20 in Book 9 of Euclid, is based on their *Prime-manufacturing Machine*. When anyone feeds primes into this machine, it spews out brand new primes—primes that are different from those fed into it. Let us first watch it in operation; we will explain afterward why it always works.

Let us feed the machine the primes 3, 5, and 11, which we have chosen at random. The machine multiplies them all together, getting 165. Then it adds 1 to this result, getting 166. Then it produces all the primes that divide 166; namely, 2 and 83. We see that 2 and 83 are indeed brand new, being different from the three primes we fed the machine.

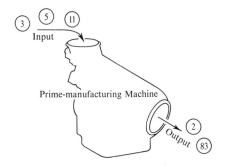

The reader may check that when he feeds the machine the primes 2, 3, and 5 it produces just the one new prime, 31. That the primes manufactured by the machine are indeed different from those fed into it is the content of Theorem 1, which depends on the first part of this

LEMMA. *If a natural number D divides each of two natural numbers A and B, then D divides their difference, A − B, and their sum, A + B.*

PROOF. By the definition of "divides," we know that there are integers $Q_1$ and $Q_2$ such that

$$A = Q_1 D \quad \text{and} \quad B = Q_2 D.$$

Thus

$$A - B = Q_1 D - Q_2 D = (Q_1 - Q_2)D.$$

Since $Q_1 - Q_2$ is an integer and $A - B = (Q_1 - Q_2)D$, we conclude that $D$ divides $A - B$. (A similar argument shows that $D$ divides $A + B$.)

THEOREM 1. *Any prime dividing the natural number that is 1 larger than the product of several primes is different from any of those primes.*

PROOF. Let us, like Adam naming the creatures, give names to the several primes in question so that we can speak of them. If we called them $A$, $B$,

$C, \ldots$, then we would limit ourselves to feeding the machine at most twenty-six primes. We can avoid this artificial limit by naming the first prime we feed the machine $F_1$ (the letter $F$ standing for Feed), the second prime we feed the machine $F_2$, and so on. When we feed the machine twenty-seven primes we name them

$$F_1, F_2, \ldots, F_{27}.$$

Now let us feed the machine $N$ primes,

$$F_1, F_2, \ldots, F_N.$$

The machine forms their product $F_1 F_2 \cdots F_N$ and then adds 1 to this product to obtain the natural number

$$1 + F_1 F_2 \cdots F_N,$$

which we will call $M$. Thus

$$M = 1 + F_1 F_2 \cdots F_N.$$

We must prove that none of the primes $F_1, F_2, \ldots, F_N$ divides $M$. To do this, we first observe that no natural number larger than 1 can divide two natural numbers that differ by 1, for as the lemma shows, when a natural number divides two natural numbers it also divides their difference. Now, each of the $F_1, F_2, \ldots, F_N$ clearly divides the product $F_1 F_2 \cdots F_N$. Hence none of the $F$'s can divide the natural number $M$, which is one larger than $F_1 F_2 \cdots F_N$. Thus the machine produces only new primes. Our proof is complete.

The wary reader might say, "Almost complete. After all, how can we be sure that $M$ always has a prime divisor?" The answer to this question is not hard. Think of the *smallest* divisor of $M$, *other than* the number 1. This smallest divisor must be a prime, for any divisor of it is also a divisor of $M$.

With the help of the machine, we can now provide an answer to "How many primes are there?" in the form of

THEOREM 2. *There is no end to the primes.*

PROOF. Let us assume that there is an end to the primes. Then we can make a list of them and denote them by $F_1, F_2, \ldots, F_N$, where $F_N$ is the largest prime. We then feed this list of all the primes into the machine. The machine, operating on them as it does with any list of primes, spews out primes different from all we fed into it. But our list was supposed to be a complete inventory of all the primes. Our assumption must be mistaken, for we have reached a contradiction—a list that is both complete and incomplete. Therefore, there is no end to the primes. This proves Theorem 2.

The proof itself deserves some comment. It is an illustration of a frequently used technique in mathematics, called "proof by contradiction." Rather than proving an assertion directly we show that the denial of the assertion leads to nonsense. We used this same technique in proving that certain Fifteen puzzle problems have no solution. We assumed that they did have a solution and deduced from this assumption that a certain number must be both odd and even, which is surely nonsense.

It is also interesting to note that, though our definition of prime involves only the notion of multiplication, our proof that there is no end to the primes also involves addition (we added 1). The reason for this is that, although "divisor" was defined only in terms of multiplication, it happens to behave nicely with respect to addition: a divisor of two natural numbers is also a divisor of their sum and of their difference.

We have already met another proof like that of Theorem 2, similar in that it goes outside of the structure in which the definitions are made and the theorems stated. In the weaver's problem we went outside of odd and even and used the idea of one number being larger or smaller than another. We will meet more proofs like these in Chapters 5 and 6.

The machine helped us answer our first question. Now let us turn to our second. "How far apart can adjacent primes be?" We answer this with

THEOREM 3. *There are gaps as big as we please between adjacent primes.*

PROOF. We illustrate the idea by proving that there are 999 consecutive natural numbers, none of which is prime. To do this, we first multiply together all the natural numbers from 1 through 1000. Call their product 1000 factorial (abbreviated 1000!). Then we will show that none of the following 999 consecutive natural numbers is prime:

$$1000! + 2, \qquad 1000! + 3, \qquad 1000! + 4, \qquad \ldots, \qquad 1000! + 1000.$$

The first is divisible by 2, since 2 divides both 2 and 1000!. The second is divisible by 3, the third by 4, and so on. Thus we have produced 999 consecutive natural numbers, none of which is prime. The same idea can, of course, be used to produce larger gaps between primes, and our proof is done.

It would seem more natural to have started with 1000! instead of 1000! + 2, but 1000! + 1 might be prime. After all, 3! + 1 is prime. This raises another question: How many natural numbers of the form $N! + 1$ are prime? No one knows. Nor is it known whether 1000! + 1 is prime.

But let us return to the next question we asked at the beginning of the chapter, "How many pairs of primes differ by 1?" This is an easy one. Clearly, 2 is the only even prime. Any two other primes, being odd, differ by an even number. Thus, 2 and 3 are the only primes differing by 1. Similarly, 2 and 5 are the only primes differing by 3. And going on, we see that 2 and 7 are the only primes differing by 5 and that there are no primes differing by 7.

But what about pairs of primes differing by 2? Clearly the two primes in such a pair must be odd. Such pairs, called "twin primes," have been the object of some of the deepest research in number theory for half a century. Yet nobody knows whether they end. A long list of primes suggests that they go on and on, but become rarer and rarer as we go farther in the list. Primes differing by 4 have also concealed their secrets well, though they have yielded some clues. For instance, they seem to be about as frequent as twin primes.

The study of the gaps between primes is intimately connected with the way in which the primes increase in size as we go farther and farther in the list of primes. If the gaps tend to grow larger and larger, then the primes grow more quickly; and we have seen that there are gaps as large as we please between primes. But if the small gaps, like 2 or 4, occurred very frequently, then the primes would not grow very quickly. Inspection of our own table of primes shows that the gaps fluctuate quite chaotically.

Yet even within chaos, mathematicians can discover order. We might even say that when things become totally chaotic order reappears. When a penny is flipped with absolute randomness, we find in the midst of that chaos this law: in the long run heads will appear half the time. In the midst of the chaos of primes there is order. In the long run, the gaps between the primes also exhibit a certain orderliness.

To discuss how the primes grow as we go farther and farther in the list, let us introduce a notation for them. Let us call the first prime $P_1$; the second prime, $P_2$; and so on. Thus, for example, $P_{10}$, the *tenth* prime, is 29. Since the gap between primes is always at least 1, we can say that the $N$th prime, $P_N$, is at least as large as $N$. But we could be more precise. Since the gaps are at least 2 (except between 2 and 3) once we go beyond the fourth prime, 7, we can say that $P_N$ is at least as large as $N \times 2$. As a matter of fact, $P_{10}$ is 29, which is $10 \times 2.9$. Now, $P_{10}$ can be thought of as the sum of ten gaps: from 0 to $P_1$, from $P_1$ to $P_2$, from $P_2$ to $P_3$, . . . , from $P_9$ to $P_{10}$. Hence we can say that the average of the first ten gaps is 2.9, since $P_{10} = 10 \times 2.9$; moreover, we see the intimate relation between the average gap up to the $N$th prime and the size of the $N$th prime. If we know one, we know the other, for $P_N$ is simply $N$ times the average of the first $N$ gaps. By the time we reach the 20th prime, 71, the average gap has grown to 3.55, since we have $P_{20} = 71 = 20 \times 3.55$. Thus, any estimate we have of the average of the first $N$ gaps gives us a way of estimating $P_N$ without our having to list all the primes from $P_1$ to $P_N$.

In 1791 Gauss noticed by inspection of tables of primes that, though the gaps behave wildly, the average gap grows slowly and is closely approximated by this formula: the average of the first $N$ gaps is approximately

$$\frac{1}{1} + \frac{1}{2} + \frac{1}{3} + \frac{1}{4} + \cdots + \frac{1}{N}.$$

He conjectured but did not prove

THE PRIME NUMBER THEOREM. $P_N$ *is approximately*

$$N \times \left( \frac{1}{1} + \frac{1}{2} + \frac{1}{3} + \cdots + \frac{1}{N} \right).$$

*That is, the difference between $P_N$ and $N \times \left( \frac{1}{1} + \frac{1}{2} + \frac{1}{3} + \cdots + \frac{1}{N} \right)$ is small in comparison with $P_N$ when $N$ is large.*

The accuracy of Gauss's estimate of $P_N$ is shown by this table:

| $N$ | $\left( \frac{1}{1} + \frac{1}{2} + \frac{1}{3} + \cdots + \frac{1}{N} \right)$ | $N \times \left( \frac{1}{1} + \frac{1}{2} + \frac{1}{3} + \cdots + \frac{1}{N} \right)$ | $P_N$ |
|---|---|---|---|
| 1 | 1 | 1 | 2 |
| 2 | 1.5 | 3 | 3 |
| 3 | $1.833\cdots$ | 5.5 | 5 |
| 4 | $2.083\cdots$ | 8.3 | 7 |
| 5 | $2.283\cdots$ | $11.4\cdots$ | 11 |
| 10 | $2.928\cdots$ | $29.2\cdots$ | 29 |
| 20 | $3.597\cdots$ | $71.9\cdots$ | 71 |
| 100 | $5.187\cdots$ | $518.7\cdots$ | 541 |
| 1000 | $7.485\cdots$ | 7485 | 7919 |

But it was not until 1896 that a proof was discovered for Gauss's conjecture. In that year Hadamard and de la Vallée Poussin, working independently of each other, proved the Prime Number Theorem. Their proofs made use of the complex numbers and the calculus. Finally, in 1948, Selberg and Erdös independently discovered a proof that requires neither the complex numbers nor the calculus. This came as a surprise to the mathematical world, which for over a century and a half had searched in vain for such a proof.

It can be shown that the Prime Number Theorem is equivalent to the assertion that *the number of primes less than the natural number $X$ is roughly*

$$X / \left( \frac{1}{1} + \frac{1}{2} + \frac{1}{3} + \cdots + \frac{1}{X} \right).$$

As $X$ increases, the numerator grows quickly, whereas the denominator grows very slowly. (See Appendix E for a discussion of the growth of the denominator.) Thus the quotient becomes very large, which tells us again that there is no end to the primes. From the very deep Prime Number Theorem we obtain a result for which we already had a simple proof.

The situation with twin primes is quite different. Although no one knows whether the twin primes end, there is much experimental evidence in favor of

THE TWIN PRIME CONJECTURE. *The number of pairs of twin primes less than the natural number X is roughly*

$$1.32X/\left(\frac{1}{1} + \frac{1}{2} + \frac{1}{3} + \cdots + \frac{1}{X}\right)^2.$$

This conjecture, made in 1923 by Hardy and Littlewood, would in particular imply that there is no end to the twin primes (for once again, the denominator is the tortoise, and the numerator is the hare). When he conjectured the Prime Number Theorem, Gauss had at least been assured by the Greeks that there is no end to the primes. It might turn out to be easier to prove the Twin Prime Conjecture than to prove that there is no end to the twin primes. We have met such a peculiar situation before, in Chapter 2, where it was easier to prove that there is an odd number of complete triangles than it was to prove that there is at least one. But the Twin Prime Conjecture, so tantalizing in its simplicity and so strongly supported by experimental evidence, remains unproved. (For more in this direction see R 4.)

The twin primes illustrate a very dependable recipe for inventing difficult questions about the primes: ask something about the relation between primes and addition. Since the primes are defined using only multiplication, it would seem fortuitous if they behaved decently with respect to addition. After all, as we saw with the Prime-manufacturing Machine, if we add 1 to a natural number we completely change its prime divisors. And the twin primes involve the addition of 2.

We must be especially grateful for any result that manages to relate the primes to addition. Let us cite, without proof, a theorem that was proved by Fermat in 1641: *Every prime that is one more than a multiple of 4 can be written as the sum of two squares in precisely one way.* (A "square" is a number of the form $N^2$ for some natural number $N$.) For example, 5 is prime, and is one more than 4, and can be written $1^2 + 2^2$. (We regard $2^2 + 1^2$ as the same representation.) Similarly, 13 is prime, and is one more than 12, a multiple of 4, and can be written $2^2 + 3^2$.

Fermat's theorem concerns writing primes in terms of other natural numbers. In the opposite direction, there are many questions concerning writing natural numbers in terms of primes. Before we can phrase some of these questions we will need the notion of a power of a natural number. *The* **powers** *of a natural number N are those natural numbers N, $N \cdot N$, $N \cdot N \cdot N$, . . . that are the product of N times itself several times.* The first few powers of 2, for example, are 2,

4, 8, 16; the first few powers of 3 are 3, 9, 27, 81. The number $N \cdot N$ is called the *square* of $N$; $N \cdot N \cdot N$ is called the *cube* of $N$. Rather than write the cumbersome $3 \cdot 3 \cdot 3 \cdot 3$, we write simply $3^4$; similarly, we write $2 \cdot 2 \cdot 2 \cdot 2 \cdot 2$ simply as $2^5$.

Now we can raise the questions mentioned in the preceding paragraph. Can every odd natural number larger than 3 be written as the sum of a prime and a power of 2? Can every odd natural number larger than 1 be written as the sum of a prime and twice a square? Can every even natural number larger than 2 be written as the sum of two primes?

A few computations suggest that the answer to all three is "yes." Examining the first question, for example, we have $5 = 3 + 2$, $7 = 5 + 2$, $9 = 7 + 2$, $11 = 7 + 4$, $77 = 13 + 64$, and so forth. If the reader extends the list he will observe that sometimes there may be more than one way of expressing a natural number as such a sum; for example, $7 = 5 + 2$, and $7 = 3 + 4$. Illustrating the second question we have $3 = 3 + 2(0)^2$, $5 = 3 + 2(1)^2$, $7 = 5 + 2(1)^2$, $9 = 7 + 2(1)^2$, $11 = 3 + 2(2)^2$, $57 = 7 + 2(5)^2$. For the third question we have $4 = 2 + 2$, $6 = 3 + 3$, $8 = 3 + 5$, $10 = 3 + 7$ or $5 + 5$, $100 = 89 + 11$, and so on.

These three questions, so similar in spirit, have not had the same fate. The first breaks down at 127—the reader is invited to check this; the second breaks down at 5777; but the third, in spite of the profound research it has inspired, still remains unanswered. Every even natural number that has been tested can be written as the sum of two primes, but no one has proved that every even natural number can be written that way. The second and third questions were raised by Goldbach, the second in 1752, the third in 1742. Clearly, it is not always possible to judge the depth of a question at first glance, though the brevity of the third has an immediate appeal that the first two lack.

Returning to the natural habitat of the primes, multiplication, we raise another question in the spirit of the last three: Can every natural number larger than 1 and not itself a prime be written as the product of primes? A list of experiments begins as follows:

| Natural Number | Remark | Natural Number | Remark |
|---|---|---|---|
| 2 | prime | 12 | $2 \cdot 2 \cdot 3$ |
| 3 | prime | 13 | prime |
| 4 | $2 \cdot 2$ | 14 | $2 \cdot 7$ |
| 5 | prime | 15 | $3 \cdot 5$ |
| 6 | $2 \cdot 3$ | 16 | $2 \cdot 2 \cdot 2 \cdot 2$ |
| 7 | prime | 17 | prime |
| 8 | $2 \cdot 2 \cdot 2$ | 18 | $2 \cdot 3 \cdot 3$ |
| 9 | $3 \cdot 3$ | 19 | prime |
| 10 | $2 \cdot 5$ | 20 | $2 \cdot 2 \cdot 5$ |
| 11 | prime | | |

*Calling the natural numbers larger than 1 and not prime* **composite,** we see that all composite natural numbers up to 20 are indeed the product of primes. Jumping ahead, let us see if 100 is the product of primes. Writing first $100 = 10 \cdot 10$ and then using the fact that $10 = 2 \cdot 5$, we have $100 = 2 \cdot 5 \cdot 2 \cdot 5$. This computation suggests a way of proving

THEOREM 4. *Every composite natural number is the product of primes.*

PROOF. If $N$ is composite, it can be written as the product of two smaller natural numbers. If those two natural numbers are prime, we have proved the theorem. If either of the two is composite, it is in turn the product of smaller natural numbers. Continuing this process until we meet only primes, we eventually have $N$ written as the product of primes.

Theorem 4 indicates how the primes earned their name. They can be thought of as the primitive bricks out of which all natural numbers can be built by multiplication, yet they themselves cannot be further reduced by multiplication. The primes are the atoms of multiplication, just as 1 is the only atom of addition, each natural number being a sum of 1's.

The reader, in checking the computations through 20, will observe that each composite natural number through 20 can be written in only one way as the product of primes. Of course, $6 = 2 \cdot 3 = 3 \cdot 2$, and so, strictly speaking, 6 can be written as a product of primes in more than one way. To avoid this trivial and awkward detail, let us agree that from now on, when we write a natural number as the product of primes, we will record the primes with their size increasing from left to right. For example, we will write $6 = 2 \cdot 3$ or $100 = 2 \cdot 2 \cdot 5 \cdot 5$.

Let us see what happens with a large natural number, well beyond our list. Say that we want to write 2000 as the product of primes. We might notice first that $2000 = 20 \cdot 100$; then $20 = 4 \cdot 5$, and $100 = 10 \cdot 10$. But $4 = 2 \cdot 2$, and $10 = 2 \cdot 5$. Thus $2000 = 2 \cdot 2 \cdot 2 \cdot 2 \cdot 5 \cdot 5 \cdot 5$. Or we might have noticed first that $2000 = 40 \cdot 50$, and then written $40 = 5 \cdot 8$ and $50 = 5 \cdot 10$. But $8 = 2 \cdot 2 \cdot 2$ and $10 = 2 \cdot 5$. Assembling these computations, we have $2000 = 2 \cdot 2 \cdot 2 \cdot 2 \cdot 5 \cdot 5 \cdot 5$. Both times we end up with four 2's and three 5's in the expression of 2000 as the product of primes. The reader is encouraged to carry out his own experiments on the number of ways he can write a composite natural number as the product of primes.

Recall that an even natural number can sometimes be written as the sum of two primes in different ways; for example, $10 = 3 + 7 = 5 + 5$. It certainly is not obvious that the situation should be any different when it concerns natural numbers written as the product of several primes.

Indeed, if the mathematicians of the great academy of Lagado were with us today, they would be shocked if we were to think, even for a moment, that all natural numbers could be written as the product of primes in only one way. In case you have forgotten, this was the academy where, according to Lemuel Gulliver, the teaching of mathematics was attempted in the following manner: "The proposition and demonstration were fairly written on a thin wafer, with ink composed of cephalic tincture. This the student was to swallow upon a fasting stomach, and for three days following eat nothing but bread and water. As the wafer digested, the tincture mounted to the brain, bearing the proposition with it."

Gulliver, for lack of space, failed to report that the King of Laputa, under whom the academy flourished, had noticed that 1 is good, for there is one sun, 4 is good, for there are four seasons, 7 is good, for there are seven days in the week, and 10 is good, for he had ten fingers. His mathematicians replied, "Yes, your majesty, we understand. All natural numbers obtained from 1 by adding a multiple of 3 are good." "A profound insight," commented the King, "hence all other natural numbers must be bad, and I hereby exile them from my kingdom." "But," pleaded the horrified mathematicians, "can't we at least use the bad natural numbers to write down the digits of the good? Otherwise, we can't even speak of 13, which is good, though 3 is bad." "Henceforth," decreed the King, "the bad natural numbers can be used only in the notation for the good ones." Thus, in Lagado, they too spoke of prime and composite natural numbers. A good natural number is a Lagado prime if it has exactly two good divisors. In the Lagado system, 4 is prime (recall that 2 was exiled), and 10 is prime (for 2 and 5 were exiled). But 16, a good natural number, is composite, even in the Lagado system, for $16 = 4 \cdot 4$. In fact, 16 can be written as the product of Lagado primes in precisely one way. The professors of Lagado made this table of all the ways in which the Lagado natural numbers up to 100 could be written as the product of Lagado primes.

| Lagado Natural Numbers | Remarks | Lagado Natural Numbers | Remarks |
|---|---|---|---|
| 1 | Neither prime nor composite | 28 | $4 \cdot 7$ |
| 4 | Lagado prime | 31 | Lagado prime |
| 7 | Lagado prime | 34 | Lagado prime |
| 10 | Lagado prime | 37 | Lagado prime |
| 13 | Lagado prime | 40 | $4 \cdot 10$ |
| 16 | $4 \cdot 4$ | 43 | Lagado prime |
| 19 | Lagado prime | 46 | Lagado prime |
| 22 | Lagado prime | 49 | $7 \cdot 7$ |
| 25 | Lagado prime | 52 | $4 \cdot 13$ |

| Lagado Natural Numbers | Remarks | Lagado Natural Numbers | Remarks |
|---|---|---|---|
| 55 | Lagado prime | 79 | Lagado prime |
| 58 | Lagado prime | 82 | Lagado prime |
| 61 | Lagado prime | 85 | Lagado prime |
| 64 | 4·4·4 | 88 | 4·22 |
| 67 | Lagado prime | 91 | 7·13 |
| 70 | 7·10 | 94 | Lagado prime |
| 73 | Lagado prime | 97 | Lagado prime |
| 76 | 4·19 | 100 | 4·25 or 10·10 |

All goes well until we reach 100. In the Lagado system, 100 can be written in two different ways as the product of Lagado primes.

Perhaps if we return to our own arithmetic and continue our own list (from which no numbers have been exiled) we too might eventually run into the same phenomenon as did the mathematicians of Lagado when they reached 100. And if we do not, then our good fortune will undoubtedly have something to do with the fact that we have banned no natural number.

Our next chapter will be devoted to proving that there is only one way to write each composite natural number as the product of primes. Not only are the primes the building blocks for all natural numbers, but they can be assembled in only one way to form each natural number. From this will flow a variety of implications, which we will examine in Chapters 6 and 7.

### Exercises

1. When there were 48 States in the U.S.A., the stars in the flag were arranged in six rows of eight each. By a *rectangular flag* we will mean an arrangement of stars in several rows, each containing the same number of stars. Thus the 48 stars could have been arranged into a 1 by 48, 2 by 24, 3 by 16, 4 by 12, 6 by 8, 8 by 6, 12 by 4, 16 by 3, 24 by 2, or 48 by 1 rectangular flag. What rectangular flags can be made with (a) 36 stars, (b) 37 stars, (c) 38 stars, (d) 39 stars, (e) 49 stars, (f) 50 stars?
   (g) How are the 50 stars of the U.S. flag arranged?
   (h) Complete this statement: If the number of stars is prime, then a rectangular flag _____.

2. In how many ways can 24 be expressed as the sum of two primes?

3. Some primes can be written in the form $1 + N^2$ for some natural number $N$. For example, $5 = 1 + 2^2$, and $17 = 1 + 4^2$.
   (a) Find three more primes of the form $1 + N^2$.
   (b) Do you think that there is an end to such primes? (No one knows.)

4. Some primes are 1 less than a square; that is, they can be written in the form $N^2 - 1$ for some natural number $N$. For example, $3 = 2^2 - 1$.
   (a) Can you find other primes of the form $N^2 - 1$?

(b) Make a conjecture.

(c) Prove it.

5. Some primes are 1 more than a power of 2; for example, $5 = 1 + 2^2$, and $17 = 1 + 2^4$.

(a) Find another such prime.

(b) Do you think that there is an end to such primes? (No one knows.)

6. Some primes are 1 less than a power of 2; for example, $3 = 2^2 - 1$, and $7 = 2^3 - 1$.

(a) Find two more such primes.

(b) Do you think that there is an end to such primes? (No one knows.)

7. If $N$ is a natural number larger than 1, is there always a prime between $N$ and $2N$?

(a) Experiment for all $N$ less than 50.

(b) Make a conjecture.

8. If $N$ is a natural number, is there a prime between $N^2$ and $(N + 1)^2$?

(a) Experiment for all $N$ less than 20.

(b) Make a conjecture. (Though it is known that the answer to E 7 is "yes," the answer to E 8 is not known.)

9. If you feed the Prime-manufacturing Machine 11 and 13, what primes does it produce?

10. If you feed the Prime-manufacturing Machine 2, 11, and 13, what primes does it produce?

11. Prove that the Greeks could have designed their Prime-manufacturing Machine by forming the number $(F_1F_2\cdots F_N) - 1$ instead of the number $M$ that they used, as long as they didn't put in only the prime 2.

12. (a) Find six primes of the form $4N + 1$ (one more than a multiple of 4).

(b) Check that each is expressible as the sum of two squares in exactly one way.

13. Let us so modify the Prime-manufacturing Machine that instead of forming $1 + F_1F_2\cdots F_n$, it forms $1 + F_1^2F_2^2\cdots F_n^2$, but otherwise operates like the original machine.

(a) What primes does this new machine produce when we feed it 3 and 5?

(b) Will the primes produced always be different from those fed in?

(c) Prove your answer to (b).

14. Let us so modify the Prime-manufacturing Machine that instead of forming $1 + F_1F_2\cdots F_n$, it forms $3 + F_1F_2\cdots F_n$, but otherwise operates like the original machine.

(a) What primes does this new machine produce when we feed it 5 and 7?

(b) Will the primes that come out always be different from those that are inserted?

(c) If your answer is "always different," prove it. If only "sometimes different," explain when, and prove your answer.

15. Some primes are four less than a square; that is, they can be written in the form $N^2 - 4$ for some natural number $N$. For instance, $5 = 3^2 - 4$.

(a) Can you find other primes of this form?

(b) Make a conjecture.

(c) Prove your conjecture.

16. (a) Compute 6!
    (b) Check that 6! + 2, 6! + 3, 6! + 4, 6! + 5, 6! + 6 are not prime.
    (c) Is 6! + 1 prime?

17. If $D$ is a divisor of $A$ and of $B$, is $D$ necessarily a divisor of (a) $A + B$, (b) of $AB$?

18. Let $N$ be a natural number larger than 1, and let $D$ be the smallest divisor of $N$ other than 1. Explain why $D$ is prime.

19. (a) Write 1500 as the product of primes.
    (b) In how many ways do you think 1500 can be written as the product of primes?

20. In how many ways can you write 1728 as the product of primes?

21. Write 57 and 117 as the product of primes.

22. Show that 6 is equal to the sum of its divisors other than itself. Find another natural number (less than 30) that is equal to the sum of its divisors other than itself. Such natural numbers are called *perfect*. It is not known whether there are any odd perfect numbers; nor is it known whether there is an end to the even perfect numbers. (See R 1 and R 5.)

23. Compare the number of primes less than $N$ to the quotient $N/(\frac{1}{1} + \frac{1}{2} + \cdots + \frac{1}{N})$ for (a) $N = 6$, (b) $N = 10$, (c) $N = 20$.

24. Which odd natural numbers can be written as the sum of three primes (allowing a prime to be duplicated)? Experiment, and make a conjecture. (See R 1.)

25. Prove that even in the Lagado system every composite natural number is the product of primes.

26. Is it true that whenever 10 divides the product of two natural numbers it must divide at least one of them?

27. (a) Show that the last digit of any even number is 0, 2, 4, 6, or 8.
    (b) Using the information in (a) prove that if 2 divides the product of two natural numbers, then it must divide at least one of them.

28. (a) Show that the last digit of a multiple of 5 is either 0 or 5.
    (b) Using (a) prove that if 5 divides the product of two natural numbers, then it must divide at least one of them.

29. Do you think that whenever 3 divides the product of two numbers it must divide at least one of them? Experiment, and make a conjecture.

30. (a) Show that 220 is a good natural number in the Lagado system and can be written in more than one way as the product of Lagado primes.
    (b) Show that 484 is a good natural number in the Lagado system and can be written as the product of Lagado primes in more than one way.

31. If a prime divides $A + B$, then must it divide at least one of $A$ and $B$?

32. (a) Prove that if every even natural number greater than 2 is the sum of two primes, then every odd natural number greater than 5 is the sum of three primes.
    (b) Give an example of an odd natural number that is the sum of three primes but not the sum of two primes. [In 1937 Vinogradov proved that every odd natural number beyond some point (which he did not determine) is the sum

of three primes. It is also known that each odd natural number less than 100,000 is the sum of three primes.] (See R 1 and p. 32 of R 10.)

33. In the next chapter we will prove that any composite number is the product of primes in essentially only one way. Many feel that this is "obvious" and requires no proof. If you feel that it is obvious, write a short essay to persuade a skeptic that it is so.

34. In the next chapter we will prove that whenever a prime $P$ divides the product of two natural numbers, then $P$ divides at least one of those two numbers. Many feel that this is "obvious" and requires no proof. If you feel that it is obvious, write a short essay to persuade a skeptic that it is so.

●

35. A prime *triplet* consists of three primes of the form $N, N + 2, N + 4$. For instance 3, 5, 7 form a prime triplet. Examine the question, "Are there more prime triplets?" If you find another such triplet, record it. If you do not, prove that there are no more.

36. The interval consisting of the six consecutive natural numbers 2, 3, 4, 5, 6, 7 contains four primes. Can any other interval of six consecutive natural numbers contain four primes?

37. Can you find an interval of twelve consecutive natural numbers that contains more primes than do the 12 natural numbers beginning with 2?

38. After several experiments fill in the blanks with the largest natural numbers, independent of the value of $N$, for which you think the statement is true.
    (a) For any even number $N$, $N^2$ is divisible by ___.
    (b) For any odd number $N$, $N^2 - 1$ is divisible by ___.

39. (a) Prove the assertion made in E 38(a).
    (b) Prove the assertion made in E 38(b).

40. Some primes are two less than a square; that is, they can be written in the form $N^2 - 2$ for some natural number $N$. For example, $7 = 3^2 - 2$.
    (a) Can you find other primes of this form?
    (b) Make a conjecture.

41. To determine whether 197 is prime why does it suffice to check whether any of the primes 2, 3, 5, 7, 11, 13 divide 197?

42. Is 223 prime? Is 221 prime? (Why does it suffice to check merely whether there is a prime less than 15 that divides these numbers?)

43. If $A$ is a multiple of $D$ and also a multiple of $E$, is $A$ necessarily a multiple of $DE$?

44. (a) Prove that if $D$ is a divisor of $N$, so is $N/D$.
    (b) Prove that if a number is not a square, then it has an even number of divisors.
    (c) Prove that if a number is a square, then it has an odd number of divisors.

45. We defined $a^n$, for any natural numbers $a$ and $n$ larger than 0, as $a \times a \times \cdots \times a$ (where $a$ appears $n$ times).
    (a) Verify that $2^3 \times 2^5 = 2^8$.

(b) Verify that $5^1 \times 5^2 = 5^3$.

(c) Prove that $a^m \times a^n = a^{m+n}$ for any natural numbers $m$ and $n$ larger than 0.

46. See E 45. We did not give the symbol $a^0$ any meaning. Show that if we want the rule $a^m \times a^n = a^{m+n}$ to hold even when $m$ or $n$ is 0, then we must assign to $a^0$ the value 1. For this reason, we (the masters) *define* $a^0$ to be 1.

● ●

47. The primes 17, 29, 41, 53 form an arithmetic progression (that is, each differs from the one before by the same amount—in this case, 12). (That there are other primes between them, such as 19, does not matter.)

(a) Find an arithmetic progression formed of five primes.

(b) Do you think that there is an arithmetic progression of 14 primes? (The longest known list has only 13 primes in it. See pp. 47–48 of R 10.)

48. Let $N!$ denote the product of all the natural numbers from 1 through $N$. For example, $6! = 1 \cdot 2 \cdot 3 \cdot 4 \cdot 5 \cdot 6 = 720$.

(a) Prove that when $N$ is composite and larger than 4, then $N$ is a divisor of $(N-1)!$.

(b) Prove that when $N$ is composite, $N$ is not a divisor of $(N-1)! + 1$.

(c) If $N$ is prime, do you think that $N$ divides $(N-1)! + 1$? (Experiment with at least four primes.) We return to this question in E 60, 61, 62 of Chapter 11.

49. Examine this question: Are there numbers of the form $4N + 3$ that can be expressed as the sum of two squares?

50. This triangular array of natural numbers

```
2   3   5   7  11  13  17
  1   2   2   4   2   4
    1   0   2   2   2
      1   2   0   0
        1   2   0
          1   2
            1
```

is formed as follows.

  (i) All primes from 2 through 17 form the first line.

 (ii) Their gaps form the second line.

(iii) Differences of adjacent numbers in the second line form the third line.

(iv) Differences of adjacent numbers in the third line form the fourth line.

 (v) The process continues until the triangle is formed.

(a) Do the same for all primes from 2 through 37 (instead of from 2 through 17).

(b) Do the same for all primes from 2 through 59.

(c) What do you notice about the entries forming the left side of the triangle in each case? (What you notice has been checked for all primes less than 792,722. Whether it always holds is not known. See R 7.)

The number 1.32 in the Twin Prime Conjecture is just a two-decimal estimate of a number whose precise definition is quite fascinating. Exercises 51–54 are devoted to a description of this number.

51. For each natural number $N$ that is greater than 1, we can form the product

$$A_N = \left(1 - \frac{1}{2^2}\right)\left(1 - \frac{1}{3^2}\right)\left(1 - \frac{1}{4^2}\right)\cdots\left(1 - \frac{1}{N^2}\right).$$

For example,

$$A_2 = \left(1 - \frac{1}{2^2}\right),$$

which is simply $\frac{3}{4}$, or decimally, 0.75. As another example,

$$A_3 = \left(1 - \frac{1}{2^2}\right)\left(1 - \frac{1}{3^2}\right),$$

which is just $\frac{3}{4} \times \frac{8}{9}$. Note that this is equal to $\frac{2}{3}$, or decimally, 0.666⋯. Compute, to three decimals, (a) $A_4$, (b) $A_5$, and (c) $A_6$.

52. See E 51. Observe that

$$1 - \frac{1}{2^2} = \frac{2^2 - 1}{2^2} = \frac{(2 - 1)(2 + 1)}{2 \cdot 2} = \frac{1 \cdot 3}{2 \cdot 2}.$$

In a similar manner show that

(a) $1 - \dfrac{1}{3^2} = \dfrac{2 \cdot 4}{3 \cdot 3}$;     (b) $1 - \dfrac{1}{4^2} = \dfrac{3 \cdot 5}{4 \cdot 4}$;     (c) $1 - \dfrac{1}{5^2} = \dfrac{4 \cdot 6}{5 \cdot 5}$.

53. Continuing E 52. (a) What equations, analogous to those in E 52(a), (b), and (c) hold for

$$1 - \frac{1}{6^2};\qquad 1 - \frac{1}{7^2}?$$

(b) Show that $A_4$ can be rewritten

$$\frac{1 \cdot 3}{2 \cdot 2} \times \frac{2 \cdot 4}{3 \cdot 3} \times \frac{3 \cdot 5}{4 \cdot 4}.$$

Therefore, after the cancellations

$$\frac{1}{2} \cdot \frac{\cancel{3}}{\cancel{2}} \cdot \frac{\cancel{2}}{\cancel{3}} \cdot \frac{\cancel{4}}{\cancel{3}} \cdot \frac{\cancel{3}}{\cancel{4}} \cdot \frac{5}{4},$$

$A_4$ is seen to be equal to $\frac{1}{2} \times \frac{5}{4}$.

(c) In a similar manner show that $A_5$ is equal to $\frac{1}{2} \times \frac{6}{5}$.

(d) Show that as $N$ grows larger and larger, $A_N$ shrinks and gets closer and closer to $\frac{1}{2}$.

54. For each natural number $N$ that is larger than 1, we can form the product

$$B_N = \left(1 - \frac{1}{(P_2 - 1)^2}\right)\left(1 - \frac{1}{(P_3 - 1)^2}\right)\cdots\left(1 - \frac{1}{(P_N - 1)^2}\right),$$

where, as in the text, $P_N$ is the $N$th prime; $P_2 = 3$, $P_3 = 5$, and so on. Thus

$$B_2 = 1 - \frac{1}{(3 - 1)^2},$$

which, as the reader may check, is 0.75. Also,

$$B_3 = \left(1 - \frac{1}{(3-1)^2}\right)\left(1 - \frac{1}{(5-1)^2}\right),$$

which reduces to $\frac{3}{4} \cdot \frac{15}{16}$, or, to three decimals, 0.703.

(a) Compute $B_N$ to 3 decimals for $N = 4$, 5, and 6. (As a check, $B_4 = 0.683\cdots$.)

(b) Why does $B_N$ shrink as $N$ increases?

(c) Why is $B_N$ always larger than the $A_N$ of E 51?

(d) Prove that for any $N$, $B_N$ is larger than $\frac{1}{2}$. [*Hint:* E 53(d) and E 54(c).]

$B_N$ shrinks as $N$ increases, but stays above $\frac{1}{2}$. With the aid of an electronic computer, it can be shown that the $B_N$ is approaching a number $C$ that is approximately 0.660. Instead of 1.32 in the Twin Prime Conjecture, we should have written $2C$. (See R 4 for more details.)

55. (a) Is the product of two Lagado numbers always a Lagado number? Prove your answer.

(b) Is the sum of two Lagado numbers ever a Lagado number? Prove your answer.

56. Prove that if $N$ is a natural number greater than 2, then there is at least one prime between $N$ and $N! - 1$.

## References

1. O. Ore, *Number Theory and Its History*, McGraw-Hill, New York, 1948. (Pages 81, 84, and 85 discuss the representation of natural numbers as the sum of primes; pp. 91–94 discuss perfect numbers.)

2. R. D. James, Recent progress in the Goldbach problem, *Bulletin of the American Mathematical Society*, vol. 55, 1949, pp. 246–260.

3. W. Sierpinski, On some unsolved problems of arithmetics, *Scripta Mathematica*, vol. 25, 1960, pp. 125–136. (This expository article presents many unsolved problems concerning primes and rational numbers.)

4. G. Polya, Heuristic reasoning in the theory of numbers, *American Mathematical Monthly*, vol. 66, 1959, pp. 375–384. [An exploration, based on electronic computers and shrewd guessing, of the differences between (not necessarily consecutive) primes.]

5. P. J. McCarthy, Odd perfect numbers, *Scripta Mathematica*, vol. 23, 1957, pp. 43–47.

6. E. T. Bell, *Men of Mathematics*, Simon and Schuster, New York, 1937. (Contains chapters devoted to Gauss and Fermat.)

7. *R. B. Killgrove and K. E. Ralston, On a conjecture concerning primes, *Mathematical Tables and Other Aids to Computing*, National Research Council, Washington, D.C., vol. 13, 1959, pp. 121–122.

8. T. Dantzig, *Number, the Language of Science*, Macmillan, New York, 1954. (Chapter 3 discusses the primes.)

9. R. G. Archibald, Goldbach's theorem, *Scripta Mathematica*, vol. 3, 1935, pp. 44–50 and pp. 153–161. (This examines the number of ways an even natural number can be written as the sum of two primes. The footnote on p. 153 notes misprints on pp. 46–48.)

10. W. Sierpinski, *A Selection of Problems in the Theory of Numbers*, Pergamon, New York, 1964.

*Chapter* **5**

# THE FUNDAMENTAL
# THEOREM
# OF ARITHMETIC

As we proved easily in the last chapter, every composite natural number is the product of primes. We approach now the question: In how many ways can a composite natural number be written as the product of primes? The few experiments we made in Chapter 4 suggest that the answer is: Precisely one way. (Recall that we have agreed to write the primes in a factorization in order of increasing size; for instance, we write $12 = 2 \cdot 2 \cdot 3$, not $12 = 2 \cdot 3 \cdot 2$.) The proof that this answer is correct is far more interesting than the proof that every composite natural number can be written as the product of primes, and will occupy this whole chapter.

Here is what we want to prove. If we have managed by some means or other to express a natural number as the product of primes (for example, as the product of $n$ primes, $p_1 p_2 \cdots p_n$), and if someone else manages to express the same number as the product of primes (for example, as the product of $m$ primes, $q_1 q_2 \cdots q_m$), then $m$ must be the same as $n$, and the $q$'s coincide with the $p$'s. (Do not confuse $p_1$, $p_2$, . . . with $P_1 = 2$, $P_2 = 3$, $P_3 = 5$, . . . .) In other words, what we want to prove is this theorem concerning "unique factorization into primes," usually called

THE FUNDAMENTAL THEOREM OF ARITHMETIC. *If* $p_1$, $p_2$, . . . , $p_n$ *and* $q_1$, $q_2$, . . . , $q_m$ *are primes (both groups written in order of increasing size) and if*

(1) $$p_1 p_2 \cdots p_n = q_1 q_2 \cdots q_m,$$

*then* $p_1 = q_1$, $p_2 = q_2$, . . . , *and hence* $m = n$.

It would be reasonable to prove first that $p_1 = q_1$. Let us try. From (1) we see that $p_1$ is a divisor of the product of all the $p$'s and hence is a divisor of the product of all the $q$'s. *If we knew that this forces* $p_1$ *to divide at least one of the* $q$'s, we could then follow this line of reasoning.

Since $p_1$ divides one of the $q$'s, and since the $q$'s are prime, $p_1$ must equal the $q$ that it divides. Thus $p_1$ occurs among the $q$'s, and hence must be at least as large as $q_1$, the smallest of the $q$'s. Identical reasoning concerning $q_1$ shows that $q_1$ is at least as large as $p_1$. Thus $p_1 = q_1$.

Since $p_1 = q_1$, we may rewrite (1) as

$$p_1 p_2 \cdots p_n = p_1 q_2 \cdots q_m$$

and divide both sides by $p_1$, obtaining

(2) $$p_2 \cdots p_n = q_2 \cdots q_m.$$

Equation (2) is just like equation (1), but with one less prime on each side. The same kind of reasoning that proved $p_1 = q_1$ also shows that $p_2 = q_2$.

Step by step we peel away the primes from each side, establishing that $p_1 = q_1$, $p_2 = q_2$, $p_3 = q_3$, and so on. Since we pair off the $p$'s with the $q$'s, we also see that $m = n$. This concludes the argument.

There is only one gap in this proof of the Fundamental Theorem of Arithmetic: we are not sure at this point that whenever a prime divides the product of several natural numbers it must divide at least one of them. After all, our definition of a prime depends on the way natural numbers divide it, not on the way it divides other natural numbers.

To check whether a natural number is prime, we only have to look at the numbers less than that number and check whether they divide it. To check the way a given natural number divides products of numbers would seem a superhuman task. Yet we must undertake it, and begin by introducing an important concept.

*Let us call a natural number larger than 1* **special** *if,* **whenever** *it divides the product of two natural numbers, it divides at least one of them.* For example, let us prove that 2 is special. We must show that whenever 2 divides $AB$, then 2 must divide at least one of $A$ and $B$. In the language of even and odd, we wish to prove that if $AB$ is even, then at least one of $A$ and $B$ is even. To prove this it is enough to verify that the product of two odd numbers is odd. This we will now do.

If $A$ is odd, it can be written in the form $2a + 1$ for some natural number $a$. If $B$ is odd, it can be written in the form $2b + 1$ for some natural number $b$. Let us examine the product $AB$ and show that it also is odd. We have

$$AB = (2a + 1)(2b + 1) = 4ab + 2a + 2b + 1,$$

which is 1 more than the number $4ab + 2a + 2b$, which is even, since the three numbers $4ab$, $2a$, and $2b$ are even. Thus since $AB$ is one more than an even number, it is odd. We have shown, therefore, that 2 is special. (Another proof is to be found in E 27 of Chapter 4.)

Clearly it is more difficult to prove that 2 is special than to prove that 2 is prime. The reader might like to prove that 3 is special (see E 45) and that 5 is special (see E 28 of Chapter 4). It is easier to show that a natural number is not special than to show that it is. For example, since 6 divides $2 \cdot 3$, yet divides neither 2 nor 3, we see that 6 is not special. This argument suggests that no composite natural number can be special. We will prove

THEOREM 1. *Every special number must be prime.*

PROOF. We must show that no special number is situated among the composite numbers. Or, to put it another way, we must show that a composite number is not special.

To do this, let $N$ be a composite number. Then $N$ is the product of two smaller natural numbers, $A$ and $B$:

$$N = A \cdot B.$$

Now, $N$ divides the product of $A$ and $B$ (for this product is just $N$). But because $A$ and $B$ are smaller than $N$, $N$ can divide neither $A$ nor $B$. Thus $N$ is not special. This proves the theorem.

Let us not be too hasty in jumping to the conclusion that every prime is special. Recall that it took some effort just to prove that 2 is special, and nowhere in that proof did we make use of the fact that 2 is prime. Moreover, recall that in the Lagado system, 10 is a prime, but not special, since it divides $4 \cdot 25$ and yet divides neither 4 nor 25. What Theorem 1 tells us is that the special numbers are to be found only among the primes.

But it is quite another matter to show that *every* prime is special. In fact, the assertions "every prime is special" and "factorization into primes is unique" are of the same depth, as we will now demonstrate.

We have already seen that *if every prime is special, then factorization into primes is unique.* We now show that *if factorization into primes is unique, then every prime is special.*

To accomplish this, let us consider a prime $P$ that divides the product, $A \cdot B$, of natural numbers $A$ and $B$. We wish to prove that $P$ divides at least one of the numbers $A$ and $B$. First of all, we know, by the definition of "divides," that there is a natural number $Q$ such that

$$A \cdot B = P \cdot Q.$$

Next, we express $Q$, $A$, and $B$ as the products of primes:

$$Q = r_1 r_2 \cdots r_s, \qquad A = p_1 p_2 \cdots p_n, \qquad B = q_1 q_2 \cdots q_m.$$

(Of course, some of $Q$, $A$, and $B$ may be prime; in that event we would have, simply, $Q = r_1$, or $A = p_1$, or $B = q_1$.) Since $A \cdot B = P \cdot Q$, we have

$$P \cdot r_1 r_2 \cdots r_s = p_1 p_2 \cdots p_n \cdot q_1 q_2 \cdots q_m.$$

Since we are assuming that factorization into primes is unique, *the prime P must occur among the p's or among the q's.* If *P* occurs among the *p*'s, then *P* divides *A*; if *P* is among the *q*'s, *P* divides *B*. This shows that *P* is special.

Thus the Fundamental Theorem of Arithmetic is equivalent to the statement "every prime is special." In this chapter we will prove that every prime is special, and thus, in a roundabout manner, prove the Fundamental Theorem of Arithmetic.

Before we turn to the problem of whether every prime is special, we ought to go back briefly to what led us to define the special numbers. We were interested in the question: If a prime divides the product of several natural numbers, must it divide at least one of them? As a small step toward answering this question we have

THEOREM 2. *If a special number divides the product of several natural numbers, then it divides at least one of them.*

PROOF. We illustrate the argument for the case in which the special number happens to divide the product of three natural numbers. Let *S* be a special number that divides the product *ABC*. We wish to prove that *S* divides at least one of *A*, *B*, and *C*.

It is clear that *S* divides the product of the two natural numbers *A* and *BC*. Since *S* is special, it must divide at least one of *A* and *BC*. If *S* divides *A*, we are done. If *S* divides *BC*, then, since *S* is special, *S* must divide at least one of *B* and *C*. Thus we see that whenever a special number divides the product of three natural numbers, then it must divide at least one of them.

The same argument can be applied to the product of more than three natural numbers. The reader can take care of the case in which a special number divides the product of four natural numbers, and also see how the proof of Theorem 2 can be carried out in general.

We have been able to go this far without getting involved with addition. Neither the notions of prime and special nor Theorems 1 and 2 and their proofs have made any mention of addition. But the proof that every prime is special leans heavily on addition. Thus the Fundamental Theorem of Arithmetic is really a theorem about multiplication and addition, not multiplication alone.

The proof that we will present is basically the one that appears in Book 7 of Euclid, published about 300 B.C., and is probably the work of a whole school of mathematicians. The reader should note carefully just where addition makes its appearance on stage and where it exits; it appears nowhere in the statement of the Fundamental Theorem of Arithmetic, but only in the proof leading up to it.

And even more important, the reader should not feel that the Fundamental Theorem of Arithmetic is obvious and that no proof is needed. The reader may have acquired the notion that this theorem is obvious from a previous course in which perhaps some numbers (surely none larger than a million) were factored into primes, and he was led to the hasty conclusion, or was told dogmatically, that the factorization is always unique. We should appreciate these words of Gauss, written in 1801 in his Disquisitiones Arithmeticae: "many modern authors have employed vague computations in place of proof or have neglected the theorem completely."

We shall break up Euclid's proof that every prime is special into several small pieces. The diagram below tells how these pieces fit together. As the reader goes through the proof, he can find where he is in the whole argument by looking back at the diagram. When he has grasped the details of the various proofs, he can then go back to the diagram to see how the separate lemmas really form one large result. The first time through, though, he may not be able to see the forest for the trees.

We shall first need the notion of a common divisor of two natural numbers. *A natural number that is a divisor of both A and B is called a* **common divisor** *of A and B.* For example, 3 is a common divisor of 9 and 12. So is 1 a common divisor of 9 and 12. Of particular interest will be the **greatest common divisor** of $A$ and $B$, which we will denote by $(A, B)$. For example, $(9, 12) = 3$. The reader may verify that $(7, 12) = 1$, $(5, 15) = 5$, and $(0, 3) = 3$. We should point out that $(0, 0)$ does not exist, since there is no largest divisor of 0.

Now we can draw the promised diagram of Euclid's proof:

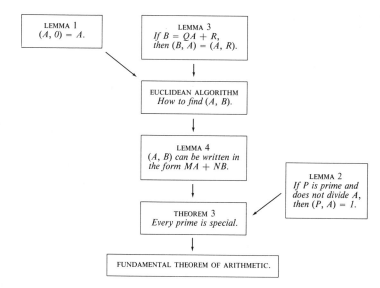

The main flow of thought is down the column; notice the presence of addition in Lemmas 3 and 4. Now let us begin the proof.

LEMMA 1. *For any natural number A, other than 0, (A, 0) = A.*

PROOF. Any natural number $D$ is a divisor of 0, since $0 = 0 \cdot D$. Thus the largest common divisor of $A$ and 0 is simply the largest divisor of $A$, namely $A$.

LEMMA 2. *If P is prime and does not divide A, then (P, A) = 1.*

PROOF. The only divisors of $P$ are 1 and $P$. Since $P$ does not divide $A$, it follows that the only common divisor of both $P$ and $A$ is 1. Thus 1 is the greatest common divisor of $P$ and $A$.

Addition appears on the scene in

LEMMA 3. *Let A and B be natural numbers, where A is not zero. When we divide A into B we obtain a quotient Q and a remainder R (hence B = QA + R). Then (B, A) = (A, R).*

As an illustration, before considering the proof, take the case $A = 12$ and $B = 57$. Since 12 "goes into" 57 four times with nine "left over," we have $Q = 4$ and $R = 9$. According to Lemma 3, we have $(57, 12) = (12, 9)$, an assertion that the reader may check. Note in general that since $R$ is less than $A$, the computation of $(A, R)$ will be easier than that of $(B, A)$.

PROOF. Though we are interested only in the greatest common divisor, we will actually prove more; namely, that the list of all common divisors of $A$ and $B$ is the same as the list of all common divisors of $A$ and $R$. From this, Lemma 3 follows easily.

Let $D$ be any natural number dividing both $A$ and $R$. Then $D$ divides $QA$, and hence also the sum $QA + R$, which is $B$. Thus $D$ divides $A$ and $B$. Hence any common divisor of $A$ and $R$ is a common divisor of $B$ and $A$.

Let us consider the opposite direction. Let $d$ be any natural number dividing both $B$ and $A$. Then $d$ divides $QA$, and hence also the difference $B - QA$, which is $R$. Thus $d$ divides $A$ and $R$. Any common divisor of $B$ and $A$ is a common divisor of $A$ and $R$.

Hence the list of common divisors of $B$ and $A$ is the same as the list of common divisors of $A$ and $R$. In particular, the greatest common divisor of $A$ and $B$ must be the same as the greatest common divisor of $A$ and $R$. This ends the proof.

A laborious way of computing $(A, B)$ consists in listing all divisors of $A$ and then all divisors of $B$, and picking out the largest divisor that appears in both of the lists. Lemma 3 is the basis for a far more efficient method of finding $(A, B)$, known as the *Euclidean Algorithm*. Let us illustrate this technique, which works for any $(A, B)$, by applying it to the case $(72, 20)$.

*Step 1.* Divide 20 into 72, and find the remainder. We have $72 = 3 \cdot 20 + 12$ (the remainder is 12). By Lemma 3, $(72, 20) = (20, 12)$.

*Step 2.* Divide 12 into 20, and find the remainder, We have $20 = 1 \cdot 12 + 8$ (the remainder is 8). By Lemma 3, $(20, 12) = (12, 8)$.

*Step 3.* Divide 8 into 12, and find the remainder. We have $12 = 1 \cdot 8 + 4$ (the remainder is 4). By Lemma 3, $(12, 8) = (8, 4)$.

*Step 4.* Divide 4 into 8, and find the remainder. We have $8 = 2 \cdot 4 + 0$ (the remainder is 0). By Lemma 3, $(8, 4) = (4, 0)$.

But Lemma 1 asserts that $(4, 0) = 4$. Combining all the steps, we conclude that $(72, 20) = 4$.

The Euclidean Algorithm *involves repeatedly dividing and finding remainders until we reach the remainder 0.* Each step produces a remainder smaller than the remainder from the previous step. Eventually we must obtain the remainder 0. The remainder before 0 is the greatest common divisor. This is the technique. The reader is invited to apply the Euclidean Algorithm to $(117, 51)$. He might also compare the ease of the Euclidean Algorithm to the labor of making lists in the computation of $(433, 144)$. (Of course, the number of steps need not be four; one keeps going till the remainder 0 is met.)

For the first time in this chapter we will need the negative integers $-1$, $-2$, $-3, \ldots$ . The term *integer* will refer to either a natural number $0, 1, 2, 3, \ldots$ or a negative integer. The arithmetic of the integers is discussed in Appendix A. In particular, Appendix A offers practical and theoretical reasons why the product of two negative numbers should be positive, for example, $(-2)(-3) = 6$. We can extend our definition of divisor to include the integers by saying that the integer $A$ divides the integer $B$ if there is an integer $Q$ such that $B = QA$. For example, since $-6 = (-3) \cdot 2$, 2 divides $-6$. We will utilize this notion further in Chapter 11.

The next lemma will be the chief tool for proving that every prime is special. Just as $(2, 3) = (-1)2 + 1 \cdot 3$, and just as $(72, 20)$, which we know to be 4, can be expressed as $2 \cdot 72 + (-7)20$, Lemma 4 asserts that $(A, B)$ for any natural numbers $A$ and $B$, where both are not zero, can be expressed similarly in terms of $A$ and $B$.

LEMMA 4. *For any natural numbers $A$ and $B$ (not both 0) there are integers $M$ and $N$ (positive or negative) such that*

$$(A, B) = MA + NB.$$

If $A$ and $B$ are small, we might find $M$ and $N$ quickly with just a little figuring. For example, if $A$ is 5 and $B$ is 8 [and $(A, B)$ is therefore 1], we can easily find these three examples:

$$1 = 5 \cdot 5 + (-3)8,$$
$$1 = (-3)5 + 2 \cdot 8,$$
$$1 = (-11)5 + 7 \cdot 8,$$

and the reader might add further illustrations.

But when $A$ and $B$ are large, it might not be at all obvious that $M$ and $N$ can be found. We would know only that if the integers $M$ and $N$ did exist, then one of them must be positive, the other negative. Now we shall turn to the proof of Lemma 4, which not only will prove that $M$ and $N$ can be found, but will provide a technique for finding them (though it may not produce the smallest possible $M$ and $N$).

The reader who has examined Chapter 3 will see that Lemma 4 is related to the problem of weighing with $A$- and $B$-ounce measures. In particular, it shows that if $(A, B) = 1$, then the two measures can weigh a 1-ounce object. More generally, the lemma asserts that $A$- and $B$-ounce measures can weigh a number of ounces as small as the greatest common divisor of $A$ and $B$. We may think of Lemma 4 as the "weighing" lemma.

In order to find $M$ and $N$ we shall unwind the Euclidean Algorithm. Let us illustrate the technique by two examples.

The computation of $(7, 29)$ by the Euclidean Algorithm involves these steps:

$$29 = 4 \cdot 7 + 1,$$
$$7 = 7 \cdot 1 + 0;$$

hence $1 = (7, 29)$. Using just the first step, we have $1 = 29 - 4 \cdot 7$, which can be written

$$1 = (-4)7 + 1 \cdot 29.$$

We have found an $M$ and an $N$ such that $1 = M \cdot 7 + N \cdot 29$. We may check that our $M$ and $N$ work by noticing that $1 \cdot 29 = 29$ and that $4 \cdot 7 = 28$.

Now for a more impressive example, say $B = 945$ and $A = 219$. We will find $(A, B)$ by the Euclidean Algorithm and then find integers $M$ and $N$ such that $(A, B) = MA + NB$. The computations for finding $(219, 945)$ appear on the left below; the underline identifies the successive $A$ and $B$ in the relation

| Euclidean Algorithm | Equations used for finding $M$ and $N$ |
|---|---|
| $945 = 4 \cdot \underline{219} + 69$ | $69 = \underline{945} - 4 \cdot \underline{219}$ |
| $\underline{219} = 3 \cdot \underline{69} + 12$ | $12 = \underline{219} - 3 \cdot \underline{69}$ |
| $\underline{69} = 5 \cdot \underline{12} + 9$ | $9 = \underline{69} - 5 \cdot \underline{12}$ |
| $\underline{12} = 1 \cdot \underline{9} + 3$ | $3 = \underline{12} - \underline{9}$ |
| $\underline{9} = 3 \cdot \underline{3} + 0$ | |

$B = QA + R$ at each stage. On the right are the equations for the remainders at each stage; it is these equations that will be used for finding $M$ and $N$.

The Euclidean Algorithm shows that $(219, 945)$ is 3. We will now use the right column, starting at the bottom and working up, to find $M$ and $N$ such that

$$3 = M \cdot 219 + N \cdot 945$$

(The reader of Chapter 3 will observe that finding $M$ and $N$ yields a method of weighing a 3-ounce potato with 219- and 945-ounce measures.)

The bottom equation on the right expresses 3 in terms of 9's and 12's (not, as desired, in terms of 219's and 945's),

$$3 = \underline{12} - \underline{9}.$$

We move up to the next equation, $9 = \underline{69} - 5 \cdot \underline{12}$, and use it to remove the $\underline{9}$ from the equation $3 = \underline{12} - \underline{9}$, as follows:

$$3 = \underline{12} - (\underline{69} - 5 \cdot \underline{12}),$$

which we rewrite as $3 = \underline{12} - \underline{69} + 5 \cdot \underline{12}$, or more simply

$$3 = 6 \cdot \underline{12} - \underline{69},$$

an equation that expresses 3 in terms of 12's and 69's. (As a check, note that $6 \cdot 12 - 69 = 72 - 69 = 3$.)

To remove the $\underline{12}$ from the equation $3 = 6 \cdot \underline{12} - \underline{69}$ we move up to the next equation, $12 = \underline{219} - 3 \cdot \underline{69}$. We obtain

$$3 = 6(\underline{219} - 3 \cdot \underline{69}) - \underline{69},$$

which we may rewrite as $3 = 6 \cdot \underline{219} - 18 \cdot \underline{69} - \underline{69}$ or

$$3 = 6 \cdot \underline{219} - 19 \cdot \underline{69},$$

an equation that expresses 3 in terms of 69's and 219's. Now, to get rid of the 69, we continue up the list of the equations on the right to the next one (which is the top one), $69 = \underline{945} - 4 \cdot \underline{219}$. Using it, we obtain

$$3 = 6 \cdot \underline{219} - 19(\underline{945} - 4 \cdot \underline{219}),$$

which simplifies to $3 = 6 \cdot \underline{219} - 19 \cdot \underline{945} + 76 \cdot \underline{219}$ or

$$3 = 82 \cdot \underline{219} - 19 \cdot \underline{945}.$$

Thus $3 = 82 \cdot 219 + (-19)945$. We have found that $M = 82$ and $N = -19$ are a pair of integers that express $(219, 945)$ in the form $M \cdot 219 + N \cdot 945$. A little arithmetic checks our answer, for $82 \cdot 219 = 17958$ and $19 \cdot 945 = 17955$; their difference is 3, as we desired.

Using the same technique, the reader might like to find $M$ and $N$ such that

$(72, 20) = M \cdot 72 + N \cdot 20$. Since the technique shown above can be applied to any $(A, B)$, we have proved Lemma 4.

With Lemma 4 in our possession, we are ready to prove

THEOREM 3. *Every prime is special.*

PROOF. Let $A$ and $B$ be natural numbers, and let $P$ be a prime that divides their product, $AB$. We wish to prove that $P$ must divide at least one of $A$ and $B$. To do this we will prove that if $P$ does not divide $A$, then $P$ must divide $B$. If $P$ does not divide $A$, we have, by Lemma 2,

$$(P, A) = 1.$$

Lemma 4 then promises us that there are $M$ and $N$ such that

$$1 = MP + NA.$$

If we multiply both sides of this equation by $B$, we obtain the new equation

$$B = MPB + NAB.$$

Now, $P$ divides $MPB$; and, since $P$ divides $AB$, it also divides $NAB$. Hence $P$ divides the sum

$$MPB + NAB.$$

But that sum is just $B$. Therefore $P$ divides $B$. Theorem 3 is proved.

This ends Euclid's proof that every prime is special. Note precisely where addition walked onto the stage; contrast this proof with the much simpler one for the converse, "every special number is prime." Moreover, having proved that every prime is special, we have also shown that the Fundamental Theorem of Arithmetic is true; for, as we observed at the beginning of this chapter, the Fundamental Theorem of Arithmetic is equivalent to the assertion that each prime is special.

In the Lagado system, the Fundamental Theorem of Arithmetic does not hold, nor are all the primes special. But there is a theorem, true both for us and for the mathematicians of the academy of Lagado, which sheds much light on our writing of natural numbers as the product of primes. Let us call this

THE CONCEALED THEOREM. *If $s_1, s_2, \ldots, s_n$ and $t_1, t_2, \ldots, t_m$ are all special (with the s's and t's increasing from left to right), and their products are equal,*

$$s_1 s_2 \cdots s_n = t_1 t_2 \cdots t_m;$$

*then $s_1 = t_1$, $s_2 = t_2$, and so on.*

The concealed Theorem, so similar to the Fundamental Theorem of Arithmetic, is quite simple to prove—so simple, in fact, that we do not need to use addition to prove it. In outline the proof goes like this. Recall that every special is prime (Theorem 1, whose proof involves no addition). Then, with the aid of Theorem 2 (whose proof involves no addition), we can carry through the identical (peeling away) argument that we provided for the Fundamental Theorem of Arithmetic.

Without using addition, we can (as could the mathematicians of Lagado) prove these theorems:

I. *Every special number is prime.*

II. *Every composite natural number can be written as the product of primes in at least one way.*

III. *Every natural number can be written as the product of special numbers in at most one way (maybe not at all).*

With the aid of addition, we (but not the mathematicians of Lagado) can then prove that in ordinary arithmetic

IV. *Every prime is special.*

In ordinary arithmetic the primes are the same as the special numbers. With this information we can merge Theorems II and III into the one elegant result: *every composite natural number is the product of primes in exactly one way.*

Not only does this result have many applications, but it also suggests new problems. These we discuss in the next chapter.

### Exercises

1. (a) List all the divisors of 100.  (b) List all the divisors of 75.
   (c) List all the common divisors of 75 and 100.  (d) What is (75, 100)?

2. Do the same as in E 1 for the numbers 90 and 76.

3. By making the complete list of divisors, find (a) (24, 51) and (b) (57, 133).

4. Using the Euclidean Algorithm compute (24, 51) and (57, 133).

5. Using the Euclidean Algorithm compute
   (a) (164, 72),  (b) (91, 39),  (c) (73, 21).

6. Find four different pairs of integers $M$ and $N$ such that $(2, 3) = M \cdot 2 + N \cdot 3$.

7. Find three different pairs of integers $M$ and $N$ such that $(3, 5) = M \cdot 3 + N \cdot 5$.

8. Using the technique given in this chapter, find a pair of integers $M$ and $N$ such that
   (a) $(24, 51) = M \cdot 24 + N \cdot 51$,
   (b) $(57, 133) = M \cdot 57 + N \cdot 133$,
   (c) $(13, 21) = M \cdot 13 + N \cdot 21$.

9. Verify that your answers in E 8 are correct by computing both sides of the equations.

10. Find an $M$ and an $N$ for each of these pairs of $A$ and $B$:
    (a) 72, 164,     (b) 39, 91,     (c) 21, 73,     (d) 255, 697.

11. Check your answers in E 10 by computing $MA + NB$ in each case and comparing it to $(A, B)$.

12. (a) Define in your own words "prime number" and "special number."
    (b) Is it easier to show that a number is prime or that it is special? Why?
    (c) Is it easier to show that every special number is prime or that every prime number is special? Why?
    (d) Are "prime" and "special" synonyms in ordinary arithmetic? In Lagado?

13. Compute
    (a) $-3(4 - 5 \cdot 6)$,     (b) $-(2 \cdot 8 - 4 \cdot 7)$,     (c) $6(28 - (-2)(-7)) - 5(-9 - 3)$.

14. (a) List all divisors of $2^3 \cdot 5^4$.
    (b) How did you use the Fundamental Theorem of Arithmetic in answering (a)?

15. Let $A = 3^3 \cdot 7^2$ and $B = 3 \cdot 5 \cdot 7^3$. Using the Fundamental Theorem of Arithmetic, show that $(A, B) = 3 \cdot 7^2$. (Explain in good English how you used the Fundamental Theorem of Arithmetic.)

16. Find (96, 144) in three ways:
    (a) by listing all the divisors of 96 and of 144 and finding which is the greatest common divisor,
    (b) by the Euclidean Algorithm,
    (c) by expressing 96 and 144 as the product of primes and using the Fundamental Theorem of Arithmetic.

17. Solve E 16(b) and (c) for (144, 196).

18. Solve E 16(b) and (c) for (120, 500).

19. Find the largest number that divides both $2^5 \cdot 7$ and $2^6 \cdot 5$.

20. Another notation for $(A, B)$ is $\text{GCD}(A, B)$ where GCD is an abbreviation for "greatest common divisor." Using the Fundamental Theorem of Arithmetic, find
    (a) $\text{GCD}(2^5, 3^7)$,     (b) $\text{GCD}(2^4 \cdot 5^3, 2^2 \cdot 5 \cdot 7)$.

21. Let $A$ and $B$ be natural numbers. Clearly the product $AB$ is both a multiple of $A$ and a multiple of $B$. But there may be a smaller "common multiple" of $A$ and $B$. Denote the least common multiple of $A$ and $B$ by $\text{LCM}(A, B)$.
    (a) Find $\text{LCM}(4, 7)$.              (b) Find $\text{LCM}(4, 6)$.
    (c) Find $\text{LCM}(2^5 3^2, 2^4 3^6)$.          (d) Find $\text{LCM}(144, 96)$.

22. This continues E 21. (a) Compute the products $AB$ and $[\text{LCM}(A, B)][\text{GCD}(A, B)]$ for $A = 4$, $B = 6$, and for $A = 2^5$, $B = 2^7$. (b) Make a conjecture. (c) Prove it.

23. (a) Show that if 1 were included among the primes, then the Fundamental Theorem of Arithmetic would not be true as stated.
    (b) If 1 were included among the primes, how would we have to reword the Fundamental Theorem of Arithmetic?

24. (a) How can you define "even number" using only addition?
    (b) What numbers have the same relation to multiplication as the even numbers do to addition?

25. If you happen to know that 12 and 16 divide $N$, what other divisors must $N$ then also have?

26. (a) If $A$ divides $B$, must $A^2$ divide $B^2$?
    (b) If $A^2$ divides $B^2$, must $A$ divide $B$? Explain your answers.

27. Prove that if a special number divides the product of four natural numbers, then it must divide at least one of them.

28. Using the Fundamental Theorem of Arithmetic, prove that every divisor of $2^{100}$ must itself be a power of 2. (Since we define $2^0$ to be equal to 1, then 1 is thus a power of 2; see E 46 of Chapter 4.)

29. See E 28. How many divisors does $2^{100}$ have?

30. Prove that every divisor of $2^5 \cdot 3^{11}$ must be a power of 2 times a power of 3. (*Hint:* Let $D$ be a divisor of $2^5 \cdot 3^{11}$ and $Q$ be the quotient: $2^5 \cdot 3^{11} = DQ$. Write $D$ and $Q$ as the product of primes, and use the Fundamental Theorem of Arithmetic.)

31. Using the Fundamental Theorem of Arithmetic, prove that
    (a) $(2^5 \cdot 3^7, 2^7 \cdot 3^2) = 2^5 \cdot 3^2$,
    (b) $(3^4 \cdot 7^6, 3^4 \cdot 11) = 3^4$.

32. Prove that no power of 3 is equal to a power of 7. (Except, of course, $3^0 = 1 = 7^0$.)

33. Prove that if $(A, B) = 1$, then $(A^2, B^2) = 1$.

34. In Chapter 3 we raised the question, "Let $A$ and $B$ be natural numbers. Which natural numbers $W$ can be expressed in the form $W = MA + NB$ for suitable integers $M$ and $N$?" Answer the question.

35. Let $A$ and $B$ be natural numbers.
    (a) If you are told that there exist integers $M$ and $N$ such that $1 = MA + NB$, what can you say about $(A, B)$?
    (b) Prove your answer.

36. Let $A$ and $B$ be natural numbers.
    (a) If you are told that there exist integers $M$ and $N$ such that $4 = MA + NB$, what can you say about $(A, B)$?
    (b) Prove your answer.

●

37. Prove that every common divisor of $A$ and $B$ is a divisor of $(A, B)$. (*Hint:* Use Lemma 4.)

38. (a) How does $(A, B)$ compare with $(A, A + B)$?
    (b) Prove your answer.
    (c) How does $(A, B)$ compare with $(A, B + NA)$, where $N$ is any natural number?
    (d) Prove your answer.

39. Show that, in the Lagado system, 100 is not the product of Lagado special numbers.

40. Prove that if $(A, B) = 1$ and $(A, C) = 1$, then $(A, BC) = 1$.

41. Is it always true that if $A$ divides $C$ and $B$ divides $C$, then $AB$ divides $C$? Prove your answer.

42. Prove that if 6 divides the square of an integer, then it divides the integer itself.

43. Prove that, in the Lagado system, $(4, 10) = 1$, but that there are no integers $M$ and $N$ for which $1 = M \cdot 4 + N \cdot 10$.

44. If the King of Lagado had decided that 1 and the even numbers other than 0 are good and that the odd numbers other than 1 are bad, he would have created a rather interesting arithmetic. (He still allows bad numbers to be used for writing good numbers.)
    (a) Find the primes of this new arithmetic.
    (b) Show that 180 can be written as the product of primes in this arithmetic in more than one way.
    (c) Prove that if $p_1, p_2, \ldots, p_n$ and $q_1, q_2, \ldots, q_m$ are primes in this arithmetic, with $p_1 p_2 \cdots p_n = q_1 q_2 \cdots q_m$, then $m = n$ (even though the $p$'s need not be the same as the $q$'s).

45. Without using the material of this chapter, prove that 3 is special. (*Hint:* Observe that every natural number, when divided by 3, leaves a remainder of 0, 1, or 2. Thus every natural number is of the form $3x$, $3x + 1$, or $3x + 2$ for some natural number $x$. Show that the product of two natural numbers of the last two types is not of the first type. Let one number be $3x + 1$ or $3x + 2$ and the other be $3y + 1$ or $3y + 2$.)

46. Using the technique of E 45, prove that 7 is special.

47. What is the smallest natural number greater than zero that can be expressed in the form $M \cdot 32 + N \cdot 14$ for some integers $M$ and $N$? Explain your answer.

48. Which natural numbers can be expressed in the form $M \cdot 39 + N \cdot 51$ for suitable integers $M$ and $N$?

49. Let $A$ and $B$ be natural numbers. If there exist integers $M$ and $N$ such that $MA + NB = 1$, are there necessarily integers $X$ and $Y$ such that $XA^2 + YB^2 = 1$? Prove your answer.

50. Let $A$ and $B$ be natural numbers. If there exist integers $M$ and $N$ such that $MA + NB = 2$, are there necessarily integers $X$ and $Y$ such that $XA^2 + YB^2 = 2$? Prove your answer.

51. Let $D = (A, B)$.
    (a) What can we say about $(A/D, B/D)$?
    (b) Prove your result with the aid of Lemma 4.
    (c) Prove the result with the aid of the Fundamental Theorem of Arithmetic.

52. Show that
    (a) $MA + NB = (M - B)A + (N + A)B$,
    (b) $MA + NB = (M + B)A + (N - A)B$.

53. See E 52. Prove that the $M$ and $N$ mentioned in Lemma 4 can be chosen in such a way that *either*
    (a) $M$ is not negative and is smaller than $B$, or
    (b) $N$ is not negative and is smaller than $A$.
    (c) Relate (a) and (b) to E 14, 19, 20, 21, 22 of Chapter 3.

54. (a) Read the jug exercises, E 30, 31, 32, 33 of Chapter 3.
    (b) Show that with an $A$- and a $B$-quart jug, such that $(A, B) = 1$, you can bring back 1 quart of water.
    (c) Generalize E 33 of Chapter 3 to arbitrary $A$-quart and $B$-quart jugs, such that $(A, B) = 1$ and solve. (*Hint:* relate to E 53).
    (d) Generalize E 34 of Chapter 3 similarly and solve.

• •

55. (a) Read the postage exercises E 23, 24, and 25 of Chapter 3.
    (b) Let the denominations available be $A$¢ and $B$¢, where $(A, B) = 1$. Study the postage problem in this general context. Consider such questions as: What amounts of postage can be obtained? How many cannot?

56. Prove that whenever a particular Lagado number is expressed as the product of Lagado primes, the number of these primes will always be the same.

57. If $(A, B) = 1$, what can be said about $(A + B, A - B)$?
    (a) Experiment,    (b) Conjecture,    (c) Prove your conjecture.

58. This exercise outlines a short proof (to be found in R 7) that every prime is special. *Assume* that there exist primes that are *not* special. We will obtain a contradiction. Let $p$ be the smallest such prime. Since $p$ is *not* special there are natural numbers $A$ and $B$ such that $p$ divides $AB$ but divides neither $A$ nor $B$. Let $m$ be the smallest such $A$, and let $n$ be a corresponding $B$. Thus $p$ divides $mn$ but divides neither $m$ nor $n$.
    (a) Show that $m$ is less than $p$. [*Hint:* Observe that $p$ divides $(m - p)n$.]
    (b) Show that $m$ is larger than 1.
    (c) Let $q$ be some prime that divides $m$. Why is $q$ special?
    (d) Since $p$ divides $mn$ there is a natural number $k$ such that $pk = mn$. Show that $q$ divides $k$.
    (e) Define $k'$ by the equation $k = k'q$ and $m'$ by the equation $m = m'q$. Show that $pk' = m'n$ and that this contradicts our definition of $m$.
    (f) Was addition used in this proof?

59. A regular $n$-gon is a polygon of $n$ equal sides and equal angles. For example, a regular 4-gon is a square. Given a regular 3-gon and regular 5-gon, inscribed in a circle, one can easily find the side of a regular 15-gon. (*Hint:* Prove that the arc $PQ$ is one-fifteenth of the circle by using the relation $\frac{1}{15} = \frac{2}{3} - \frac{3}{5}$.)

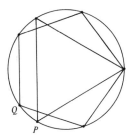

60. (a) Prove that if $(A, B) = 1$, then there are natural numbers $C$ and $D$ such that $1/AB = C/A - D/B$ or, perhaps, $C/B - D/A$.
    (b) Use (a) to show that if $(A, B) = 1$, then from a regular $A$-gon and a regular $B$-gon inscribed in a circle and sharing a vertex, it is easy to construct a regular $A \times B$-gon. (*Hint:* See E 59.)

61. With straightedge and compass construct the following regular polygons:
    (a) 3-gon,   (b) 4-gon,   (c) 6-gon,   (d) 8-gon,   (e) 12-gon.

62. (a) Show how to obtain by straightedge and compass a regular $2N$-gon from a regular $N$-gon.
    (b) Show how to obtain by straightedge and compass a regular $N$-gon from a regular $2N$-gon.
    (c) Why do (a) and (b) reduce the problem of constructing regular polygons to the construction of those with an odd number of sides?

63. See E 61 and E 62. In 1796 Gauss proved that a regular $N$-gon, with $N$ odd, can be constructed with straightedge and compass only when $N$ has the following property: when $N$ is written as the product of primes no prime appears twice, and each prime is one more than a power of 2.
    (a) Using Gauss's theorem, show that a regular 9-gon, or regular 11-gon, or regular 25-gon cannot be constructed with straightedge and compass.
    (b) Using Gauss's theorem, show that a regular 5-gon, or regular 3-gon, or regular 17-gon can be constructed with straightedge and compass.
    (c) What role does the Fundamental Theorem of Arithmetic play in the statement of Gauss's theorem?

64. (a) Find all natural numbers less than 20 that have exactly four divisors.
    (b) How many natural numbers do you think there are that have exactly four divisors? State why.

65. How could $(A, B)$ be defined *without* using the concept of "greatest"?

66. How many divisors are there of
    (a) 243,   (b) 32,   (c) (243)(32),   (d) $2^{70} \cdot 3^{20}$?

### References

1. I. Niven and H. S. Zuckerman, *An Introduction to the Theory of Numbers*, Wiley, New York, 1960. (An alternate proof of the Fundamental Theorem of Arithmetic is given on p. 14.)

2. R. Courant and H. Robbins, *What Is Mathematics?*, Oxford University Press, New York, 1960. (On p. 23 is the same proof as is given in R 1.)

3. *B. L. van der Waerden, *Modern Algebra*, vol. 1, Ungar, New York, 1949. (The theory of Euclidean rings is presented on pp. 32–62; unique factorization is discussed on pp. 55–61.)

4. *G. Birkhoff and S. MacLane, *Survey of Modern Algebra*, Macmillan, New York,

1965. (Unique factorization into primes is discussed for natural numbers on p. 20, for polynomials on pp. 74–76, and for complex integers on p. 386.)

5. H. Davenport, *The Higher Arithmetic*, Harper and Brothers, New York, 1960. (Another short proof of the Fundamental Theorem of Arithmetic is given on pp. 19–20.)

6. J. B. Roberts, *The Real Number System in an Algebraic Setting*, W. H. Freeman and Company, San Francisco, 1962. (Another proof that each prime is special is given on pp. 34–36.)

7. K. Rogers, Unique factorization, *American Mathematical Monthly*, vol. 70, 1963, pp. 547–548. (A short proof of the unique factorization of integers and polynomials is presented here.)

8. B. A. Trakhtenbrot, *Algorithms and Automatic Computing Machines*, Heath, Boston, 1963. (The Euclidean Algorithm and algorithms for winning games and searching a labyrinth are discussed on pp. 3–24. The Euclidean Algorithm serves as an example of machine coding on pp. 48 and 71.)

9. C. F. Gauss, *Disquisitiones Arithmeticae*, translated by A. A. Clarke, Yale, 1966. (The quote in this chapter is from p. 5.)

*Chapter* **6**

# RATIONALS AND IRRATIONALS

As long as we use numbers just for counting, we can get along quite well with just the natural numbers 0, 1, 2, 3, 4, 5, . . . . In Chapter 1 we counted switches and backward pairs; in Chapter 2, edges and triangles of special types. But in Chapter 3 we introduced the negative integers to simplify the statements of the weighing problems, replacing three equations by one. And in Chapter 5 the negative integers were of use in finding $M$ and $N$ such that $(A, B) = MA + NB$. For some purposes, however, such as measuring temperature or distance, we require numbers that are not positive or negative integers.

We can think of the integers as descriptions of certain regularly spaced points on a thermometer or ruler:

$$\cdots \quad -2 \quad -1 \quad 0 \quad 1 \quad 2 \quad 3 \quad 4 \quad 5 \cdots$$

Once this scale is fixed, then to every point on the line there corresponds a number, and to every number there corresponds a point on the line. [A proof of this requires a close study of the geometric notion "line" and of the algebraic notion "number," and is quite technical (see R 6, R 7, R 8)]. In view of this fact, we might refer to "the point 3," meaning, of course, the point corresponding to the number 3.

For more accurate measurements, we usually introduce a finer division than the integers by cutting each segment into 4, 8, 16, or 32 identical pieces, if we are working with inches, or into 10, 100, or 1000 pieces, if we are working with the metric system. There is nothing sacred about these particular numbers. The mathematicians of Babylonia, some 4000 years ago, chose to divide each segment into 60 pieces. If we wish, we can divide each segment into 117 pieces, but no civilization has ever chosen to do so.

Let us look at the points of division when we cut each segment into, say, five pieces:    $\cdots \quad -2 \quad -1 \quad 0 \quad 1 \quad 2 \quad 3 \quad 4 \quad 5 \cdots$

The finer unit of measure is now $\frac{1}{5}$. The points of division are labeled

$$\cdots \frac{-7}{5}, \frac{-6}{5}, \frac{-5}{5}, \frac{-4}{5}, \frac{-3}{5}, \frac{-2}{5}, \frac{-1}{5}, \frac{0}{5}, \frac{1}{5}, \frac{2}{5}, \frac{3}{5}, \frac{4}{5}, \frac{5}{5}, \frac{6}{5}, \frac{7}{5} \cdots$$

Of course, $\frac{0}{5}$ is just the point 0, and $\frac{5}{5}$ is 1. Perhaps the reader has the impulse to change $\frac{7}{5}$ to $1\frac{2}{5}$, which is short for $1 + \frac{2}{5}$, but let him restrain himself. Whether such a change complicates or simplifies depends on the application we have in mind. For the purpose of this chapter, it will turn out that to insist on writing $\frac{7}{5}$ as $1 + \frac{2}{5}$ will only complicate matters.

If we divided each segment into, say, twenty-seven pieces instead of five, then we could label all points that can be written in the form $M/27$ for some integer $M$. More generally, when we divide each segment into $N$ pieces, then the points of division will correspond to numbers of the form $M/N$.

*Numbers that can be expressed in the form $M/N$, for some integers $M$ and $N$ ($N$, of course, not $0$), we shall call* **rationals.** That is, a rational is a number describing a point that appears as one of the marks of division in some refinement of our original fixed scale 0, 1, 2, 3, $\cdots$. Observe that every integer is a rational; for example, $6 = 6/1, -7 = -7/1$, and so on; and, generally, $N = N/1$. We might think that every point on the line is described by some rational—that is, we might think that each point is the division mark of some perhaps extremely small refinement of the given scale. After all, as we draw in more and more division points of smaller and smaller refinements, we seem to be filling up the line gradually with an unbroken blur of ink. It is hard to imagine that a point could escape being a mark of at least one refining scale or, what amounts to the same thing, that there are numbers that are not rational.

Though it is hard to imagine, we must try; for, some 2400 years ago, the Pythagoreans, much to their astonishment, discovered numbers that are not rational. Let us see what kind of problems led them to their discovery of the world of numbers beyond the realm of rationals. (For the history of this discovery see R 4.)

More than 1000 years before Pythagoras, the mathematicians of Babylonia had a formula relating the two shorter sides of a right triangle to the side opposite the right angle. If the lengths of the three sides of the right triangle are given by the numbers $a$, $b$, and $c$, as in this diagram,

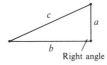

then, say the perfectly preserved Babylonian clay tablets,

$$a^2 + b^2 = c^2.$$

Though they used this formula for over a thousand years, there is no evidence that they had a proof for it. But, using it, they knew that if $a = 3$ and $b = 4$, then $c = 5$ ($3^2 + 4^2 = 5^2$) and that if $a = 8$ and $b = 15$, then $c = 17$ (as the reader may check).

Hundreds of proofs now exist for the *Pythagorean Theorem, which asserts that in all right triangles $a^2 + b^2 = c^2$*. Some of these date back to the Greeks, perhaps even to Pythagoras himself. The proof that we shall present here depends on a pictorial interpretation of the three numbers $a^2$, $b^2$, and $c^2$.

The area of a square of side $a$ is $a^2$. Similarly, $b^2$ and $c^2$ can be thought of as the areas of squares of sides $b$ and $c$, respectively. With this point of view, we are therefore concerned with the three squares shown in this figure:

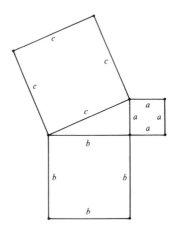

To prove the Pythagorean Theorem, we will prove that the two smaller squares together have an area equal to that of the big square. To do so, let us look closely at these two big squares (of side $a + b$):

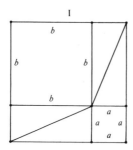 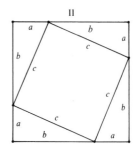

Squares I and II have equal area. If we take away from each the same area, namely, the four triangles, then what we are left with in each case must be the same. In square I we are left with the two small squares; in square II we are left with the big square of side *c*. We have proved the Pythagorean Theorem.

The reader who objected to Mackie the cat in Chapter 2 is probably quite exasperated. "Look here, you with your areas and your squares. You are proving a theorem that concerns lengths and angles. What's the idea of dragging in area?" Once again this scrupulous reader has a point. Fortunately, there is a proof involving only angle and length, and he can find it in E 49, E 50, and E 51.

Now we are ready to observe how geometry presents numbers that are not rationals. When we apply the Pythagorean Theorem to the simplest right triangle anyone might think of,

we get into trouble in trying to describe *c* numerically. Since

$$1^2 + 1^2 = c^2,$$

we know that $c^2 = 2$. But what is *c*? Clearly *c* is somewhere between 1 and 2. In fact, as this picture suggests, *c* is a little less than $\frac{3}{2}$.

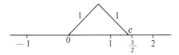

The Babylonians estimated that *c* is about $1 + 24/60 + 51/(60)^2 + 10/(60)^3$ (see R 1), which is decimally about 1.414, or, if you prefer to write our decimals, similarly, as sums of certain rationals, $1 + 4/10 + 1/(10)^2 + 4/(10)^3$. In any case, *c* is trapped between the rationals 1414/1000 and 1415/1000. Perhaps *c* is of the form *M/N*, where *N* is not a power of 10. For example, we might try to find *M* such that $(M/13)^2 = 2$. Now, as the reader may check, $(18/13)^2$ is less than 2, whereas $(19/13)^2$ is more than 2. Thus no rational with denominator equal to 13 can have its square equal to 2. Similarly, the reader can check that no rational *M/N* with *N* less than 13 has its square equal to 2. And, if he has time, he might easily eliminate $N = 14, 15, 16$, or 17, and so on (depending on the time available) as possibilities.

But the Greeks disposed of all *M/N* in one sweeping analysis, and showed that no rational has its square equal to 2. We will present a different proof, one that employs the Fundamental Theorem of Arithmetic and which will help us study far more than the single case $c^2 = 2$. (The Greeks' proof, which rests on the theory of odd and even, is sketched in E 45.)

For convenience, let us call the positive number whose square is 2 "*the square root of 2.*" Similarly, we could speak of the square root of 25, which happens to be 5. The square root of the natural number $A$ we will write simply $\sqrt{A}$. We know that $\sqrt{2}$ is a little larger than 1.414 and will show that it is not rational.

A *number that is not rational we will call* **irrational**. The rationals and irrationals together describe all points on the line. Thus there are two types of numbers, rational and irrational, which together constitute the real numbers.

We shall prove that there is no rational whose square is 2; that is, we shall prove that the number $c$ in question, which is a little bigger than 1.414, is not rational. To prove that $c$ is not rational we shall reason by contradiction, as we did with the Fifteen puzzle, where we proved that a certain assumption leads us to a natural number that is both odd and even. To begin, we assume that there are natural numbers $M$ and $N$ such that

$$\frac{M}{N} \cdot \frac{M}{N} = 2,$$

and from this assumption deduce that there is a natural number that is both odd and even. Such an assumption, leading us to nonsense, must itself be nonsense. Thus our reasoning will show that $c$ is not rational.

Multiplying both sides of the equation by $N^2$ we obtain

$$M^2 = 2N^2,$$

an equation that involves only natural numbers and which asserts that "the square of $M$ is twice the square of $N$." We now wish to prove that the square of a natural number can never be twice the square of another natural number (other than the case $0^2 = 2 \cdot 0^2$).

To do this, we are going to examine the factorization of $M^2$ and $2N^2$ into primes. Imagine that $M$ is expressed as the product of $d$ primes,

$$M = M_1 M_2 \cdots M_d$$

(if $M$ is prime, $d = 1$). Similarly express $N$ as the product of, say, $e$ primes

$$N = N_1 N_2 \cdots N_e.$$

Then $M^2 = (M_1 M_2 \cdots M_d)(M_1 M_2 \cdots M_d)$ is the product of $d + d = 2d$ primes. Similarly $N^2$ is the product of $2e$ primes. Thus $2N^2$ is the product of $2e + 1$ primes, because of the extra prime factor, 2.

The number of primes in the prime factorization of $M^2$ is even. The number of primes in the prime factorization of $2N^2$ is odd. But $M^2$ is supposed to equal $2N^2$: we have a violation of the Fundamental Theorem of Arithmetic, summarized in this diagram:

| $M$ | $M$ | $=$ | $2$ | $N$ | $N$ |
|---|---|---|---|---|---|
| Product of $d$ primes | Product of $d$ primes | One prime | | Product of $e$ primes | Product of $e$ primes |

Product of an even number of primes      Product of an odd number of primes

This contradiction proves that there is no rational number whose square is 2. The number whose square is 2 is not rational.

The same reasoning that showed that $\sqrt{2}$ is irrational would apply to the square root of any prime, since the only property we used of the number 2 is that it is prime. The Fundamental Theorem of Arithmetic has therefore provided us

THEOREM 1. *The square root of any prime is irrational.*

Let us see which natural numbers have rational square roots and which have irrational square roots. Both $\sqrt{0}$ and $\sqrt{1}$ are rational, for $\sqrt{0} = 0$ (since $0^2 = 0$) and $\sqrt{1} = 1$ (since $1^2 = 1$). Theorem 1 shows that $\sqrt{2}$ and $\sqrt{3}$ are irrational. Next, consider $\sqrt{4}$; since $2^2 = 4$, we have $\sqrt{4} = 2$, which is rational. By Theorem 1, $\sqrt{5}$ is irrational.

The first case that offers a fresh challenge is $\sqrt{6}$, for 6 is neither a prime nor is it the square of a natural number. Let us try to prove that $\sqrt{6}$ is irrational by the same argument that showed that $\sqrt{2}$ is irrational.

If $\sqrt{6} = M/N$, then, as before,

$$M^2 = 6N^2.$$

Let us see what happens if we count the number of primes in the prime factorizations of $M^2$ and $6N^2$. Both $M^2$ and $N^2$ would have an even number of primes. Since $6 = 2 \cdot 3$, the product of two primes, the number of primes in the factorization of $6N^2$ is two more than an even number, hence even. There is no contradiction this time. We simply conclude that if $M^2 = 6N^2$, then the prime factorizations of $M^2$ and $6N^2$ both involve an even number of primes. This argument does not allow us to conclude whether $\sqrt{6}$ is irrational.

| $M^2$ | $=$ | $6$ | $N^2$ |
|---|---|---|---|
| Product of an even number of primes | | Product of two primes | Product of an even number of primes |

Product of an even number of primes
(No contradiction; no information about $\sqrt{6}$.)

Imitation of our previous argument brought us to a dead end. But just a slight refinement in the reasoning will lead us to a contradiction. *Instead of counting the total number of primes*, let us *count only the number of 2's* in the prime factorizations of $M^2$ and $6N^2$. In the prime factorizations of both $M^2$ and $N^2$ the prime 2 appears an even number of times (maybe zero) but in the factorization of 6 it appears once. Thus 2 appears an *even* number of times in the factorization of $M^2$ but an *odd* number of times in the factorization of $6N^2$. The contradiction proves that $\sqrt{6}$ is irrational.

$$\underbrace{M^2}_{\substack{\text{Even number of 2's in} \\ \text{prime factorization}}} = \underbrace{\underbrace{6}_{\text{One 2}} \underbrace{N^2}_{\substack{\text{Even number of 2's in} \\ \text{prime factorization}}}}_{\substack{\text{Odd number of 2's} \\ \text{in prime factorization}}}$$

(Contradiction; hence $\sqrt{6}$ is irrational.)

THEOREM 2. *The square root of 6 is irrational.*

This refined line of reasoning is delicate enough to determine which natural numbers have rational square roots and which have irrational square roots. Consider, for instance, $\sqrt{360}$. First factor 360 into primes, $360 = 2 \cdot 2 \cdot 2 \cdot 3 \cdot 3 \cdot 5$. Our experience with $\sqrt{6}$ would suggest that we tabulate appearances of 2's, or of 3's, or of 5's. If we tabulated 3's we would not arrive at a contradiction, since there is an even number of them in the factorization of 360. But if we tabulate 2's (or 5's), we will arrive at a contradiction, for there is an odd number of 2's (and also of 5's). Thus $\sqrt{360}$ is irrational. The reader may check that the same type of reasoning establishes

THEOREM 3. *If in the prime factorization of the natural number $A$ there is at least one prime that appears an odd number of times, then $\sqrt{A}$ is irrational.*

On the other hand, what can we say if, when we write $A$ as the product of primes, each prime appears an even number of times? For example, what if $A$ is $3 \cdot 3 \cdot 5 \cdot 5 \cdot 5 \cdot 5$? In this case, we can easily compute $\sqrt{A}$; it is simply $3 \cdot 5 \cdot 5$, the natural number formed by taking each prime half as often as it appears when $A$ is written as the product of primes. That is, $A$ is a square. More generally, we obtain

THEOREM 4. *If in the prime factorization of the natural number $A$ each prime appears an even number of times, then $A$ is the square of a natural number, hence $\sqrt{A}$ is rational.*

Taken together, Theorems 3 and 4 tell us that the only natural numbers that have rational square roots are the squares.

The Fundamental Theorem of Arithmetic has shown its power. Not only did it help us prove that $\sqrt{2}$ is irrational, but it even helped us to determine all natural numbers that have irrational square roots. The Fundamental Theorem of Arithmetic, which depends on the addition and multiplication of natural numbers, has provided an insight into the irrationals.

It was geometry, through the Pythagorean Theorem, that led us to the irrational numbers. But even ordinary arithmetic can show us that there are numbers that are not rational. To see this, consider the decimal expansion of a particular rational number, say $\frac{11}{14}$. Carrying out the usual division,

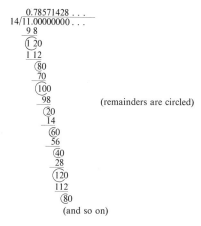

we find that the decimal representation of $\frac{11}{14}$ is quite orderly. It is simply 0.7 followed by the block 857142 repeated endlessly:

$$0.7857142857142857142\ldots,$$

which we will indicate simply as $0.7\overline{857142}$. (Similarly we would write $0.3333\ldots$ as $0.\overline{3}$.) As soon as we met the remainder 12 for the second time, our computation began to repeat exactly, step for step, remainder for remainder, an earlier part of the division.

The next theorem shows that the decimal representation of every rational resembles that for $\frac{11}{14}$ in that it goes on and on, repeating the same block. Although $\frac{1}{2} = 0.5$ is a terminating decimal, we can also consider it, as well as any terminating decimal, a repeating decimal by adding a repeating block of 0's, $\frac{1}{2} = 0.50\overline{0}$. Such a decimal representation we call **repeating.**

THEOREM 5. *If A and B are natural numbers, then the decimal representation of A/B is repeating.*

PROOF. We will show that when we carry out the division of $B$ into $A$, a remainder must be repeated. But this is easy to see, since there are only $B$ possible remainders when a number is divided by $B$; namely, 0, 1, 2, . . . , $B - 1$. By the time we have carried out $B + 1$ divisions, at least one remainder must be duplicated. Thus the decimal representation of $A/B$ is repeating, and the repeating block has at most $B$ digits.

Theorem 5 assures us that *any nonrepeating decimal* cannot be a rational number, and hence *must be an irrational number*. In particular,

$$0.101001000100001000001 \ldots,$$

formed by alternating 1's with longer and longer blocks of 0's, is an *irrational number*.

Are there any irrational numbers whose decimal representation is repeating? The answer is "no," as is shown by

THEOREM 6. *If the decimal representation of a number is repeating, then that number is rational.*

PROOF. We illustrate the reasoning by proving that the number $X$ whose decimal representation is $2.6\overline{145}$ is rational. We have

$$X = 2.61454545 \ldots$$

Now we multiply both sides of this equation by 100 (in order to slide the decimal two places to the right), obtaining

$$100X = 261.45454545 \ldots.$$

Subtracting the first equation from the second one, and noticing the cancellation of 45's, we obtain

$$99X = 258.84,$$

hence

$$9900X = 25884,$$

and finally

$$X = \frac{25884}{9900} = \frac{719}{275}.$$

Thus $X$ is a rational number. The same technique—multiplying the repeating decimal $X$ by an appropriate power of 10 (the power is the number of digits in the repeating block), subtracting the resulting two equations, and solving for $X$,—can be applied to write any repeating decimal in the form $A/B$, where $A$ and $B$ are natural numbers. Thus any repeating decimal is a rational number, and Theorem 6 is proved.

Theorems 5 and 6 show that the distinction "rational" versus "irrational" corresponds to the distinction "repeating decimal" versus "nonrepeating decimal." In particular, since $\sqrt{2}$ is irrational, we may conclude that the decimal representation of $\sqrt{2}$, which begins 1.41421 . . . , is nonrepeating.

Just as geometry raised a question about numbers and exposed the irrationals, so does the existence of irrationals in turn have an impact upon geometry. This we will examine in the next chapter, on covering rectangles with squares.

## Exercises

1. Draw the points on the number line corresponding to (a) $\frac{3}{4}$, (b) $\frac{6}{8}$, (c) $\frac{7}{5}$, (d) $\frac{14}{10}$.

2. (a) Fill in the blanks: $\frac{3}{4} = \frac{6}{8} = \frac{9}{12} =$ _____ $=$ _____ $=$ _____ $= \ldots$ .
   (b) Fill in the blanks: $\frac{2}{5} = \frac{4}{10} = \frac{6}{15} =$ _____ $=$ _____ $=$ _____ $= \ldots$ .
   (c) Use (a) and (b) to compute $\frac{3}{4} + \frac{2}{5}$.

3. Why is it easier to multiply two fractions than to add them?

4. (a) Draw a right triangle.
   (b) Measure its three sides (preferably in centimeters).
   (c) Compute $a^2 + b^2$ and $c^2$.
   (d) Compare your results in (c) with the Pythagorean Theorem.

5. The reader may make the proof of the Pythagorean Theorem more vivid by cutting out of cardboard eight right triangles of the same size, and three squares, each corresponding to one of the three sides of his triangle. With these he can then assemble the squares I and II of the proof.

6. By straightforward arithmetic, check that there is no rational number with denominator 7 whose square is 3.

7. (a) Compute $(1.7)^2$ and $(1.8)^2$.
   (b) In view of (a), what can we say about the decimal representation of $\sqrt{3}$?
   (c) Compute $(1.73)^2$ and $(1.74)^2$.
   (d) In view of (c), what can we say about the decimal representation of $\sqrt{3}$?

8. (See E 7.) Using the idea suggested in E 7 and a little arithmetic, fill in the blanks:
   $\sqrt{5} = 2.\_\_\_ \ldots$ .

9. Some students have raised the question, "How do you know what $\sqrt{2}$ is if you've never written it as a decimal?" How would you answer this question?

10. Which of these statements about $\sqrt{2}$ is the most fundamental: (a) $\sqrt{2}$ is irrational, (b) $\sqrt{2} = 1.414\cdots$, (c) $\sqrt{2} \cdot \sqrt{2} = 2$? Why?

11. (a) If the two shorter sides of a right triangle are 5 and 12, find the third side.
    (b) If the two longer sides of a right triangle are 29 and 21, find the smallest side.

12. Assume that the two short sides of a right triangle are both 30.
    (a) Show that the long side is $30\sqrt{2}$,
    (b) Give a decimal estimate of $30\sqrt{2}$,

(c) Compare this estimate to your measurement of the long side of a right triangle with its two short sides each of length 30.

13. Compare your estimate of E 12(b) to the Babylonian estimate:

$$42 + \frac{25}{60} + \frac{35}{(60)^2}.$$

14. If the two shorter sides of a right triangle are 4 and 5, find the third side to one decimal place.

15. Without using Theorem 3, prove that $\sqrt{10}$ is irrational.

16. Without using Theorem 3, prove that $\sqrt{30}$ is irrational.

17. Without using Theorem 3, prove that $\sqrt{108}$ is irrational.

18. Is $1 + \sqrt{2}$ rational or irrational? Prove your answer.

19. If $N$ is an integer, is $N + \sqrt{2}$ always irrational? Prove your answer.

20. If $R$ is a rational number, is $R + \sqrt{2}$ always irrational? Prove your answer.

21. If $R$ is a rational number, is $R\sqrt{2}$ always irrational? Prove your answer.

22. Is $\sqrt{2} + \sqrt{3}$ rational or irrational? Prove your answer.

23. What is the relation of Lemma 4 of Chapter 5 (the weighing lemma) to our proof that $\sqrt{2}$ is irrational?

24. Is there a smallest number that is greater than $0$?

25. Is $\sqrt{\frac{5}{3}}$ rational or irrational? Prove your answer.

26. If $a$ and $b$ are numbers such that $b = a^3$, we call $a$ "the cube root of $b$" and write $a = \sqrt[3]{b}$. For instance, $\sqrt[3]{8} = 2$ and $\sqrt[3]{27} = 3$. Using the Fundamental Theorem of Arithmetic, prove that $\sqrt[3]{2}$ is irrational. (*Warning:* Review the proof of Theorem 1, but be careful. "Odd and even" is not of use.)

27. See E 26. Prove that $\sqrt[3]{135}$ is irrational.

28. See E 25 and E 26. For which natural numbers $A$ is $\sqrt[3]{A}$ rational? Prove your answer.

29. Using simple arithmetic, fill in the two blanks: $\sqrt[3]{2} = 1.\_\_\_\ldots$ .

30. See E 26. Prove that $\sqrt[4]{72}$ is irrational. [Note that $(\sqrt[4]{72})^4 = 72$.]

31. (a) Express as decimals: $\frac{13}{10}, \frac{5}{3}, \frac{1}{7}$.
    (b) What is the repeating block in each case?

32. Express $\frac{3}{17}$ as a decimal.

33. Find natural numbers $M$ and $N$ such that $M/N$ is equal to (a) 1.31, (b) 2.14$\overline{57}$, (c) 5.4$\overline{17}$.

34. Find natural numbers $M$ and $N$ such that $M/N$ is equal to (a) 1.414, (b) 0.38$\overline{7}$, (c) 6.1$\overline{9}$, (d) 8.3$\overline{724}$.

35. (a) Prove that for natural numbers $A$ and $B$, we have $\sqrt{A}\sqrt{B} = \sqrt{AB}$. (*Hint:* Square both sides.)
    (b) Show by an example that $\sqrt{A} + \sqrt{B}$ need not equal $\sqrt{A + B}$.

36. (a) Using the distinction between odd and even, prove that $(\frac{3}{2})^{12}$ is not equal to $2^7$.
  (b) Using the Fundamental Theorem of Arithmetic, prove that $(\frac{5}{3})^{10}$ is not equal to $(\frac{7}{5})^{16}$.
  (c) Compute $(\frac{3}{2})^{12}$ and $2^7$ in decimals, and compare them.

37. The frequency of a note that is an octave above another is exactly twice the frequency of the lower note. The frequency of a note that is a fifth above another note is $\frac{3}{2}$ the frequency of the lower note. Starting at the lowest $C$ on the piano, you will notice that going up 7 octaves will bring you to the highest $C$; going up 12 fifths will also bring you to the top $C$. Prove that if a piano could be tuned such that fifths as well as octaves were true, then $(\frac{3}{2})^{12}$ would equal $2^7$ (but compare with E 36).

38. (a) Is the product of two rationals always rational? Prove your answer.
  (b) Is the product of two irrationals always irrational? Prove your answer.

39. (a) If you multiply two nonrepeating decimals, can you ever obtain a repeating decimal?
  (b) If you multiply two repeating decimals, can you ever obtain a nonrepeating decimal?

40. Draw as well as you can the point on the number line corresponding to 0.101001000100001 . . . .

41. Determine whether there are two different numbers such that
  (a) there is no number between them,
  (b) there is no rational number between them,
  (c) there is no irrational number between them.

42. (a) Give an example of a rational number whose decimal expansion begins with 0.712 . . . .
  (b) Give an example of an irrational number whose decimal expansion begins with 0.712 . . . .

43. (a) When you add two repeating decimals, do you necessarily obtain a repeating decimal? Prove your answer.
  (b) When you add two nonrepeating decimals do you necessarily obtain a repeating decimal? Prove your answer.

●

44. Prove that if $M$ and $N$ are natural numbers, there are natural numbers $A$ and $B$ such that $(A, B) = 1$ and $A/B = M/N$.

45. The Pythagoreans based their proof of the irrationality of $\sqrt{2}$ on the distinction between odd and even, but did not require the Fundamental Theorem of Arithmetic. Carry out their proof, which we now sketch:
  (i) If $\sqrt{2}$ is a rational number, then we may assume that $\sqrt{2} = A/B$ with $(A, B) = 1$ (see E 10).
  (ii) From $2B^2 = A^2$, we see that $A$ is even and thus can be written as $2C$ for some natural number $C$.
  (iii) From $2B^2 = 4C^2$, we see that $B$ is even.
  (iv) Thus $(A, B)$ is at least 2, a contradiction with (i).

46. Prove that between any two numbers there is an endless supply of both rationals and irrationals.

47. The three dimensions of a box are 1, 2, and 2. Find the length of the long diagonals (the diagonals not on the surface of the box).

48. The three dimensions of a box are 2, 3, and 6. Find, as in E 47, the length of the long diagonals.

49. Two triangles are similar if the three angles of one are equal to the three angles of the other. Prove that the triangles $ADC$, $ACB$, and $CDB$ of this figure are similar:

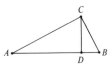

$CD$ and $AB$ are perpendicular; so are $AC$ and $CB$.

50. It is proved in geometry that if $a$, $b$, and $c$ are the lengths of the three sides of one triangle and $a'$, $b'$, $c'$ the lengths of the three corresponding sides of a similar triangle, then

$$\frac{a}{a'} = \frac{b}{b'} = \frac{c}{c'}.$$

Using this fact and E 49, prove that $b/p = c/b$ and $a/q = c/a$, where the letters are defined by this diagram:

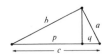

51. Continued from E 50. From the relations $p + q = c$, $b/p = c/b$ and $a/q = c/a$, prove that

(a) $b^2 = pc$,   (b) $a^2 = qc$,   (c) $a^2 + b^2 = (p + q)c$,   (d) $a^2 + b^2 = c^2$.

(E 50 and E 51 provide a proof of the Pythagorean Theorem that uses length and angle but not area.)

52. See E 45 and E 46 of Chapter 4. We have given the symbols $2^0$, $2^1$, $2^2$, ... meaning. Let us, since we are the masters, now give the symbol $2^{1/2}$ meaning.
    (a) If we want the rule $2^{m+n} = 2^m \cdot 2^n$ to hold even when $m$ and $n$ are not natural numbers, show that we would want $2 = 2^{1/2} \cdot 2^{1/2}$.
    (b) Show that (a) advises us to use $2^{1/2}$ as a new name for $\sqrt{2}$.

53. See E 52. We have not given the symbol $2^{-1}$ any meaning.
    (a) Show that if we want the rule $2^{m+n} = 2^m \cdot 2^n$ to hold even when $m$ or $n$ is negative, then we would want $2^0 = 2^1 \cdot 2^{-1}$.
    (b) Show that (a) tells that $2^{-1}$ should be a new name for $\frac{1}{2}$.

54. Using the ideas of E 52 and E 53, show that we should make (a) $25^{1/2} = 5$, (b) $3^{1/2} = \sqrt{3}$, (c) $3^{-2} = \frac{1}{9}$.

55. See E 52. Using the ideas of E 52 and E 53, show that we should make (a) $4^{3/2} = 8$, (b) $4^{-(3/2)} = \frac{1}{8}$.

56. (a) Show that $(2^3)^4 = 2^{3 \cdot 4}$.
   (b) Why does $(2^m)^n = 2^{mn}$ for natural numbers $m$ and $n$?
   (c) (See Exercise 53.) Why should $a^{-1}$ stand for the reciprocal of $a$, that is, $1/a$, for any nonzero number $a$?
   (d) Show that if we want the rule $(2^m)^n = 2^{mn}$ to hold when $m = -1$ and $n = -1$, then we must have $(-1)(-1) = 1$. (This is another plausibility argument for "negative times negative is positive.")

● ●

57. Prove that the sum of the three angles of any triangle is 180° by drawing the dotted lines indicated in the figure below and using the facts that $A = D$, $B = E$, where the letters name angles in this diagram.

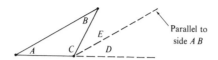
Parallel to side $A B$

58. In square II of the proof of the Pythagorean Theorem, we tacitly *assumed* that the figure whose four sides are all of length $c$ is a square. Show that it is indeed a square by proving that each of its four corners is a right angle. (*Hint:* Use E 57.)

59. Consider an endless, evenly spaced arrangement of dots in the plane. Any four adjacent dots are the vertices of a one-inch square, which we will take as our basic unit for measuring areas.
   (a) Show that the area of any triangle whose three corners are chosen among the given dots is rational.
   (b) Can three of the given dots be found that are the corners of an equilateral triangle?

60. With an unmarked straightedge and a compass can you cut a given line segment into two segments of equal length? Into three segments of equal length?

61. With the equipment described in E 60, can you cut a given angle into two angles of equal size? Into three angles of equal size?

62. (a) Show how two rulers or yardsticks can be used to make a slide rule for addition and subtraction.
   (b) To make a slide rule for multiplication and division, mark two strips of paper (or wooden slats) as follows. At the left end write the numeral 1. One inch from the left write the numeral 2. Two inches from the left write the numeral 4. Three inches from the left write the numeral 8. Continue in the same manner for several inches; that is, $m$ inches from the left write the decimal form of $2^m$. How can these two strips of paper be used to multiply powers of 2? Why is the relation $2^m \cdot 2^n = 2^{m+n}$ important here?

63. A natural number is called *square-free* if 1 is the only square that divides it. How can you decide whether a natural number is square-free if you have it written as the product of primes?

64. See E 63.
    (a) How many square-free natural numbers are there from 1 to 100?
    (b) How many square-free natural numbers are there from 101 to 200?
    (c) How many square-free natural numbers are there from 201 to 300?

65. This continues E 64.
    (a) Compute $6/\pi^2$ to two decimals.
    (b) Compute to two decimals the fraction of natural numbers from 1 to 200 that are square-free.
    (c) It is known that as we make longer and longer lists, the fraction of square-free natural numbers gets closer and closer to $6/\pi^2$. Use this fact and (b) to estimate $\pi$.

66. (a) On a piece of paper or tagboard draw a right triangle *ABC* and two squares against its shorter sides. Cut the squares into eight triangles, as shown (using diagonals of the squares and lines parallel to the longest side of the right triangle).

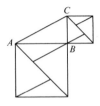

    (b) Cut out the eight triangles forming the two squares, and assemble them into a square whose side is the longest side of the right triangle *ABC*.
    (c) How is (b) related to the Pythagorean Theorem?

67. There is a very fast way of estimating the square root of a positive real number $A$; that is to say, of finding the positive root of the equation $X^2 - A = 0$. We illustrate the method by estimating $\sqrt{19}$. As a first guess we take 4. We then divide 4 into 19, obtaining 4.75. Next, we average 4 and 4.75; that is, we add 4 to 4.75 and divide the sum by 2. This average is 4.375. Now we use 4.375 as the new guess and repeat the process. That is, we divide 4.375 into 19, say, to 3 decimals, getting 4.349; and average 4.375 and 4.349, getting 4.362. If we wish, we can proceed with 4.362 as our guess. In any case, 4.362 is a close estimate for $\sqrt{19}$, as the reader may check by squaring 4.362.
    (a) Compute the square of our first guess, 4.
    (b) Compute the square of our second guess, 4.375.
    (c) Compute the square of our third guess, 4.362.
    This technique was used by the Babylonians.

68. (a) Using the method of E 67, estimate $\sqrt{5}$, starting with a guess of 2 and applying the process twice.
    (b) Compare the square of your final guess to 5.

69. The same as E 68 but estimate $\sqrt{7}$, and choose your own first guess.

70. A Pythagorean triplet $a$, $b$, $c$ consists of integers $a$, $b$, $c$ such that $a^2 + b^2 = c^2$. For instance, 3, 4, 5 and 6, 8, 10 are Pythagorean triplets.

    (a) Let $m$ and $n$ be any natural numbers. Show that

$$a = 2mn, \qquad b = m^2 - n^2, \qquad c = m^2 + n^2$$

    constitute a Pythagorean triplet. (For instance, taking $m = 3$, $n = 2$, we have $a = 12$, $b = 5$, $c = 13$; note that $12^2 + 5^2 = 13^2$.)

    (b) Use the method described in (a) to find more Pythagorean triplets. Be sure to include the case $m = 2$, $n = 1$.

71. See E 70. Let $a$, $b$, $c$ be a Pythagorean triplet such that $(a, b) = 1$.

    (a) Show that $(b, c) = 1$.

    (b) Why can't both $a$ and $b$ be odd, nor both be even?

    (c) In view of (b), let us assume that $a$ is even and $b$ is odd. Show that

$$a^2 = (c - b)(c + b).$$

    (d) Show that $(c - b, c + b) = 2$.

    (e) Combining (c) and (d) show that $c - b$ is twice the square of an odd number, and that the same is true of $c + b$.

    (f) Setting $c + b = 2m^2$ and $c - b = 2n^2$, deduce that $c = m^2 + n^2$, $b = m^2 - n^2$, and $a = 2mn$.

### References

1. B. L. van der Waerden, *Science Awakening*, Noordhoff Ltd., Groningen, Holland, 1954. (A discussion of the Babylonian and Egyptian treatments of right triangles is to be found on pp. 6, 76–77. Plate 8, facing p. 65, has a photograph of the clay tablet showing a right triangle whose two short sides have length 30.)

2. O. Ore, *Number Theory and Its History*, McGraw-Hill, New York, 1948. (The right triangle interested the ancients in the problem of finding all pairs of square natural numbers whose sum is again a square. The case of two cubes whose sum is a cube is discussed on pp. 204–207.)

3. I. Niven, *Numbers: Rational and Irrational*, Random House, New York, 1961. (The impact of the irrationals on the proof that corresponding sides of similar triangles are proportional is shown on pp. 46–49.)

4. K. von Fritz, The discovery of incommensurability by Hippasus of Metapontum, *Annals of Mathematics*, vol. 46, 1945, pp. 242–264.

5. O. Neugebauer, *The Exact Sciences in Antiquity*, Princeton University Press, 1952.

6. G. D. Birkhoff and R. Beatley, *Basic Geometry*, Chelsea, New York, 1959. (This is a specially fine text on high school geometry.)

For material on the geometry of points and the arithmetic of numbers and their relation, see R 8, R 9, and R 10 of Chapter 16.

## Chapter 7

# TILING

There is no barrier between one part of mathematics and another. In the last chapter a theorem in geometry led to a discovery in the algebra of numbers—the existence of the irrationals. To prove that, for example, $\sqrt{2}$ is irrational, we used the Fundamental Theorem of Arithmetic from the part of mathematics called number theory, which is concerned primarily with the study of the integers. The distinction between even and odd, also part of number theory, was of aid in our study of switches (part of algebra) and of labeling vertices (part of topology). We will find that the existence of irrationals has an impact on geometry; in particular, on the division of rectangles into squares.

It is a simple matter to tile a rectangle of dimensions 5 by 7 with nonoverlapping congruent squares. (Two geometric figures are *congruent* if they have the same shape and size; if they are not congruent, they are said to be *incongruent*.) The rectangle shown in this diagram

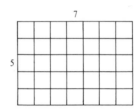

has been cut into 35 congruent squares. We could divide each of these squares into four smaller squares, all of equal size, and make a tiling of the rectangle with 140 squares.

Let us try to tile with congruent squares a more challenging rectangle, one whose dimensions are $4\frac{2}{3}$ by $8\frac{5}{7}$. We can rewrite $\frac{2}{3}$ as $\frac{14}{21}$ and $\frac{5}{7}$ as $\frac{15}{21}$; thus a square whose dimensions are $\frac{1}{21}$ by $\frac{1}{21}$ can tile this rectangle. We can even compute how many tiles are necessary. Since

$$4\frac{2}{3} = 4 + \frac{2}{3} = 4 + \frac{14}{21} = \frac{84}{21} + \frac{14}{21} = \frac{98}{21},$$

the little square whose sides are $\frac{1}{21}$ must be laid off 98 times along the width. Since

$$8\frac{5}{7} = 8 + \frac{5}{7} = 8 + \frac{15}{21} = \frac{168}{21} + \frac{15}{21} = \frac{183}{21},$$

the same little square must be laid off 183 times along the length. The total number of squares covering the rectangle is $98 \times 183 = 17,934$; but this number is of less interest than the 98 and 183 which already assure us that the tiling exists. The same reasoning proves

THEOREM 1. *A rectangle whose sides are rational can be tiled with congruent squares.*

What happens if the rectangle has irrational sides? For example, can we tile a rectangle whose dimensions are $10\sqrt{2}$ by $5\sqrt{2}$? Since this rectangle happens to be twice as long as it is wide, it can be tiled with two congruent squares, each with side $5\sqrt{2}$. Thus even this rectangle, though it has irrational sides, can be tiled with congruent squares.

Next let us consider the rectangle

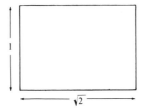

Is it possible to find a square, perhaps a very small one, that can be used to tile this rectangle of dimensions $\sqrt{2}$ by 1? That is, can we find a square that we can lay off along the width, and along the length, without having anything left over?

We can easily show that no such square can be found. Imagine, for the moment, that we have such a square and that its sides are of length $A$, perhaps some very small number. Since this square is supposed to tile the rectangle, there are natural numbers $M$ and $N$ such that

$$\sqrt{2} = MA$$

and

$$1 = NA.$$

From these two equations we deduce that

$$\frac{\sqrt{2}}{M} = A$$

and

$$\frac{1}{N} = A,$$

and hence

$$\frac{\sqrt{2}}{M} = \frac{1}{N}.$$

Thus $\sqrt{2} = M/N$. But we have already proved that $\sqrt{2}$ is irrational. From this contradiction, we gather that our assumption must be wrong. We have proved

THEOREM 2. *The rectangle of dimensions $\sqrt{2}$ by 1 cannot be tiled with congruent squares.*

In a similar manner, the reader may prove that we cannot tile the $\sqrt{2}$ by $\sqrt{3}$ rectangle with congruent squares. What is the general rule for determining which rectangles can be tiled with congruent squares? It is not simply that both sides must be rational, since the $5\sqrt{2}$ by $10\sqrt{2}$ rectangle can be tiled. What is it, then, that made this case so simple? The answer is that the quotient $(10\sqrt{2})/(5\sqrt{2})$ is 2 (a rational), even though the dimensions are not rational.

We should probably look at the quotient of the two dimensions rather than the dimensions themselves. Perhaps the reader has already guessed

THEOREM 3. *A rectangle whose dimensions are L by W, with L/W a rational, can be tiled with congruent squares.*

PROOF. Since $L/W$ is rational, we know that there are natural numbers $M$ and $N$ with

$$L/W = M/N.$$

Thus $LN = WM$. Dividing both sides by $MN$, we obtain

$$L/M = W/N.$$

This last equation tells us what size square can be used to tile the rectangle. A square of side $L/M$ fits exactly $M$ times along the length. Moreover, since $L/M$ is equal to $W/N$, this square fits exactly $N$ times along the width. Thus this little square, when laid down $MN$ times, tiles the rectangle. This proves Theorem 3.

Which rectangles cannot be tiled with congruent squares? It is reasonable to suspect those for which $L/W$ is irrational. One can prove

THEOREM 4. *A rectangle whose dimensions are L by W, with L/W irrational, cannot be tiled with congruent squares.*

The proof, similar to that of Theorem 2, is left to the reader.

Having learned which rectangles can be tiled with congruent squares, let us go to the opposite extreme and ask: Which rectangles can be tiled with squares *no two of which are the same size?* We will keep our tacit restriction; namely, that a tiling has only a finite number of tiles.

This is a question on which mathematicians have worked only in the twentieth century. But the Egyptians, forty centuries ago, were involved with a similar question about rationals. In their daily computations they represented a rational as the sum of unit fractions, no two being the same size. (A unit fraction is a rational of the form $1/N$, with $N$ a natural number.) For example, they expressed $\frac{2}{5}$ as $\frac{1}{3} + \frac{1}{15}$ and $\frac{2}{29}$ as $\frac{1}{36} + \frac{1}{236} + \frac{1}{531}$. On the other hand, it is a simple matter to represent $\frac{2}{5}$ as the sum of equal unit fractions, for example,

$$\frac{2}{5} = \frac{1}{5} + \frac{1}{5}.$$

Though their representation may strike us as peculiar, it is similar to our own decimal system. After all, when we write $\frac{5}{16} = 0.3125$ we are writing $\frac{5}{16}$ as the sum of rationals, no two the same size: $\frac{3}{10}, \frac{1}{100}, \frac{2}{1000}, \frac{5}{10000}$. When we write the numbers less than 1 decimally, we are actually expressing them as the sum of rationals of different sizes and of the form $M/N$, where $M$ is at most 9 and $N$ is a power of 10. (This is discussed in more detail in Chapter 15.)

But, after this detour, let us return to the problem of tiling a rectangle* with incongruent squares, a problem remotely resembling the Egyptians' problem of writing rationals. There are two ways of approaching this question. We may either try to piece together squares and hope that after a while we succeed in forming a rectangle, or we may start with a rectangle and try to cut it into squares, no two of the same size. Let us first try to assemble a rectangle out of incongruent squares. Using just two squares we would have either

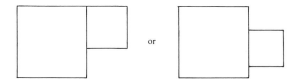

or

which surely are not rectangles. If we now add one square we get many possibilities, two of which are

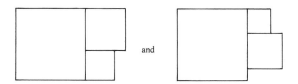

and

---

* If the rectangle happens to be a square, then it can of course be covered with one square. Such a trivial situation we will not call a tiling.

but neither of which forms a rectangle. Of course, if we allow two of the squares to be congruent, we could have this tiling of a rectangle:

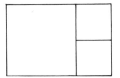

The reader, in assembling four squares to form a rectangle, will see that there are five substantially different ways

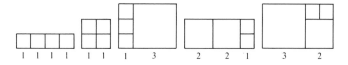

and that each has at least two congruent tiles. With five squares we can assemble a rectangle in many more ways, of which these are three samples:

Observe that in each rectangle there are congruent squares. The reader may add to the list and also go on to assemble six squares to form a rectangle. He will find, as we have seen so far, that each such tiling will have at least two congruent squares.

At this point we can either go on, trying to assemble seven incongruent squares into a rectangle, or we can seek a proof showing that it is impossible to tile a rectangle with incongruent squares. Or we could take the attitude of some research mathematicians who devote Monday to trying to prove a conjecture, Tuesday to trying to disprove it, Wednesday to trying to prove it, and so on, until either the conjecture is resolved or the mathematician is exhausted.

Already with five or six squares it is difficult to keep an accurate list of all possibilities and to be sure we have overlooked none. Going on to seven, therefore, is not promising. Thus, like the "alternating" mathematician, let us try to prove that it is impossible to tile a rectangle with incongruent squares.

In all our examples the smallest square is duplicated. This suggests that we should look closely at the smallest square and its neighboring squares. Imagine, for the moment, that the smallest square, $S$, lies on the border of the rectangle and that no two squares are congruent. We might have either

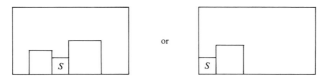

In either example, a square placed above *S* would have to be smaller than *S*. Thus we have

THEOREM 5. *If a rectangle is tiled with incongruent squares, then the smallest square does not touch the border of the rectangle.*

Now all we must do is prove that the smallest square must touch the border, and we will have a proof that a rectangle cannot be tiled with incongruent squares. Let us try.

Let us assume that the smallest square does not touch the border. Looking closely at this square *S* and at its surrounding neighbors, we see that if *S* had two neighboring squares touching one of its edges, for example,

then adding another square, *T*, above *S*, would give us something like this (since *T* must be larger than *S*):

Thus another square, *U*, placed to the left of *S*, would give

And similarly, any square placed below *S* would have to be smaller than *S*. Thus there can be only one square bordering *S* on its right (and only one on each of its other three sides), and the squares surrounding *S* will form a windmill.

If, as the other initial possibility, $S$ has only one larger neighboring square touching one of its edges, as in this figure

a similar analysis, as the reader may check, shows that there would be a square bordering $S$ and congruent to it. We have proved

THEOREM 6. *If a rectangle is tiled with incongruent squares, then the smallest square is surrounded by four squares.*

Instead of obtaining a contradiction to Theorem 5, we just learned a little more about the smallest square. Perhaps, like our "alternating" mathematician, we should try again to tile a rectangle with incongruent squares.

Here is a way that might work. Draw a rectangle cut up into rectangles in such a way that there is a chance of distorting all the small rectangles into squares (and, incidentally, the big rectangle into one of different shape). In view of Theorem 6, we should be sure to include a rectangle surrounded by four rectangles. One candidate, with nine rectangles, looks like this:

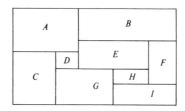

Either $D$ or $H$ might become the smallest square.

To see what stretching and shrinking might accomplish, let us use a little algebra. Imagine that rectangle $E$ becomes a square of side $x$ and that rectangle $H$ becomes a square of side $y$ and that all the other rectangles become squares. Square $F$ will then have side $x + y$. Thus $I$ will have side $y + (x + y)$, or simply $x + 2y$. Then $G$ will have side $y + (x + 2y)$, or simply $x + 3y$. If we call $z$ the side that $D$ must have, then we see that $z + x = (x + 3y) + y$. Thus $z = 4y$, and hence $D$ will have side $4y$. Square $C$ will have side $4y + (x + 3y)$, or simply $x + 7y$. Hence $A$ will have side $(x + 7y) + 4y$, or simply $x + 11y$.

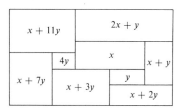

To obtain an equation relating $x$ and $y$, we must use an important piece of information: the total length of the right edges of squares $A$ and $D$ is the same as the total length of the left edges of squares $B$ and $E$. In algebraic terms we have $(x + 11y) + 4y = (2x + y) + x$. Collecting terms on each side of the equation, we get

$$x + 15y = 3x + y.$$

Subtracting $x$ from both sides yields

$$15y = 2x + y.$$

Subtracting $y$ from both sides yields

$$14y = 2x.$$

Dividing both sides by 2 yields

$$7y = x,$$

a relation between $x$ and $y$. If we choose $y = 100$, then $x$ must be 700. For convenience of arithmetic let us choose $y = 1$. Then $x = 7$, $x + y = 7 + 1 = 8$, and $F$ becomes a square of side 8. Similarly, we can compute the sides of the other six squares. Luckily, none of their sides turns out to be zero or negative; moreover, no two are equal. We have finally obtained a partition of a rectangle into (nine) incongruent squares.

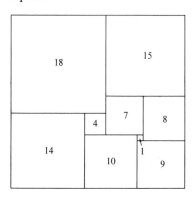

If we were to place a 33 by 33 square above this 32 by 33 rectangle, we would have ten incongruent squares tiling a 33 by 65 rectangle. Such a tiling, depending so much on a previous tiling, is not very appealing. Let us call a tiling of a rectangle by squares *simple* if no subcollection of the squares forms a rectangle. We see that a 32 by 33 rectangle has a simple tiling by incongruent squares. Even a square has a simple tiling by incongruent squares, as is shown by this tiling of a 503 by 503 square

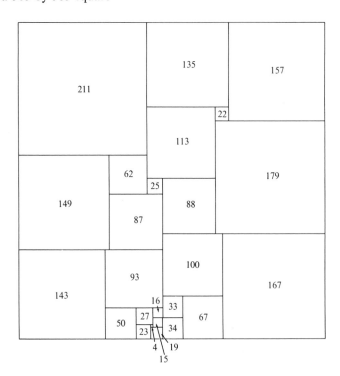

The mathematician J. C. Wilson reported this tiling by 25 squares in 1964. Whether the number of incongruent squares forming a simple tiling of a square can be reduced to less than 25 is not known. (It is known that any tiling of a square, simple or not, by less than 20 squares has a pair of congruent squares.) But no one knows which rectangles have simple tilings; for example, it is not known whether the 1 by 2 rectangle has a simple tiling.

Still keeping the restriction of incongruent tiles, we can go in various directions. We might ask: Can an equilateral triangle be tiled with incongruent equilateral triangles? Or we might move into space and ask: Can a box be filled with incongruent cubes?

Before he feels he understands the specific case of tiling rectangles, a mathematician will automatically examine such closely related questions. Perhaps he will find a general theorem to be true for tilings of many different objects. Such a theorem is more satisfying than knowledge of one special case, such as tiling by squares. And if he can find no all-inclusive theorem, no insight into a wide class of structures, then at least he knows the extent of his theory, and understands just what it is he is proving theorems about.

As it turns out, the case of an equilateral triangle, or of a box, is quite different from that for the rectangle. In R 1, Tutte proves

THEOREM 7. *It is impossible to tile an equilateral triangle with incongruent equilateral triangles.*

His proof is quite indirect and involved, and will be omitted; in E 31 we sketch what might be made the basis of a more direct and elementary proof.

The equilateral triangle thus behaves quite differently from the square. We should feel very fortunate that the square can be tiled with incongruent squares, yet we are disappointed that the theory is not so general as we hoped it would be. When we consider the problem of filling a box, we meet

THEOREM 8. *It is impossible to fill a rectangular box with incongruent cubes.*

PROOF (By contradiction). We will assume that it is possible to fill a box with incongruent cubes and deduce that there is an *endless* tower of cubes within the given *limited* supply of cubes. This contradiction will suffice to prove the theorem.

Let us look first only at those cubes that rest on the bottom of the box. Consider the smallest of these. We know from Theorem 5 (which states that when a rectangle is tiled with incongruent squares, the smallest square does not touch the border) that this cube—call it *C*—does not touch the border of the bottom; hence it will have a position such as this:

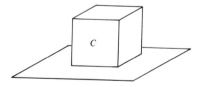

Thus the other cubes resting on the bottom of the box, and in touch with *C*, enclose *C* with a wall, since their tops all extend above *C*. Hence the cubes resting on *C* are all smaller than *C*, and their bases tile the top of *C* with incongruent squares. Now if we look at the smallest of the cubes resting on $C$—give it the name $D$—we see that we have started a tower of cubes:

It can easily be seen that we can continue the tower endlessly (by looking at the smallest cube resting on $D$, and so on). Thus the alleged tiling must have an endless supply of cubes. Since we have a limited number of cubes at our disposal, we have reached a contradiction: there can be no endless tower. Thus it is impossible to fill a box with incongruent cubes.

The tiling of rectangles with incongruent squares is, in a sense, a piece of luck. It is a theory that exists for the plane, but not for space; for rectangles, but not for equilateral triangles.

However, within the theory of tiling rectangles is a result that is not only a vast generalization of Theorem 4, but will illustrate how the machinery developed in one branch of mathematics can be applied to another, seemingly unrelated, branch.

In Chapter 8 we will use the theory of electrical networks to prove that it is impossible to tile the 1 by $\sqrt{2}$ rectangle with squares, even if we remove all restrictions on their relative size.

### Exercises

1. Tile a rectangle $21\frac{5}{7}$ by $17\frac{2}{5}$ with congruent square tiles.

2. What is the smallest number of congruent square tiles that will tile the rectangle of E 1?

3. Tile a rectangle $10\frac{3}{4}$ by $5\frac{1}{2}$ with congruent square tiles.

4. What is the smallest number of congruent square tiles that will tile the rectangle of E 3?

5. Can the following two rectangles be tiled with congruent square tiles?
   (a) $5\sqrt{3}$ by $7\sqrt{3}$,            (b) $7\sqrt{2}$ by $11\sqrt{3}$. Explain.

6. Can the following two rectangles be tiled with congruent square tiles? Explain.
   (a) $10\sqrt{5}$ by $15\sqrt{5}$,            (b) 10 by $12\sqrt{2}$.

7. (a) A box has dimensions $L$, $W$, and $H$. If the box can be filled with congruent cubes, what relations must hold between these numbers? Prove why.
   (b) If those relations hold, can the box necessarily be filled with congruent cubes? Explain.

8. In finding a tiling of a rectangle by incongruent squares (see p. 98), we set the side of *E* equal to *x* and the side of *H* equal to *y*. But we need not have started with *E* and *H*. Set the side of *B* equal to *x* and the side of *E* equal to *y*. Then proceed as we did on p. 98.

9. See E 8. Set the side of *E* equal to *x* and the side of *F* equal to *y*. Then proceed as we did on p. 98.

10. See E 8. Set the side of *C* equal to *x* and the side of *D* equal to *y*. Then proceed as we did on p. 98. It might be necessary to introduce more unknowns than *x* and *y* in the process.

11. Distort this tiling of a rectangle into a tiling of a rectangle with incongruent squares.

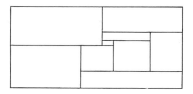

12. (a) In how many ways can 36 congruent squares be put together to form a rectangle?
    (b) In how many ways can 37 congruent squares be put together to form a rectangle?

13. Can this tiling of a rectangle by rectangles be distorted into (a) a tiling of a rectangle by squares? (b) a tiling of a square by squares? Explain.

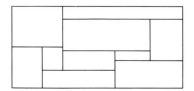

14. What is the largest square tile that can be used to tile a 151 by 721 rectangle with congruent squares? Why?

15. What is the largest square tile that can be used to tile a 27 by 52 rectangle with congruent squares? Why?

●

16. Complete the unfinished case of the analysis leading to Theorem 6.

17. A triangle has angles 30°, 60°, and 90°. Prove that it can be tiled with incongruent triangles having the angles 30°, 60°, and 90°.

18. Prove that a right triangle whose two smaller angles, *A* and *B*, are not 45° can be tiled with incongruent triangles similar to the given triangle (that is, with triangles whose angles are *A*, *B*, and 90°).

19. Show that a rectangle 1 by $\sqrt{2}$ can be tiled with incongruent square tiles if we permit ourselves to use an endless supply of tiles.

20. Show that any rectangle can be tiled with square tiles if we permit ourselves an endless supply of tiles.

21. Prove that for any natural number $N$, from 9 on, it is possible to put $N$ incongruent squares together to form a rectangle.

22. Two squares are "adjacent" if they touch at more than just a corner. Prove that if a rectangle is tiled with squares in such a way that no two adjacent squares are congruent, then the smallest square does not touch the border of the rectangle.

23. Two cubes are "adjacent" if they touch at more than an edge. Prove that if a box is filled with cubes, then there exist two adjacent cubes that are congruent. In fact, they will share one of their six faces. (*Hint:* Use E 22.)

24. A rectangle is tiled with squares in such a way that no two adjacent squares are congruent. Prove that any smallest square of the tiling is surrounded by precisely four squares.

25. (a) Show all ways that you can find to fit five squares (of various or equal sizes) together to form a rectangle.
    (b) On the basis of (a) which rectangles do you conclude can be tiled with five squares? (There is no restriction on the size of the squares; they may be all of the same size or of various sizes.)

26. Cut out of paper or cardboard a quadrilateral having no two sides parallel, no two sides of equal length, and no indentations, such as

    Show that an endless floor can be tiled with copies of such a figure.

27. (a) Prove that the sum of the four inside angles of any four-sided figure such as the one of E 26 is 360°. See E 57 of Chapter 6.
    (b) Why is part (a) of importance in the tiling requested in E 26?

● ●

28. The two end squares of a diagonal of a checkerboard are removed. Can the remaining 62 squares be covered (tiled) with 31 dominoes, each of which can cover two squares?

29. **R. G. Taylor** devised a tiling of a triangle whose angles are 90°, 45°, and 45° by six incongruent triangles with the same angles. Is six the smallest number of tiles required?

30. (a) Cut 13 squares of cardboard to fit together as in this tiling

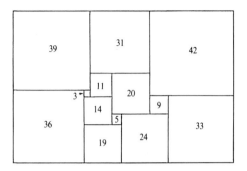

(b) Reassemble the 13 squares in a different way, again forming a rectangle. (Don't just simply interchange top and bottom or right and left. The new assembly should be substantially different from the first.)

31. Consider a tiling of an equilateral triangle with small incongruent equilateral triangles.
   (a) Prove that the smallest tile, *S*, touching the base of the big triangle meets the base at just one point:

   (b) Prove that the smallest tile, *T*, touching the top side of *S* meets that side at just one point:

   (c) Can this method be continued to produce an endless tower of little triangles and thus an elementary proof of Theorem 7? No one has yet found an elementary proof of that theorem.

32. (a) Cut out of thick paper or tagboard at least eight copies of the first polygon and four copies of each of the other six polygonal tiles shown (each differs from a square by four notches or prongs):

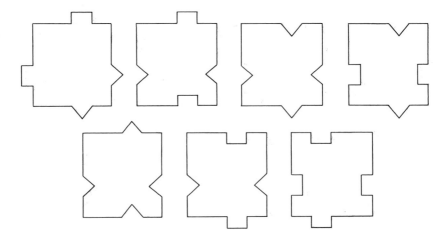

(b) Fit four of your tiles together in the form of a 2 by 2 square. (You may turn the tiles over and rotate them.)

(c) Fit six of your tiles together in the form of a 2 by 3 rectangle.

(d) Fit nine of your tiles together in the form of a 3 by 3 square.

(e) Fit sixteen of your tiles together in the form of a 4 by 4 square.

R. M. Robinson announced in 1967 that if you make an infinite number of copies of the seven tiles shown, then you can tile the entire plane with them. Moreover, he proved that no such tiling consists of one pattern repeated at equal intervals. His work is related to R 3 and R 4.

### References

1. W. T. Tutte, The dissection of equilateral triangles into equilateral triangles, *Proceedings of the Cambridge Philosophical Society*, vol. 44, 1948, pp. 463–482.

2. _____, The quest of the perfect square, *American Mathematical Monthly*, vol. 72, 1965, pp. 29–35. (This is a survey of recent results.)

3. H. Wang, Games, logic, and computers, *Scientific American*, vol. 213, 1965, pp. 98–106. (The Euclidean Algorithm, tiling by colored squares, and computing machines are shown to be related to fundamental problems in logic.)

4. R. Berger, The undecidability of the domino problem, *Memoirs of the American Mathematical Society*, no. 66, 1966. (This paper proves that there is no general algorithm for deciding whether copies of a finite set of polygons can tile the entire plane.)

See also references of Chapter 8.

# TILING AND
# ELECTRICITY

In the last chapter we saw that rectangles whose dimensions are $L$ by $W$, with $L/W$ irrational, cannot be tiled with congruent squares. We will now prove much more: *such rectangles cannot be tiled with squares in any way whatsoever.* To carry out the proof, we will show that tiling a rectangle with rectangles is intimately related to the flow of electricity through wires. Then we will be able to use for our purposes the mathematics of electrical networks, which was developed a century ago. We will first present the electrical background.

Since it will be several pages before we return to the tiling of rectangles by rectangles, the reader is asked to be patient. After we have developed the electrical theory it will be a very simple matter to prove that a 1 by $\sqrt{2}$ rectangle cannot be tiled with squares in any way at all.

*A collection of wires we will call a* **network.** *A point where two or more wires meet we will call a* **vertex.** Here, for example, is a network with eight wires and six vertices.

We will assume that our networks are **connected**; that is, that a bug can get from any vertex to any other by traveling along the wires.

Each wire has the ability to conduct electricity. This ability is influenced by the material of which the wire is made (silver conducts electricity better than copper); the length of the wire (the shorter the wire, the shorter the distance the electricity must travel); and the radius of the wire (the bigger the cross section, the easier it is for electricity to pass through). With this in mind, let us cause electricity to flow in our network by hooking up a battery.

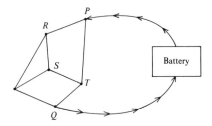

The electrons enter the network at $P$ and leave it at $Q$. Some of the electrons will choose to flow through the wire $PR$; and others, through the wire $PT$. If $PR$ conducts electricity better than $PT$, we might expect all the electrons to choose $PR$. But if they do, $PR$ may get so crowded that some will choose $PT$. At $R$, $S$, and $T$, the electrons have more choices. For example, some electrons leaving $T$ may choose to move toward $S$. Other electrons might move from $S$ toward $T$. Nobody has developed a microscopic psychology of electrons. Yet it is important to be able to predict the current in each wire; too much current will blow a fuse, too little will result in dim lights.

After many experiments, physicists came to two simple conclusions about the flow of electrons through a network. With these two rules, they can predict the mass behavior of the electrons. They observed first that in the simplest net-work—a single wire—with both ends attached to a battery,

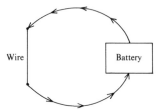

the current flowing in the wire is proportional to the strength of the battery. That is, if they doubled the strength of the battery (say, by hooking a second battery to the first) then the flow of electricity in the wire doubled; similarly, if they tripled the strength of the battery, the flow of electricity in the wire tripled. In short, if the strength of a battery was measured in volts, $V$, and the flow of electricity in amperes, $A$, it was observed that, *for each wire*, $A/V$ was always a fixed number. *This number—called the* **conductance** *of the wire—describes the ease with which the wire lets electricity pass through it*, and is denoted by the letter $C$. Thus $A/V = C$, or, avoiding division, we have

$$A = CV.$$

Here $V$ records the voltage and $A$ the current. We may assume that the con-ductance of each wire is known.

Voltage can also be thought of as a pressure. With a voltmeter, physicists can measure the difference in pressure between any two vertices of a network. The difference in voltage causes the current, just as a difference in air pressure causes the wind. If we set the voltage at the bottom vertex of the network equal to zero (the "sea level" above which voltages are measured), then we can assign to each vertex a voltage equal to the reading of a voltmeter attached to that vertex and to the bottom vertex. Now, without a private interview with each electron, we are ready to state the first rule of electrical networks.

I. *If a wire in the network has conductance C and voltages V and V' at its two vertices, then the current A in the wire flows from the larger voltage (say, V) to the smaller (V'), and*

$$A = C(V - V').$$

We may read the preceding equation as: current is equal to conductance times the difference in pressure. (The difference in pressure, $V - V'$, is what we call the voltage drop in the wire.) In the psychology of the individual wire, we would say that the wire thinks it is hooked up to a battery with voltage $V - V'$ and knows nothing about the other wires of the network.

The second rule concerns only the currents.

II. *The total current flowing into each vertex equals the total current leaving that vertex.*

Rule II simply says that electrons do not accumulate at any vertex. Before discussing these two rules further, let us see how we can use them to predict the flow of electrons through a network. Take, for example, this network of two wires and two vertices.

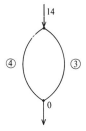

The conductance of each wire is circled. The voltage at the bottom vertex is 0. The current flowing into the network is 14 amperes (we will no longer bother drawing the battery). We want to know the voltage at the top vertex and the current in each of the two wires. We would also like to know the current leaving the bottom vertex (we suspect that it will be 14 amperes).

The unknown voltage we will call $V_1$. The unknown currents will be $A_1$, $A_2$, $A_3$, as in this picture:

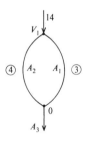

Rule I, applied to each of the two wires, gives

$$A_1 = 3(V_1 - 0),$$
$$A_2 = 4(V_1 - 0).$$

Rule II, applied to each of the two vertices, gives

$$14 = A_1 + A_2,$$
$$A_3 = A_1 + A_2.$$

The last two equations inform us that $A_3 = 14$, as we hoped. Thus we may dispense with the fourth equation. The first three are simply

$$A_1 = 3V_1, \qquad A_2 = 4V_1, \qquad 14 = A_1 + A_2.$$

These three equations yield

$$14 = 3V_1 + 4V_1.$$

Thus $14 = 7V_1$, and hence $V_1 = 2$. Having found $V_1$ we then find $A_1$ and $A_2$: $A_1 = 3 \cdot 2 = 6$, $A_2 = 4 \cdot 2 = 8$.

Are the answers $A_1 = 6$ and $A_2 = 8$ reasonable? Yes, for we would expect more current in the wire with the larger conductance. This example indicates that rules I and II seem to be adequate for predicting the currents and the voltages. That is, rules I and II seem to contain enough information to enable us to pinpoint all the unknowns.

Let us see if rules I and II are adequate for this fancier network.

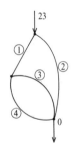

Here 23 amperes are flowing into a network with four wires and three vertices. We would like to know the current in each of the four wires and the voltage at each of the two vertices (the voltage at the bottom vertex is always taken as 0). As a check on our arithmetic, we might also compute the current leaving the bottom vertex (we suspect that it should be 23). We have seven unknowns, $A_1$, $A_2$, $A_3$, $A_4$, $A_5$, $V_1$, and $V_2$, as indicated in this figure.

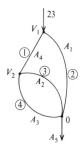

Rule I, applied to each of the four wires, gives

$$A_4 = 1(V_1 - V_2),$$
$$A_1 = 2(V_1 - 0),$$
$$A_2 = 3(V_2 - 0),$$
$$A_3 = 4(V_2 - 0).$$

Rule II, applied to each of the three vertices, gives

$$23 = A_1 + A_4,$$
$$A_4 = A_2 + A_3,$$
$$A_5 = A_1 + A_2 + A_3.$$

We have seven equations in seven unknowns. Let us omit the last equation and use it only as a check after we have found $A_1$, $A_2$, and $A_3$ ($A_5$ should be 23). The six equations in six unknowns, slightly simplified, are:

$$A_4 = V_1 - V_2, \qquad A_1 = 2V_1, \qquad A_2 = 3V_2, \qquad A_3 = 4V_2,$$
$$23 = A_1 + A_4, \qquad A_4 = A_2 + A_3.$$

Now we use the information given by rule I to simplify the equations given by rule II. We combine $23 = A_1 + A_4$ with $A_4 = V_1 - V_2$ and $A_1 = 2V_1$ to deduce

$$23 = 2V_1 + (V_1 - V_2) = 3V_1 - V_2.$$

Similarly, from

$$A_4 = A_2 + A_3, \qquad A_2 = 3V_2, \qquad A_3 = 4V_2, \qquad A_4 = V_1 - V_2,$$

we find that

$$V_1 - V_2 = 3V_2 + 4V_2.$$

Therefore we now have only two equations in two unknowns:

$$23 = 3V_1 - V_2,$$
$$V_1 - V_2 = 3V_2 + 4V_2.$$

The second of these equations can be shortened. Since $3V_2 + 4V_2 = 7V_2$, we have $V_1 - V_2 = 7V_2$. Adding $V_2$ to both sides yields $V_1 = 8V_2$. Thus we have the two equations

$$23 = 3V_1 - V_2,$$

and

$$V_1 = 8V_2.$$

We can now substitute $8V_2$ for $V_1$ into the first to obtain

$$23 = 3(8V_2) - V_2 = 24V_2 - V_2 = 23V_2.$$

Hence we have just one equation in one unknown,

$$23 = 23V_2.$$

Dividing both sides by 23 yields $V_2 = 1$. Now that we have found $V_2$ we can immediately find $V_1$, since $V_1 = 8V_2$. In fact, $V_1 = 8 \times 1 = 8$. Having found all the voltages, we can now compute all the currents:

$$A_4 = 1(V_1 - V_2) = 1(8 - 1) = 7,$$
$$A_1 = 2(V_1 - 0) = 2(8) = 16,$$
$$A_2 = 3(V_2 - 0) = 3(1) = 3,$$
$$A_3 = 4(V_2 - 0) = 4(1) = 4.$$

Finally, as a check, let us compute $A_5$. We obtain

$$A_5 = A_1 + A_2 + A_3 = 16 + 3 + 4 = 23,$$

as we suspected.

So, again, rules I and II were adequate for pinpointing all the unknown currents and voltages. So far, the answers have all been integers. This need not always be the case, of course, as the following example shows. This third example we will work out in detail also, to illustrate how rules I and II are sufficient to deal with an even more complicated network. After we finish this example, we will be ready to return to tilings of rectangles.

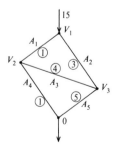

Here we have three voltages and five currents to find. Let us not bother with the current leaving the network; it will equal $A_4 + A_5$ and will serve as a check, since it should equal 15.

Rule 1 gives

$$A_1 = 1(V_1 - V_2), \quad A_2 = 3(V_1 - V_3), \quad A_3 = 4(V_2 - V_3),$$
$$A_4 = 1(V_2 - 0), \quad A_5 = 5(V_3 - 0).$$

Rule II gives $15 = A_1 + A_2$, $A_1 = A_3 + A_4$, $A_2 + A_3 = A_5$.

As before, we use the equations given by rule I to eliminate the $A$'s. We end up with three equations involving only the $V$'s:

$$15 = 1(V_1 - V_2) + 3(V_1 - V_3),$$
$$1(V_1 - V_2) = 4(V_2 - V_3) + 1(V_2 - 0),$$
$$3(V_1 - V_3) + 4(V_2 - V_3) = 5(V_3 - 0).$$

Simplifying these three equations slightly, we have

$$15 = V_1 - V_2 + 3V_1 - 3V_3,$$
$$V_1 - V_2 = 4V_2 - 4V_3 + V_2,$$
$$3V_1 - 3V_3 + 4V_2 - 4V_3 = 5V_3.$$

Moving all unknowns to the right side,* we obtain

$$15 = 4V_1 - V_2 - 3V_3,$$
$$0 = -V_1 + 6V_2 - 4V_3,$$
$$0 = -3V_1 - 4V_2 + 12V_3.$$

This time we have three equations and three unknown voltages. What should we do with these three equations in order to find $V_1$, $V_2$, and $V_3$?

One possibility is this. Get rid of $V_1$ and hope we can find two equations in the two unknowns $V_2$ and $V_3$. That will be a step in the right direction.

From $15 = 4V_1 - V_2 - 3V_3$ and $0 = -V_1 + 6V_2 - 4V_3$ we can obtain an equation involving just $V_2$ and $V_3$. To do this, multiply the second equation by 4, which gives

$$0 = -4V_1 + 24V_2 - 16V_3,$$

and add the equation just obtained to

$$15 = 4V_1 - V_2 - 3V_3;$$

we thus have

$$15 = 23V_2 - 19V_3.$$

To find another relation between $V_2$ and $V_3$ we might get rid of $V_1$ between the two equations

$$0 = -V_1 + 6V_2 - 4V_3$$

and

$$0 = -3V_1 - 4V_2 + 12V_3.$$

---

* How this is done is shown in slow motion in Example 17 of Appendix C.

To do so, multiply $0 = -V_1 + 6V_2 - 4V_3$ by $-3$, and add the result to $0 = -3V_1 - 4V_2 + 12V_3$. Thus we will be adding

$$0 = 3V_1 - 18V_2 + 12V_3$$

to

$$0 = -3V_1 - 4V_2 + 12V_3.$$

The $V_1$'s cancel when we add these two equations; their sum is

$$0 = -22V_2 + 24V_3.$$

As a result we have two equations that involve only $V_2$ and $V_3$:

$$15 = 23V_2 - 19V_3,$$
$$0 = -22V_2 + 24V_3.$$

We want just one equation in one unknown. We can devise such an equation by eliminating, say, $V_2$, from the last two equations. Using just the second of these, we have

$$22V_2 = 24V_3.$$

Dividing both sides by 22 we obtain

$$V_2 = \frac{24}{22} V_3,$$

or, if we prefer,

$$V_2 = \frac{12}{11} V_3.$$

Combining this relation with $15 = 23V_2 - 19V_3$ gives

$$15 = 23 \left(\frac{12}{11} V_3\right) - 19V_3.$$

Thus

$$15 = \frac{276}{11} V_3 - 19V_3.$$

Multiplying both sides by 11 we obtain

$$165 = 276V_3 - 209V_3,$$

or simply

$$165 = 67V_3.$$

Dividing both sides by 67 we find

$$V_3 = \frac{165}{67}.$$

Having found $V_3$ it is a simple matter to reverse our steps and find $V_2$, then $V_1$, and finally the $A$'s.

From $V_2 = \frac{12}{11}V_3$ we learn that $V_2 = \frac{12}{11} \cdot \frac{165}{67}$. Since $\frac{165}{11} = 15$, we see that

$$V_2 = (12 \cdot 15)/67.$$

Thus

$$V_2 = \frac{180}{67}.$$

Any of the equations involving $V_1$, $V_2$, and $V_3$ can now tell us what $V_1$ must be. Using, say,

$$0 = -V_1 + 6V_2 - 4V_3,$$

we find

$$0 = -V_1 + 6\left(\frac{180}{67}\right) - 4\left(\frac{165}{67}\right).$$

Thus

$$V_1 = 6\left(\frac{180}{67}\right) - 4\left(\frac{165}{67}\right),$$

or

$$V_1 = \frac{1080}{67} - \frac{660}{67}.$$

Hence

$$V_1 = \frac{420}{67}.$$

Once these voltages are known, the equations given by rule I tell us the currents. For example,

$$A_1 = 1(V_1 - V_2) = 1\left(\frac{420}{67} - \frac{180}{67}\right) = \frac{240}{67}.$$

The reader may easily compute $A_2$, $A_3$, $A_4$, and $A_5$.

In our third example, the input information (conductance and current) consisted entirely of natural numbers. The unknowns, as it turned out, were not natural numbers, but at least they were rationals. Even if the input information had consisted of rationals, the unknowns would still have been rationals, for all we did was multiply, divide, add, or subtract the numbers we had to work with. These operations, applied to rationals, yield only rationals, never an irrational such as $\sqrt{2}$.

In the three examples, rules I and II were sufficient to predict all the unknowns. It can be proved by the theory of determinants that rules I and II are sufficient to predict all the unknowns in any network. Rules I and II make it unnecessary to psychoanalyze myriads of individual electrons.

In particular, our technique of first writing down all the equations provided by rules I and II, finding equations involving only the $V$'s, and then gradually getting rid of the $V$'s until we have a single equation involving only one of the $V$'s, will always work. All that we will require from the mathematics of electrical networks is this conclusion:

THEOREM 1. *If the conductances in a network and the current entering the network are all rationals, then so are the voltages at all the vertices and the currents in all the wires.*

We are finally prepared to show the intimate relation between electrical networks and the tiling of a rectangle. We will show that, from any tiling of a rectangle with rectangles (in particular, with squares) a network of wires, conductances, currents, and voltages can be derived that will satisfy rules I and II. Though we will illustrate the process with a particular example, we will describe it in terms general enough to be applied to any tiling of any rectangle by rectangles. Let us start with a rectangle tiled with rectangles as in this diagram:

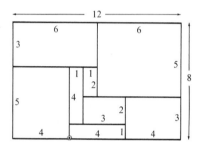

The horizontal lines have been drawn heavier for emphasis. Observe that the sum of the horizontal dimensions of the rectangles hanging from a given horizontal line is equal to the sum of the horizontal dimensions of the rectangles resting on that horizontal line. This sounds a good deal like rule II and suggests that the horizontal dimension of each rectangle could be a current in a wire and that each darkened horizontal line could somehow be a vertex. This observation suggests the following steps—to be applied to any tiling of a rectangle with rectangular tiles—for the construction of an electrical network.

*Step 1. Darken each horizontal line.*

*Step 2. The midpoint* (or any other point on that line, if you prefer) *of each horizontal line shall be a vertex* (for the network being constructed).

*Step 3. With each rectangle of the tiling associate a wire joining the midpoints of the two darkened lines that the rectangle touches on its top and bottom.*

*Step 4. The current in a wire shall be the horizontal dimension of the rectangle with which it is associated.*

*Step 5. The voltage at each vertex shall be the height of this vertex above the lowest horizontal line.*

Thus the number of wires equals the number of rectangles; the number of vertices equals the number of horizontal lines.

We next define the conductance in each wire such that rule I is satisfied. Rule I reads

$$Current = Conductance \times Difference\ in\ voltages.$$

Now, the difference in voltages between the ends of a wire is just the vertical dimension of the associated rectangle. The current is the horizontal dimension of that rectangle. So we must define the conductance $C$ in each wire such that

$$\frac{Horizontal\ dimension\ of}{associated\ rectangle} = C \times \frac{Vertical\ dimension\ of}{associated\ rectangle}.$$

This brings us to

*Step 6. Set the conductance of each wire equal to $h/v$, where $h$ is the horizontal dimension and $v$ the vertical dimension of the associated rectangle.*

*Step 7. The current flowing into the top vertex and out of the bottom vertex will equal the horizontal dimension of the tiled rectangle.*

(Step 7 guarantees that rule I holds at the top and bottom vertices.) When applied to our example, these seven steps create this electrical network:

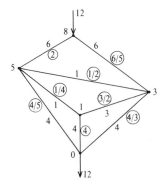

which satisfies rules I and II. The reader might check a few vertices and wires to assure himself that this electrical network, derived from a rectangle tiled with rectangles, satisfies rules I and II.

Now we are ready to prove a theorem that the reader should contrast with the weaker Theorem 2 of Chapter 7.

THEOREM 2. *The rectangle of dimensions $\sqrt{2}$ by 1 cannot be tiled with squares in any way whatsoever.*

PROOF. Place the rectangle such that its horizontal dimension is 1 and its vertical dimension is $\sqrt{2}$.

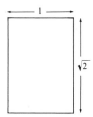

If it could be tiled with squares, then we would have an electrical network with input current equal to 1 and all conductances equal to 1. By Theorem 1, all the currents and voltages must be rational. In particular, the voltage at the highest vertex must be rational. But $\sqrt{2}$ is not rational. Thus the 1 by $\sqrt{2}$ rectangle cannot be tiled with squares in any way whatsoever.

With a slight change in technique we can prove that a rectangle $\sqrt{2}$ by $\sqrt{3}$ cannot be tiled with squares. Were we simply to take $\sqrt{2}$ as the input current, we would not be able to conclude that all voltages are rational. But we notice that, by shrinking each dimension by the factor $\sqrt{2}$, we obtain a smaller rectangle of dimensions 1 by $\sqrt{3}/\sqrt{2}$,

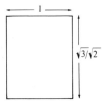

which is of the same shape as the larger rectangle

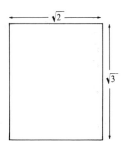

To any tiling of the larger rectangle by squares there corresponds a tiling of the smaller rectangle by squares (obtained by shrinking all dimensions by the factor

$\sqrt{2}$). Since $\sqrt{3}/\sqrt{2}$ is not rational, the smaller rectangle cannot be tiled with squares, and hence the larger cannot be.

Similar reasoning proves

THEOREM 3. *A rectangle of dimensions L by W, such that the ratio L/W is irrational, cannot be tiled with any finite collection of squares.*

Combining Theorem 3 of Chapter 7 with Theorem 3, just obtained, we arrive at

THEOREM 4. *If a rectangle cannot be tiled with congruent squares, then it cannot be tiled with any finite collection of squares.*

The arguments leading to Theorems 2, 3, and 4 needed a device for recording how the squares fit together to form a rectangle. This was provided by the electrical network diagrams, which described in a somewhat skeletal fashion the relative positions of the tiles. What we knew about networks we then applied to tilings, which are just networks in disguise.

When we phrased rules I and II we were thinking of electrical networks. But it turns out that, without intending to, we were simultaneously talking about tilings. Of course, we must then reinterpret the key words in the two rules. For "vertex" read "horizontal line"; for "wire" read "rectangle"; for "current" read "horizontal dimension"; for "voltage" read "height."

We thus have two quite different interpretations of rules I and II: the electrical, dating back to the middle of the nineteenth century, and the tiling, introduced in the middle of the twentieth century. Will someone in the twenty-first century find a third interpretation? We do not know. But we can say that rules I and II are important and deserve to be studied for their own sake, and in a manner not limited by a particular application—either electrical or tiling. When divorced from any concrete interpretation, or, as mathematicians prefer to say, model, rules I and II would be called "Axioms I and II." Whatever mathematicians learn about "Axioms I and II" will be applicable to electricity, tiling, or to any model satisfying these two axioms that may be devised in future centuries.

To a mathematician, then, networks and tilings of a rectangle are just two examples of structures that satisfy rules I and II. Their main difference lies in our feeling about them; their underlying form is the same. And it is the form that a mathematician prefers to study. For this reason he strips away the specific setting as a camouflage, and, by abstracting, obtains a keener insight into the structure. The deeper his understanding of the pure structure, the freer will be his imagination to recognize, through the clutter of reality, an example, or model, of this structure.

### Exercises

1. The conductance of a certain wire is 8. How many amperes flow in this wire if the difference in voltage between its two ends is (a) 7 volts, (b) 14 volts, (c) 10 volts?

2. The conductance of a certain wire is 5. How strong a battery is needed to push through this wire a current of (a) 10 amperes, (b) 20 amperes, (c) 12 amperes?

3. Find the unknown voltage and the three unknown currents for this network:

4. Check that the voltages and currents of E 3 satisfy rules I and II.

5. Find the unknown voltage and the three unknown currents for this network:

6. Check that the voltages and currents of E 5 satisfy rules I and II.

7. Find all currents and voltages in this network:

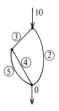

8. Check your answers to E 7 by showing that they satisfy rules I and II.

9. Find all currents and voltages in this network:

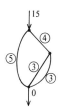

10. Check your answers to E 9 by showing that they satisfy rules I and II.

11. Draw the network, with its conductances, currents, and voltages, that is associated with the tiling at the bottom of p. 99.

12. Draw the network, with its conductances, currents, and voltages, that is associated with the tiling of E 30 of Chapter 7.

13. (a) Draw the electrical network that is associated with the tiling of a 5 by 7 rectangle into 35 congruent squares. Assume that the long sides of the rectangle are horizontal.
    (b) Like (a), but first rotate the rectangle 90°.

14. Draw the network, with its conductances, currents, and voltages, that corresponds to this tiling of a rectangle by rectangles:

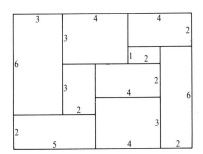

15. Draw the tiling of a rectangle by squares that corresponds to each of these electrical networks:

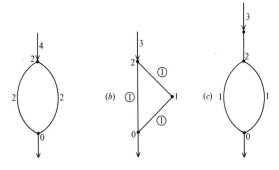

16. Draw the tiling by rectangles that corresponds to the following network, currents, and voltages:

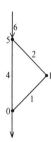

17. Find $x$ and $y$ if $2x + 3y = 7$ and $4x - y = 11$. Check your answers by showing that they satisfy the two equations.

18. Find $s$ and $t$ if $4s - 9t = 5$ and $3s + 5t = 6$. Check your answers by showing that they satisfy the two equations.

●

19. In our networks we have an unknown voltage at each vertex except the bottom one. We also have an unknown current in each wire and an unknown current leaving the bottom vertex. Prove that the number of equations provided by rules I and II is equal to the total number of unknowns.

20. Find the unknown voltages and currents in this network:

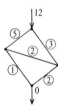

21. Check your answers to E 20 by determining whether the current leaving the network equals the current entering.

22. Find the unknown voltages and currents in this network:

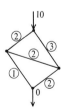

23. Check your answers to E 22 by determining whether the current leaving the network equals the current entering.

24. Find the unknown voltages and currents in this network:

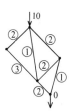

25. Draw the tiling of a rectangle into five rectangles that corresponds to this network:

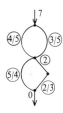

26. Draw the tiling of a rectangle into six rectangles that corresponds to this network:

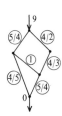

27. If we rotate through 90° a rectangle tiled with rectangles, the horizontal and vertical lines change roles. Thus the electrical network associated with the rotated rectangle is different from that associated with the original rectangle. Find the network associated with the rectangle of p. 116 rotated 90°.

28. The same as E 27, but for the rectangle of E 11.

29. The same as E 27, but for the rectangle of E 12.

30. A driver knows that his car gets 10 miles per gallon in town and 15 miles per gallon on the highway. In a trip of 220 miles his car used 18 gallons. How much of the trip was in town and how much on the highway?

31. How can we derive an electrical network from a box filled with smaller boxes? (Think first of satisfying rule II.)

32. Show that further examples of systems satisfying rule II are:
    (a) water flowing through narrow pipes,
    (b) traffic on a system of one-way roads, alleys, and boulevards.

33. Call a tiling of a rectangle "rational" if for each of the tiles the ratio of length to width is rational. Prove the following generalization of Theorem 2: *If a rectangle has dimensions L by W, with L/W irrational, then it has no rational tiling whatsoever.*

### References

1. M. Gardner, Mathematical Games, *Scientific American*, November, 1958, pp. 136–142. (Contains further examples of tiling by squares.)

2. ———, *The Second Scientific American Book of Mathematical Puzzles and Diversions*, Simon and Schuster, New York, 1961. (This includes R 1 and some new results obtained with the aid of electronic computers. See pp. 186–209.)

3. ———, *Martin Gardner's New Mathematical Diversions from Scientific American*, Simon and Schuster, New York, 1966.

## Chapter 9

# THE HIGHWAY
# INSPECTOR AND
# THE SALESMAN

If we stare at a diagram of an electrical network long enough, it might, like an optical illusion, turn into a map of highways and towns. This diagram, for example, formed of ten wires and six vertices where the wires are soldered together,

might turn into a map of ten sections of highway joining six towns. Even if we lose all the numbers, the voltages, the currents, and the conductances, such a map still raises interesting questions, which concern not electrons, but a highway inspector and a salesman.

A thrifty highway inspector has this problem. He would like to find a route that takes him over each section of highway exactly once. After all, he does not want to waste the taxpayers' money on unnecessary inspections.

A traveling salesman has another problem. He would like a route that would take him through each town exactly once. He is not at all interested in covering each section of highway.

Let us try to find the inspector and the salesman their desired routes on this simple highway system of six towns and nine highway sections:

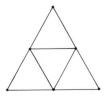

After a few tries, the reader might come up with this route for the inspector,

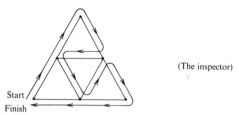

(The inspector)

and this for the salesman:

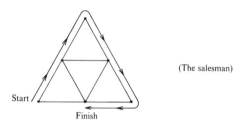

(The salesman)

Notice that the inspector has the added convenience of ending his inspection in the town where he started; he will not need two headquarters. The salesman happened to finish his journey in a town adjacent to the one where he started.

Let us try another highway system; for example, this one with five towns and five highway sections:

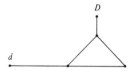

Working first on the inspector's problem, we notice that the terminal points of his route must be the towns $d$ and $D$. As the reader can easily verify, each of the two routes from $D$ to $d$ fails to cover all the sections of highway. We must inform the inspector that for this highway there is no route that takes him exactly once over each section.

Working on the salesman's problem for the same highway system, we see that his route must also have its ends at $d$ and $D$. Of the two routes from $d$ to $D$, one of them, fortunately, takes the salesman through each town:

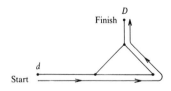

Here is a highway system of four towns and three sections, which, as the reader can check, makes both the inspector and the salesman unhappy, for neither has his desired route:

Finally, here is a system of five towns and five sections, which pleases the inspector but not the salesman, as the reader may check:

But the mere accumulation of examples is not mathematics any more than a dictionary is a novel. What we want is an insight into highway systems that will enable us to decide which highway systems have a route for the inspector or the salesman and which do not. Is there a way of knowing whether such routes can be found in a particular highway system without going through perhaps thousands of trials? We will consider the inspector's problem first.

Let us probe more deeply the inspector's problem by a close examination of this system:

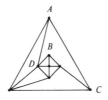

We first concentrate our attention on town *A*. Assume, for the moment, that the inspector does not begin his tour at town *A*. When he drives through *A* he inspects two of the sections at *A*; one as he approaches *A* and the other as he leaves *A*. This leaves him one section at *A* that he still must inspect. When he does inspect this third section he will find himself at *A*, with no exit. If he is to inspect each section once, he must do so before he gets stuck in *A*—that is, his tour must finish at *A*. To reach this conclusion, we assumed, let us recall, that he did not start at *A*. We see, then, that if the inspector does have a route, it must either begin or end at *A*.

By similar reasoning (concentrating his attention on one town at a time), the reader can easily show that the inspector's route must also begin or end at *B*; that it must begin or end at *C*; and that it must begin or end at *D*. Thus if the highway inspector does have a route, then that route must have four ends. But, as we all know, a route cannot have more than two ends. With this analysis, we know that the highway inspector can find no route for this system.

Our insight suggests that the number of sections joining one town to other towns is quite important in studying the inspector's problem. *We will call the* **degree** *of a town the number of sections having an end at that town.* For example, a town at a dead end has degree 1. The same reasoning that we applied to town *A* in the system just considered proves

THEOREM 1. *If a town has odd degree, then the inspector must either begin or end his tour of inspection at that town.*

And, in the same way that we showed the above system to have no route for the inspector, we can prove

THEOREM 2. *If there are more than two towns of odd degree in a system, then the inspector has no route (taking him over each section exactly once).*

Theorem 2 is not one to please the inspector. There may still be a chance to please him, however, if we look at highway systems containing no more than two towns of odd degree. There are three cases to consider: systems with no town of odd degree; systems with one town of odd degree; and systems with two towns of odd degree. (The reader might try some examples of his own and guess what we will discover for the inspector.)

We will first prove that the second of the three cases cannot occur.

THEOREM 3. *There is no highway system with exactly one town of odd degree.*

PROOF. We shall prove that a system containing a town of odd degree must contain at least one other town of odd degree. To do so, let *T* be a town of odd degree, and imagine starting a vacation trip at *T*. We shall travel quite at random, except that we agree *never to use the same section of highway twice.* Each time we pass through a town we are free to choose as an exit any of the sections over which we have not already traveled.

Let us now make such a journey, beginning at *T*. If we are unlucky, the journey could be quite short. With luck, we could have a long journey. But since there is a limit to the number of sections, our vacation must end in some town, which we will call *E*, for "end."

Could *E* simply be *T*? Observe that, since we started at *T*, an even number of sections at *T* remained unused; hence each time we entered *T* subsequently, we had an escape route. Thus *E* is not *T*.

If *E* were of even degree, then each time we entered it we could have found an exit. But since our trip ended at $E$, the degree of $E$ must be odd. This proves the theorem.

Only two cases remain before we have a complete solution to the inspector's problem: systems with no town of odd degree, and systems with two towns of odd degree. Treating first the case in which there is no town of odd degree, we will prove

THEOREM 4. *If a highway system has no town of odd degree, then the inspector can find a route that will take him over each section exactly once. (Moreover, he can begin his inspection at any town; his tour must end where it starts.)*

Before we begin the proof, the reader should draw a highway system with no town of odd degree and show that it has a route for the inspector and that it ends at the town where it begins.

We should also mention something that we have up to now tacitly assumed in all our highway systems; namely, that they are connected. *A system is* **connected** *if it is possible to travel on its highways from any town to any other town.* For example, the highway system of the United States is, in all likelihood, connected.

Now we are ready for the

PROOF OF THEOREM 4. Not only will we prove that the inspector can find a route through any (connected) highway system with no town of odd degree, but we will show him how to find it.

Select any town $T$ in the system. We will prove that the inspector has a route that begins at $T$ and covers all the sections (clearly this route must end at $T$).

Here is a recipe for discovering such a route. Get into a car at $T$, and travel about at random as in the proof of Theorem 3. That is, go over no section twice, and end your journey in a town from which you cannot escape.

Since the highway system has a limited supply of sections, this random trip must end. Where must it end? Since all towns are of even degree, it must end at $T$. This whimsical trip probably fails to cover all the sections. If so, we will enlarge the trip with a vacation excursion in the following manner.

If the trip did not pass through all the sections, then there must be a town on the trip that is a terminus of one of the sections not covered (since the system is connected). Call this town $U$. Start a random side trip at $U$ that takes us over no section we have already covered and that, in addition, covers no section twice. Such a side trip will have to end at $U$. We can combine the first trip and this side trip into one journey by starting at $T$, traveling along the first trip until reaching $U$, taking the side trip from $U$ to $U$, and then continuing on our original trip back to $T$.

We can continue adding more side trips, forming them into bigger and bigger routes, until finally we obtain a route that uses all the sections. Such a route begins at $T$, takes the inspector over each section once, and deposits him at $T$. The inspector has a route in any highway system with no town of odd degree, and he has a way of finding it—by piecing together side trips. Thus theorem 4 is proved.

The method provided in the proof of Theorem 4 is quite easy to use. For example, let us watch the inspector find a route over this highway system, which has no town of odd degree:

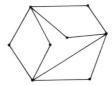

First he chooses a town for his headquarters and takes a random journey until he is blocked. Perhaps this is the journey:

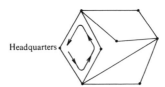

Then he tosses in an excursion from one of the towns he passed through on his first journey:

These two trips he can combine into one trip:

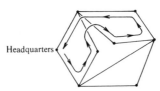

Since he has still not covered all the sections, he makes another excursion over sections he has not covered:

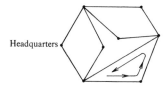

Now he has covered all the sections and can combine his first trip and the two side trips into one route that carries him over each section once:

Of course on a more complicated system the inspector might have to use more side trips before he reaches a route that covers the whole system.

The only case remaining is covered by

THEOREM 5. *If a highway system has two towns of odd degree, then the inspector can find a route taking him over each section exactly once. (Moreover, he must begin his inspection at one of the two towns of odd degree and must end it at the other.)*

The proof of this, similar to that of Theorem 4, is left to the reader.

We have completely solved the inspector's problem. With a quick count of the towns of odd degree we can tell whether there is a route; if there is, we have a way of finding it. But the inspector is still not happy.

"On these narrow roads I can inspect both lanes at the same time. But soon I will be promoted and will have to inspect the superhighways, which sometimes have the two directions of traffic far apart. How will I be able to tell whether I can find a route then?"

The inspector has raised a new problem. After all, so far we have not bothered about the direction he chose to travel in covering a section. We have been able to think of a section as a line rather than as a one-way arrow. Now we must replace each section by two sections, each furnished with an arrow indicating the permitted direction of travel. For example, instead of a highway system such as this,

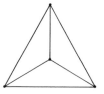

we have this one, which consists of one-way roads:

Though we might expect the new problem to be harder than the one we already solved, it turns out to be quite simple. Observe that the number of one-way roads entering a particular town is equal to the number of one-way roads leaving that town. Not only is every town of even degree, but every time the inspector enters a town (other than his headquarters) he can leave it (legally). The reader may use the same kind of reasoning that proved Theorem 4 to prove

THEOREM 6. *If for each town in a system of towns and one-way roads there are as many roads entering as there are leaving, then the inspector has a route that enables him to inspect each one-way road exactly once. (Moreover, he can choose any town as his headquarters. His tour must end where it begins.)*

The proof of this is almost identical to the proof of Theorem 4, and is left to the reader. Actually, Theorem 6 is more than the inspector requires. It includes a system such as:

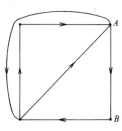

(Note, for example, that the towns labeled *A* and *B* are joined by a single one-way road.)

Theorem 6 would also apply to systems with loops, such as:

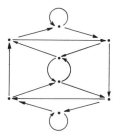

Note that as many one-way routes enter a town as leave it.

In the next chapter we will make use of Theorem 6. In particular, we will be using systems with loops, such as the one just drawn, to analyze an ancient word used in India as an aid for remembering certain rhythmic patterns.

Though we have satisfied the inspector, we have done nothing for the salesman, who wants to know when he can find a route that will take him through each town exactly once. This problem, proposed by W. R. Hamilton in 1859, is still unsolved. Perhaps there is no general way of deciding whether a highway system has a salesman's route other than by simply trying to find one. The inspector will just have to count the towns of odd degree. If there is no such town, or if there are two such towns, he will then know that he has a route; in all other cases, he will know that he has no route. If there is a solution to the salesman's problem, it will probably be much more complicated.

We have previously met pairs of problems that are similar in their formulation but quite different in their difficulty. It was easy to show that there is only one prime that is one less than a square, but nobody knows how many primes are one more than a square (see E 3 and E 4 of Chapter 4). It was fairly easy to prove that there is no end to the primes; but nobody knows whether there is an end to the twin primes. It is very easy to see that if a triangle is cut into triangles and the dots labeled *A*, *B*, and *C* with an odd number of *AA* edges on the border, then there must be a triangle with duplicated *A*'s (simply take any triangle with an *AA* edge on the border); but it was much harder to prove, as we did in Chapter 2, that if there is an odd number of *AB* edges on the border, then there must be at least one *ABC* triangle. It was easy to prove that every special number is prime; but it was not easy to prove the converse—that every prime is special. The contrast between the problems posed by the inspector and the salesman will play a role in the next chapter.

### Exercises

1. Does the salesman have a route in this system? Does the inspector?

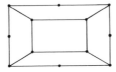

2. Does the salesman have a route in this system? Does the inspector?

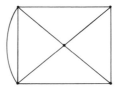

3. (a) Count the towns of odd degree in several highway systems.
   (b) Count the towns of even degree in several highway systems.
   (c) Make a conjecture.

4. Using the method we used to prove that there cannot be just one town of odd degree, prove that in an ordinary highway system (with a limited supply of sections) the number of odd towns is even.

4. Prove Theorem 5.

5. Prove Theorem 6.

6. The problem of the inspector was solved in 1735 by Euler, who wrote, "In the town of Koenigsberg there is an island called Kneiphof, with two branches of the river Pregel flowing around it. There are seven bridges crossing the two branches. The question is whether a person can plan a walk in such a way that he will cross each of these bridges once but not more than once. . . . On the basis of the above I formulated the following very general problem for myself: Given any configuration of the river and the branches into which it may divide, as well as any number of bridges, to determine whether or not it is possible to cross each bridge exactly once."

   This is a diagram of the seven bridges of Koenigsberg:

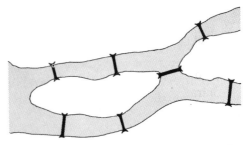

   Can a person plan a walk that will take him across each bridge exactly once?

7. Can a person find a path in this house that will take him through each door exactly once?

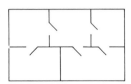

8. Can a person find a path in this house that will take him through each door exactly once?

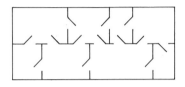

9. Can a salesman visit each of the towns of this system exactly once?

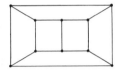

10. Show that a salesman can visit each of the towns of this system exactly once and return to his starting point.

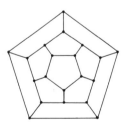

11. Using the technique provided by the proof of Theorem 4, find a route for the inspector of this system:

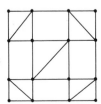

12. Using the technique provided in the proof of Theorem 4. find a route for the inspector of this system:

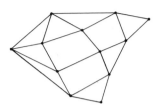

13. Using the method we used to prove that there cannot be just one town of odd degree, prove that in an ordinary highway system (with a limited supply of sections) the number of odd towns is even.

14. Devise an example of a highway system with exactly one town of odd degree (but with an *unlimited* collection of sections and towns of even degree).

15. (a) Prove that when a section is deleted from a highway system the number of towns of odd degree changes by an even number.
    (b) Use (a) to prove that the number of towns of odd degree is even.

16. In a given highway system let $T(d)$ be the number of towns of degree $d$. Thus, for example, $T(1)$ is the number of dead-end towns.
    (a) By a bookkeeping of the way in which sections are contributed by towns, prove that

$$1T(1) + 2T(2) + 3T(3) + 4T(4) + \cdots$$

    is equal to twice the number of sections.
    (b) Use (a) to prove that the number of towns of odd degree is even.

17. How many inspectors are necessary to inspect a highway system with four towns of odd degree if none of them is to inspect a section more than once? There is no restriction on the number of towns of even degree.
    (a) Carry out three experiments.
    (b) Make a conjecture.
    (c) Prove your conjecture.

18. A highway system consists of three towns, each two of which are joined by a single one-way road.
    (a) There are essentially two such systems. Draw them.
    (b) Show that in both of them the salesman has a route.

19. Using E 18, prove that in any system of four towns in which a one-way road connects each two towns, there is a route for the salesman.

20. (a) Using E 19, prove that any system of five towns, with a one-way road between each two towns, has a route for the salesman.
    (b) Extend the result to any number of towns.

21. In a round robin volleyball tournament, each of 673 teams played each other once. There were no tie games. Prove that it is possible to label the teams $T_1, T_2, T_3, \ldots,$ $T_{672}, T_{673}$ in such a way that $T_1$ defeated $T_2$, $T_2$ defeated $T_3$, $T_3$ defeated $T_4$, and so on. *Hint:* Relate to E 20(b).

22. A "circuit" in a highway system is a route that begins at some town and, without using any section twice, returns to that town. A "tree" is a highway system without circuits. (The system need not be connected.)
    (a) Draw a tree.
    (b) Remove a path joining two dead ends.
    (c) From the tree that now remains, remove a path joining two of its dead ends.
    (d) Continue till all the sections are removed.
    (e) Count the number of paths you removed.
    (f) Repeat the process on the same tree (possibly using different paths), and count the number of paths you removed.
    (g) Count the number of odd towns in your tree.
    (h) Make a conjecture.
    (i) Prove your conjecture.

23. A tree was defined in E 22. Let $t$ be the number of towns and $s$ the number of sections in a connected tree.
    (a) Find $s$ and $t$ for three connected trees.
    (b) Conjecture a relation between $s$ and $t$ for any connected tree.

24. Prove your conjecture of E 23.

●

25. Draw a connected highway system that is not a tree (one that has circuits). The police want to put up roadblocks in the middle of some sections in such a way that racing around any circuit will be impossible but it will still be possible to travel from any town to any other town without passing a roadblock. In short, the police want to make some cuts in the system to turn it into a connected tree.
    (a) Show (in the highway system you drew) some of the various ways in which the police can set up their roadblocks.
    (b) Count the number of roadblocks in each case.
    (c) Make a conjecture.
    (d) Test your conjecture on another highway system.

26. (a) In a highway system, set up roadblocks as in E 25. Count $t$, the number of towns; $s$, the number of sections; and $b$, the number of blocks.
    (b) Do this for three more highway systems.
    (c) Make a conjecture linking $s$, $t$, and $b$.

27. Prove the conjecture of E 26(c). (Use E 23 and E 24.)

28. The manager of a metropolitan transportation system wants to route his buses on a given street layout in such a way that (i) each route is a loop, ending where it begins and passing through no intersection more than once, and (ii) each section of street is on exactly one route.
    (a) Give an example of street layout without dead-ends for which such routes cannot be found.
    (b) For which street layouts can such routes be found?

●  ●

29. A standard domino set consists of 28 pieces, from double-zero to double-six.
    (a) Is it possible to arrange all these pieces in a straight line in such a way that the dots of any pair of adjacent pieces match?
    (b) Is it possible to arrange them in a circle and still meet the condition in (a)?

30. Let us call the standard domino set, which goes up to double-six, a double-six set. More generally, a double-$n$ set consists of all dominoes from double-zero to double-$n$. For instance, a double-four set consists of the 15 dominoes (0, 0), (0, 1), (0, 2), (0, 3), (0, 4), (1, 1), (1, 2), (1, 3), (1, 4), (2, 2), (2, 3), (2, 4), (3, 3), (3, 4), (4, 4).
    (a) Like E 29(a), (b), but for a double-four set.
    (b) Like E 29(a), (b), but for a double-three set.
    (c) Like E 29(a), (b), but for a double-$n$ set.

31. Were a chemist to look at our maps, he might think of the towns as atoms, the sections as bonds between atoms, and the whole highway system as a molecule. What we call the "degree of a town" he would call "the valence of an atom." The valence of hydrogen is 1; the valence of oxygen is 2. Thus the water molecule $H_2O$ corresponds, in our language, to the tree H—O—H.

    The valence of a carbon atom is 4. Let a molecule have $c$ carbon atoms, $h$ hydrogen atoms, $a$ oxygen atoms, and $b$ bonds.
    (a) Prove that $4c + 2a + h = 2b$.
    (b) If the molecule is a tree, prove, using E 23, that
    $$c + a + h - 1 = b.$$
    ("Tree" was defined in E 22.)
    (c) Combining (a) and (b), prove that in a tree of carbon, hydrogen, and oxygen atoms,
    $$h = 2c + 2.$$
    (d) The chemical formula of glucose is $C_6H_{12}O_6$. Can the glucose molecule be a tree? Draw a possible arrangement of the atoms.

32. Prove that, in any connected system in which all the towns are of degree 2, both the inspector and the salesman have a route.

33. From a highway system in which each town is of even degree we can obtain a system—not necessarily a connected system—in which each town is of degree 2, by replacing every town of degree $2N$ by $N$ nearby towns of degree 2 (for example, a town of degree 6, such as ✳, would be replaced by ⫲ ).

    Combine this with E 32 to obtain a new proof of the fact that the inspector has a route over a system in which all of the towns are of even degree.

34. Using E 22, prove that a tree with $s$ sections and $t$ towns of odd degree must have a path with at least $2s/t$ sections. (Of course the number of sections in a path is always a natural number. Perhaps it would be better to say: "There is a path whose number of sections is at least as large as $2s/t$.")

### References

1. J. R. Newman, *The World of Mathematics*, vol. 1, Simon and Schuster, New York, 1956. (Euler's solution of the inspector's problem is presented on pp. 573–580.)

2. O. Ore, *Graphs and Their Uses*, Random House, New York, 1963. ("Linear graph" is the technical term for "highway system.")

3. B. A. Trakhtenbrot, *Algorithms and Automatic Computing Machines*, Heath, Boston, 1963. (Highway systems are applied to game theory on pp. 8–24.)

*Chapter* **10**

# MEMORY WHEELS

Mathematics, like every branch of knowledge, is the product of the interplay between past and present, between accumulated knowledge and curiosity, between an autonomous structure and the tastes and needs of the time. What one age considers a pressing question, another may not ask at all. The pure mathematics of one era may be applied in another, perhaps centuries later. Pushing into the unknown, the mathematician is an explorer who is likely to find what he did not seek and who cannot predict how others will use his discoveries.

A problem on which I once worked illustrates these aspects of the growth of knowledge, and its solution captures the flavor of mathematical research. This particular adventure began when the composer George Perle told me about an elaborate theory of rhythm that had been developed in India perhaps some thousand years ago. "While reading about this theory," he said, "I learned my one and only Sanskrit word: *yamátárájabhánasalagám.*" I asked him what it meant.

"It's just a nonsense word invented as an aid in remembering the names of certain rhythms." *

"If a person can remember that," I replied, "he can remember anything."

"There is a lot in those ten syllables," said Perle. "As you pronounce the word you sweep out all possible triplets of short and long beats. The first three syllables, *ya má tá* have the rhythm short, long, long. The second through the fourth are *má tá rá:* long, long, long. Then you have *tá rá ja:* long, long, short. Next there are *rá ja bhá:* long, short, long. And so on."

I wrote down the word and saw that what Perle said is true. Each successive triplet of syllables displays a different pattern, and the whole word displays all eight possible patterns, giving each once and only once. As a mathematician I was fascinated to find that such a sequence could exist.

That night I returned to the ancient word. To strip it of irrelevancies I replaced

---

* S. V. Arya of Kharagpur informs me that this word is still used in the study of medieval Indian poetry. Perle learned of the word in R 2.

the syllables with digits, letting 0 stand for a short beat and 1 for a long beat. In this notation *yamátárájabhánasalagám* became 0111010001.

After staring at the simplified string for a while, I noticed a lovely thing. The first two digits are the same as the last two; if I bent the string into a loop, it would look like a snake swallowing its own tail. That is, the last 01 could be placed over the first 01, and the two pairs of digits would merge into a single pair. Instead of a line of ten digits I now saw a circle of eight.

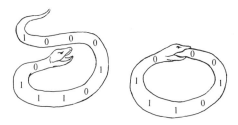

I could begin anywhere on this "memory wheel" and move around it in either direction, sweeping out triplets of 0's and 1's. Starting at the top and reading counterclockwise, for example, gave 011, 111, 110, 101, 010, 100, 000, and 001.

The next thing that occurred to me, as it would automatically to any mathematician, was to generalize what I had found. Is there a "word" for listing all quadruplets of 0's and 1's once and only once? For quintuplets? For groups of any size? And if so, does the snake always swallow its tail?

Before attacking the problem for quadruplets I decided to go back and look at couplets. Is there a word that lists each of the four couplets 00, 01, 10, and 11 exactly once? Does it close up on itself? Writing down the sequence 0011, I saw that I already had the three couplets 00, 01, and 11. Adding one more 0, to form 00110, gave the final couplet: 10. I noted that, since the first and last digits are 0's, the snake does swallow its tail. The five-digit word could be bent into this four-digit wheel containing each couplet once and only once.

Now I was ready to take on the quadruplets. I began methodically by listing all the possible groups of four digits composed of 0's and 1's. To do this I first wrote down the eight triplets and placed a 0 in front of each. This gave me all the quadruplets beginning with 0. Then I repeated the triplets and placed a 1 before each, obtaining the quadruplets beginning with 1.

| | |
|------|------|
| 0000 | 1000 |
| 0001 | 1001 |
| 0010 | 1010 |
| 0011 | 1011 |
| 0100 | 1100 |
| 0101 | 1101 |
| 0110 | 1110 |
| 0111 | 1111 |

Since the list contained 16 quadruplets, I saw that any memory word that would sweep out each of them once (if any exists) would have to have 19 digits. The first four digits make one quadruplet, and each of the next 15 numbers completes another.

Somewhere in the word I was seeking, I knew there would have to be a string of four 1's and, somewhere, four 0's. Why not put them together and see what would happen? I wrote down the eight symbols 11110000. Then I checked off my list the five quadruplets it contained: 1111, 1110, 1100, 1000, and 0000. So far so good. To avoid getting 0000 twice, I next had to add a 1, which gave 111100001. I checked off 0001 on the list. Adding 010 produced three more quadruplets, all new, and brought the sequence to 111100001010.

From here I proceeded one digit at a time, checking the list to make sure there were no duplications. At each of the next three positions the choice was clear: one of the digits would form a quadruplet already checked; the other would not. I had then reached the 15-symbol word: 111100001010011.

Only four to go. Considering the next digit, I found that neither 1 nor 0 provided a duplication. But a 1 would lead to trouble in the next position, where either 0 or 1 would produce a duplication. I was afraid I might not be able to reach 19 symbols after all. It turned out, however, that a 0 in the sixteenth position involved no such difficulty, and the last three positions presented unambiguous choices. I had found my word of 19 symbols—a word containing each of the 16 quadruplets precisely once: 1111000010100110111.

As soon as I had finished, I looked at the first three symbols and the last three. They were the same. This snake, too, could swallow its tail. This 19-symbol word could be bent into a wheel of 16 symbols:

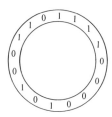

Inspired by this success, I decided that there must be memory words for quintuplets, sextuplets, and so on. Furthermore, I felt sure they would all close up into wheels. It was time to stop experimenting, however, and look for a proof of the conjecture.

Grappling with the problem, I began to look at the Indian memory word in a slightly different way, concentrating on the eight overlapping triplets it contains:

011

111

110

101

010

100

000

001

In this light, the word appeared as a means of arranging the triplets such that the last two symbols of one are the same as the first two of the next. Suppose the word had not been invented. How would one have gone about finding it? I decided to spread the triplets over a piece of paper and connect the appropriate pairs with arrows. That is, I drew an arrow from one triplet to another whenever the last two symbols of the former were the same as the first two symbols of the latter; for example, $001 \rightarrow 010$ and $001 \rightarrow 011$. After moving the triplets and arrows around a little to make a simple pattern, I obtained this diagram:

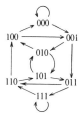

As I gazed at the configuration, I suddenly saw it as a map in which the arrows were one-way roads and the triplets were towns. The problem of arranging the eight triplets into a memory word could now be stated in the terms of our traveling salesman of Chapter 9, who is looking for a route over the one-way roads that will take him through each town just once. With the help of the Indian memory word, I traced one possible route. As this illustration shows, the "town" in which the journey ends is adjacent to the one in which it starts.

```
              000
       100 ←       ↘ 001    Finish
              010
              ⤷
                101
       110 ←      ↗ 011     Start
              111 ←
```

There is a section of road that will take the salesman from the finishing point back to the start. This, of course, reflects the fact that the memory word closes into a wheel.

Clearly the same scheme would apply to overlapping couplets, to quadruplets, or to groups of any size. The general problem had been translated into a new language. The question "Is there always a memory word?" now reads "Is there always a route for the salesman?" The question "Does every memory word close up on itself?" now reads "Does the salesman always finish his trip in a town adjacent to the town in which his trip began?"

Unhappily, the translation did me no good. I had absolutely no luck in solving either of the salesman problems. There was (and still is) no known general technique, other than trying all paths, to tell whether a highway system has a route passing just once through every town (see Chapter 9). And I did not see why these special highway systems arising from overlapping strings of 0's and 1's should have a route for the salesman. After several days without progress, I put the matter aside. A few months later, as I was leafing through the 1946 volume of *The Journal of the London Mathematical Society*, I saw a diagram that resembled my highway system. It appeared in a paper entitled "Normal Recurring Decimals," by I. J. Good. Quite a different topic had brought Good to a consideration of the memory wheel problem—and he had solved it. He was chiefly interested in showing how to produce an endless string of 0's and 1's in which each of the possible sextuplets of 0's and 1's appears with equal frequency. His solution was perfectly general, and applied to groups of any size. He noted further that ". . . the result has an application to the construction of teleprinters that use alphabets whose letters consist of a finite number (usually five or six) of 0's and 1's."

Good had also recognized that the problem is related to a highway system, but his system was different from mine. I had seen the triplets (or groups of any other size) as towns, and the roads between them as overlaps. Good saw the problem the other way around: he saw the triplets as roads, and the couplets—the overlapping portions of the triplets—as towns.

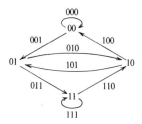

For example, the highway 011 runs from the town 01 to the town 11. To put it another way, there are four towns: 00, 01, 10, and 11. One of these is joined to another if the last digit of the former is the same as the first digit of the latter (01 is joined to 11 by 011).

Now each road represents a triplet, and every incoming triplet overlaps every outgoing triplet for every town. This means that if a person—say, an efficient highway inspector—could find a route taking him over each road just once, he would, in effect, trace a memory wheel. For example, suppose the inspector began at town 01. He could follow a route consisting of highways 011, 111, 110, 101, 010, 100, 000 and 001. The trip is, of course, exactly the one given by *yamátárájabhánasalagám*. Notice also that the last highway, 001, leads into the first one, 011.

A glance at the diagram makes it clear that two roads lead into, and two lead out of, every town. Thus, as we saw in the last chapter, the highway inspector can always find a route over such a highway system (Theorem 6 of Chapter 9). Hence it was no accident that he could find a route over the diagram constructed to find a memory word for triplets.

The same reasoning works just as well on Good's highway system associated with quadruplets. In this case a triplet is a town, and a quadruplet is a one-way road that goes from the first triplet to the last triplet of the town:

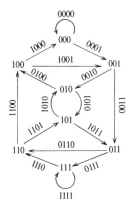

Again, notice that two roads enter each town and that two roads leave. Thus, by Theorem 6 of Chapter 9, the inspector has a route; that is, there is a memory wheel for quadruplets. Similarly, there is a memory wheel for quintuplets, and so on.

Since the salesman's problem for my highway systems is equivalent to the inspector's problem for Good's systems, the salesman also has a route, and his journey must end in a town adjacent to the one in which he starts. As it turns out, I had quite unintentionally come upon a special class of highway systems for which the salesman has a route.

The story does not end here. Through *Mathematical Reviews* I learned that other mathematicians in various parts of the world had been involved with the memory-wheel problem in one guise or another. Twelve years before Good had published his paper, M. H. Martin had solved the problem in a completely different way; he had been led to it by a problem in dynamics. Another mathematician, D. Rees, seeking a more constructive technique than Good's, solved the problem by considering certain divisors of polynomials of the type $X^n - 1$. In 1951, N. M. Korobov, who was studying the fractional part of multiples of numbers, described an entirely different technique for constructing memory wheels. A Dutch engineer, K. Posthumus, working on the theory of telephone circuits, posed another question about memory wheels: How many different wheels are there for couplets, triplets, quadruplets, and so on? He found that there is one for couplets and that there are two for triplets, 16 for quadruplets, 2048 for quintuplets; and he guessed at a general formula for groups of any size ($n$-tuplets): 2 raised to the power $2^{n-1} - n$. In 1946, N. G. de Bruijn proved the conjecture.

I learned that memory wheels have found technological applications. A computer made in Czechoslovakia employs a memory wheel of 1024 symbols to search through an equal number of storage locations on a memory drum. And in the Damodar Valley near Calcutta a memory wheel in a central office scans automatically transmitted reports from 64 rain gauges. Memory wheels are also used in constructing error-correcting codes for communication theory.

Only after publishing an article on memory wheels (based on an early version of this chapter) in the *Scientific American* of May 1961, did I learn that the modern history of these wheels goes back to the French telegraph engineer Émile Baudot, who in 1882 used quintuplets of 0 and 1 to transmit information. In a letter appearing in the July, 1961, issue of the *Scientific American*, D. A. Kerr, an engineer working on printing telegraphy, writes, " . . . (It) appears in a Western Electric tape teletypewriter used very early in this century . . . , and it is known here as the combination-wheel and seeker mechanism. The combination wheel is, of course, a mechanically coded memory wheel, and the seekers

HISTORY OF MEMORY WHEELS

| *Who* | *Where* | *When* | *Why* |
|---|---|---|---|
| Poets | India | A.D. 1000 (?) | Used memory word for triplets of short and long beats. |
| Émile Baudot | France | 1882 | Used memory wheel for the 32 quintuplets of telegraphy (for the 26 letters, idle condition, space, line feed, carriage return, shift to regular alphabet, shift to numbers). |
| M. H. Martin | U.S.A. | 1934 | Interested in a problem in dynamics; provided simple algorithm for making memory wheels. |
| K. R. Popper | Austria | 1934 | Constructed random sequences for probability theory with memory wheels. |
| I. J. Good | Great Britain | 1946 | Proved existence of memory wheels in answering a question concerning decimal representation of numbers. |
| N. G. de Bruijn | Holland | 1946 | Answered question raised in telephone engineering; found number of memory wheels. |
| D. Rees | Great Britain | 1946 | Stimulated by Good's work, proved existence of memory wheels by using divisors of the polynomials $X^n - 1$. |
| N. M. Korobov | Russia | 1951 | Interested in properties of the decimal representation of numbers, devised way of constructing all memory wheels (generalizing Martin's technique). |
| E. N. Gilbert | U.S.A. | 1953 | Constructed memory wheels for Bell Telephone Laboratories using methods illustrated in E 12, 13, 14, 15, 16. |
| R. L. Goodstein | Great Britain | 1956 | Interested in servomechanisms; independently rediscovered Martin's technique, for which he gave a proof using congruences. |
| L. R. Ford, Jr. | U.S.A. | 1957 | Constructed memory wheels to imitate the binomial distribution of probability theory; independently rediscovered Martin's technique, for which he gave a new proof. |
| S. Golomb and L. Welch | U.S.A. | 1957 | Extended Gilbert's technique, examined various formulas for producing memory wheels, gave new proof of de Bruijn's theorem. |
| T. E. Stern and B. Friedland | U.S.A. | 1959 | Interested in error-correcting codes for communication theory; independently rediscovered Rees's technique. |
| E. B. Leach | U.S.A. | 1960 | Constructed certain sequences for probability theory by memory wheels, which he devised in a new way (out of a wheel for $n$-tuplets he constructed a wheel for $(n + 1)$-tuplets). |

form a device for locating the desired quintuplet on its periphery, thus indexing the desired character for printing."

This has been a fable with many morals. It deals with a question that could have been raised a thousand years ago in India, but was not. And little did Euler dream that his work on a puzzle would be of practical use in the twentieth century. Moreover, the variety of applications reflects both the mathematical mood of our time and the fantastic amount of mathematics being created all over the world. The fact that so many of the workers were ignorant of the others' results, and therefore duplicated much effort, points to another problem: how to record the myriad discoveries constantly flowing into man's store of knowledge.

**Exercises**

1. Is the resemblance between the highway systems of pp. 143 and 145 accidental? If not, explain their relationship.

2. Check that each of the 64 6-tuplets of 0's and 1's appears exactly once in this memory word:

    0000010000110001011111101100111100100110111010101101001010001110000000.

3. Show that there is only one memory wheel for couplets of 0's and 1's.

4. Show that there are exactly two memory wheels for triplets formed of 0's and 1's, and that one of them is just the other read backwards.

5. (a) Count the number of 0's and the number of 1's in three memory wheels.
   (b) Make a conjecture.

6. See E 5.
   (a) Prove that the number of 0's is the same as the number of 1's in any memory wheel for quintuplets. (*Hint:* Do bookkeeping with respect to contributions of 0's and 1's by quintuplets. Observe, for example, that each 0 and each 1 in the memory wheel is part of 5 quintuplets.)
   (b) Do the same for 6-tuplets.

7. If you write down the string 000 and then build up a word by writing down in succession the larger of the digits 0 and 1 that does not lead to a duplicated triplet, show that you obtain a memory word for all triplets.

8. If you write down the string 0000 and then build up a word by writing down in succession the larger of the digits 0 and 1 that does not lead to a duplicated quadruplet, show that you obtain a memory word for all quadruplets.

9. If you write down the string 00000 and then build up a word by writing down in succession the larger of the digits 0 and 1 that does not lead to a duplicated quintuplet, show that you obtain a memory word for quintuplets. (The method of E 7, E 8, and E 9 is due to M. H. Martin, who in R 1 proved that it always works.)

10. Is it necessary in E 7 to start with 000? Could we have started with the triplet 010 or the triplet 111?

●

11. Prove directly, without using the notion of highway systems, that each memory word for 6-tuplets of 0's and 1's must close up into a memory wheel. (Your argument should apply to couplets, triplets, and so on.)

12. Another method of producing a string of quadruplets that fit together to form a memory wheel is used by the Bell Telephone Laboratories. Given a quadruplet $Q$, form the next quadruplet this way. If the first two digits of $Q$ are the same, write a 0 to the right of $Q$. If the first two digits of $Q$ are not the same (that is, if one of them is 1 and the other 0), write a 1 to the right of $Q$. Erase the first digit of $Q$ from the quintuplet just formed. We are left with a quadruplet that overlaps $Q$ in such a way that the two quadruplets might possibly be part of a memory wheel for quadruplets. (This method applied to 1111 gives 1110. When applied to 1101, it gives 1010. When applied to 1010, it gives 0101.)
    (a) Applying this technique repeatedly, manufacture all quadruplets except 0000. (You may start with any quadruplet other than 0000.)
    (b) Using (a) build a memory wheel of 15 digits for the 15 quadruplets other than 0000.
    (c) Insert a 0 in the wheel built in (b) so as to obtain a wheel for the 16 quadruplets.

13. Show that the method of E 12, when applied to the triplet 111, produces all triplets except 000. Use it to form a memory wheel for triplets.

14. Show that the method of E 12, when applied to 11111 (or any other quintuplet except 00000) does *not* produce the 31 quintuplets other than 00000.

15. Show that the method of E 12, when applied to 111111, produces the 63 6-tuplets other than 000000.

16. Show that the method of E 12, when applied to the 7-tuplet 1111111, produces the 127 7-tuplets other than 0000000. (Combining E 12, E 13, E 14, E 15, and E 16, we see that this method can be used to make memory wheels for $N$-tuplets, $N = 3, 4, 6$, and 7 but not for $N = 5$. It also works for $N = 15$. It is not known, in general, for which $N$ it works.)

17. Is it necessary to start with 111111 in E 15?

●●

18. This is how E. B. Leach, in R 13, devises a memory wheel for quadruplets out of one for triplets: He writes out the wheel of 8 symbols in a line; for example, 11010001. He then forms a string of 8 symbols by first writing below the left digit of the given string the digit 0. Below the second digit he writes a 0 if the first digit of the given string and the string he is building are the same; otherwise he writes a 1. In the example, he writes 1. Below the third digit he writes a 0 if the second digit of the string he is forming and the second digit of the given string are the same; otherwise he writes a 1. Continuing in this manner, he builds the string

01001111. Then he obtains a second string by interchanging 0's with 1's in the string he has just formed. This gives him 10110000.

To assemble the two strings

$$01001111 \quad \text{and} \quad 10110000$$

into a string of length 16, first form the second string into a wheel. Then read this wheel, beginning with the first triplet, 010, of the first string, and write out the result, 01011000. Place the latter string to the right of the string 01001111, and obtain a string of length 16,

$$0100111101011000,$$

which, when placed on a wheel, forms a memory wheel for quadruplets. Check that it does.

19. Using the method of E 18, construct a memory wheel for quintuplets.

20. Check that this wheel lists each of the nine possible couplets formed of 0, 1, and 2:

21. Devise a memory word for listing without duplication each of the 27 triplets formed of 0, 1, and 2. Show that it can be pulled around to form a memory wheel.

22. Prove that there is a memory wheel for quadruplets formed of 0, 1, and 2. (*Hint:* Use the same reasoning as Good did, and notice that three roads enter, and that three leave, each town.)

23. Draw the highway system of 9 towns and 27 one-way roads that would be of use in proving that a memory wheel for triplets of 0, 1, and 2 exists. (The towns are the couplets formed with 0, 1, and 2; the roads are the triplets formed with 0, 1, and 2.)

### References

1. *M. H. Martin, A problem in arrangements, *Bulletin of the American Mathematical Society*, vol. 40, 1934, pp. 859–864.

2. C. Sachs, *Rhythm and Tempo*, Norton, New York, 1953, pp. 98–101.

3. K. R. Popper, *The Logic of Scientific Discovery*, Basic Books, New York, 1959. (In particular, pp. 162–163, 292.)

4. *N. G. De Bruijn, A combinatorial problem, *Akademie van wetenschappen, Amsterdam*, vol. 8, 1946, pp. 461–467.

5. *I. J. Good, Normal recurring decimals, *Journal of the London Mathematical Society*, vol. 21, 1946, pp. 167–169.

6. *D. Rees, Note on a paper by I. J. Good, *ibid.*, pp. 169–172.

7. *N. M. Korobov, Normal periodic sequences, *American Mathematical Society, Translations*, series 2, vol. 4, pp. 31–58. (Translation of an article published in Russian in 1951.)

8. R. L. Goodstein, A permutation problem, *Mathematical Gazette*, vol. 40, 1956, pp. 46–47. (This uses the language of congruences, discussed in Chapter 11.)

9. *L. R. Ford, Jr., A cyclic arrangement of *n*-tuplets, The Rand Corporation, 1957.

10. *S. Golomb and L. Welch, Nonlinear shift registers, Jet Propulsion Laboratory, California Institute of Technology, Pasadena, 1957.

11. (a) N. M. Blachman, Central European computers, *Association for Computing Machinery* (Communications), vol. 2, 1959, pp. 14–18.
    (b) M. Nadler and A. Sengupta, Unusual applications, *ibid.*, pp. 40–43.

12. *B. Friedland and T. Stern, Applications of modular sequential circuits to single error-correcting P-Nary codes, *IRE Transactions*, vol. IT-5, 1959, pp. 114–123.

13. *E. B. Leach, Regular sequences and frequency distributions, *Proceedings of the American Mathematical Society*, vol. 11, 1960, pp. 566–574.

14. *W. W. Peterson, *Error-correcting Codes*, M.I.T. Press and John Wiley, New York, 1961.

15. S. Stein, The mathematician as an explorer, *Scientific American*, May, 1961, pp. 149–158. (This is essentially the same as the present chapter.)

## Chapter 11

# CONGRUENCE

This chapter provides a language or viewpoint in terms of which much of our work of the first five chapters can be stated more compactly, and, in addition, develops machinery that we will need in the next two chapters.

This viewpoint may have occurred to anyone looking at a calendar, such as this one

September

| Sun | M | T | W | Th | F | Sat |
|---|---|---|---|---|---|---|
|  | 1 | 2 | 3 | 4 | 5 | 6 |
| 7 | 8 | 9 | 10 | 11 | 12 | 13 |
| 14 | 15 | 16 | 17 | 18 | 19 | 20 |
| 21 | 22 | 23 | 24 | 25 | 26 | 27 |
| 28 | 29 | 30 |  |  |  |  |

As the eye travels down a column corresponding to a day of the week, it sees several numbers, each 7 larger than the one directly above it. For instance, the Tuesdays have the dates 2, 9, 16, 23, and 30. Any two of these Tuesday dates differ from each other by a multiple of 7. The same is true for any of the Wednesday dates, 3, 10, 17, 24; any two of these dates differ by a multiple of 7. Thus, just as the number 7 breaks the days into seven groups, so does it break the integers into seven groups. Of course, we need not restrict ourselves to the number 7, chosen long ago as a unit of time; any natural number from 1 on might be treated as a measure. This brings us to the key notion of this chapter, which we phrase for any "measuring number," not just for the number 7.

Definition of **congruence**.* If the integers $A$ and $B$ differ by a multiple of the natural number $M$ ($M$ not zero), then we will say "$A$ is congruent to $B$ modulo $M$" and write

$$A \equiv B \;(\mathrm{mod}\; M).$$

We will call $M$ the modulus (from the Latin for "little measure").

---

* Congruence of integers is not related to congruence of geometric figures.

The phrase "*A* is congruent to *B* modulo *M*" is simply shorthand for "*M* divides *A* − *B*" or "There is an integer *Q* such that *A* − *B* = *QM*." Returning to the calendar and its modulus, 7, we have, for instance, 3 ≡ 17 (mod 7), for 3 − 17 is a multiple of 7 [3 − 17 = (−2)7]. The reader may check that 2 ≡ 23 (mod 7), 23 ≡ 2 (mod 7), 5 ≡ 19 (mod 7) by a little arithmetic or a glance at the calendar: two September dates are congruent modulo 7 if they correspond to the same day of the week.

The question, "Do the 326th and the 111th days of the year fall on the same day of the week?" reads, in the language of congruences, "Is 326 congruent to 111 modulo 7?" or, more briefly,

$$\text{Is } 326 \equiv 111 \text{ (mod 7)?}$$

To answer the question we determine whether 7 divides 326 − 111 = 215. It does not, so the answer is "no."

Let us turn our attention to the modulus 3. Suppose we were to ask, "Which integers are congruent to 1 modulo 3?" We have, as examples, 4 ≡ 1 (mod 3), −2 ≡ 1 (mod 3), and 7 ≡ 1 (mod 3). More generally, all integers we obtain from 1 by adding a multiple of 3 are congruent to 1 modulo 3. We sketch them on the number line:

The labeled integers are congruent to 1 modulo 3

Which integers are congruent to 2 modulo 3? We have 2 ≡ 2 (mod 3), since 3 divides 2 − 2 (2 − 2 = 0·3); 5 ≡ 2 (mod 3); 8 ≡ 2 (mod 3); −1 ≡ 2 (mod 3). We sketch some of the endless list of integers congruent to 2 modulo 3:

The labeled integers are congruent to 2 modulo 3

The above two sketches list all integers except 0, 3, −3, 6, −6, . . . . We may describe them as the integers congruent to 0 modulo 3. (Some may prefer to describe them as the integers congruent to 15 modulo 3; others, as the integers congruent to 3 modulo 3.) We sketch them:

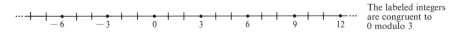

The labeled integers are congruent to 0 modulo 3

Next let us consider the smallest interesting modulus, 2. How does the assertion "*A* ≡ *B* (mod 2)" translate into words? First of all, it says that *A* and *B* differ by a multiple of 2, that is, by an even number. So the congruence "*A* ≡ *B* (mod 2)" says, "Either *A* and *B* are both odd or they are both even." Thus, while the modulus 7 separates the integers into seven classes, the modulus 2 separates the integers into only two classes, the even numbers, . . . , $-4$, $-2$, 0, 2, 4, 6, . . . , and the odd numbers, . . . , $-3$, $-1$, 1, 3, 5, . . . .

The duality of even and odd is the basis of Chapters 1 and 2, the main results of which can be stated quite compactly in the congruence notation. In Chapter 1 we proved that the number of switches, $S$, in a seamless hatband is even. This now reads: $S \equiv 0 \pmod 2$. In the same chapter we also observed that if $B$ is the backward number of one arrangement, and if $B'$ is the backward number of another arrangement, then $S$, the number of switches in going from the first to the second arrangement, is always even if $B - B'$ is even; odd if $B - B'$ is odd. These two assertions are both included in the statement

$$S \equiv B - B' \pmod 2.$$

In Chapter 2 (Theorem 2) we proved that if the number of $AB$ edges on the border of a triangle is odd, then so is the number of $ABC$ triangles. We also proved that if the number of $AB$ edges on the border of the triangle is even, then so is the number of $ABC$ triangles. Both of these statements are included in this assertion:

number of $AB$ edges on the border $\equiv$ number of $ABC$ triangles (mod 2).

Definitions and results in other chapters can also be cast into the language of congruences. In Chapter 4 our statement that $A$ is a divisor of $B$ can be translated as $B \equiv 0 \pmod A$. The basic idea behind the Prime-manufacturing Machine can be expressed (for $M$ larger than 1) as follows: if $A \equiv 0 \pmod M$, then $A + 1$ is not congruent to 0 (mod $M$).

In Chapter 5 we defined a special natural number as one that, whenever it divides the product of two natural numbers, divides at least one of them. In the language of congruences, the natural number $S$ is special if

$$AB \equiv 0 \pmod S, \text{ then } A \equiv 0 \pmod S \text{ or } B \equiv 0 \pmod S.$$

In Chapter 5 we also met the important theorem: if $A$ and $B$ are natural numbers, then there are integers $M$ and $N$ (one positive and one negative) such that

$$(A, B) = MA + NB.$$

This can be translated in two different ways into the language of congruences. First, using the modulus $A$, we could assert: for any natural numbers $A$, $B$, there is an integer $N$ such that

$$(A, B) \equiv NB \pmod A.$$

Secondly, using the modulus $B$, we could assert: for any natural numbers $A$, $B$, there is an integer $M$ such that

$$(A, B) \equiv MA \pmod B.$$

There is another way of looking at the congruence $A \equiv B$ (mod $M$) when $A$ and $B$ are both natural numbers. Consider, for instance, those natural numbers $A$ such that $A \equiv 1$ (mod 3). They are 1, 4, 7, 10, . . . , the Lagado numbers; that is, those numbers that when divided by 3 leave the remainder 1. This may suggest that the congruence "$A \equiv B$ (mod $M$)" may be interpreted as "$A$ and $B$ have the same remainder when divided by $M$." That this is so is the substance of

THEOREM 1. (a) *If the natural numbers $A$ and $B$ have the same remainders when divided by $M$, then $A \equiv B$ (mod $M$).*

(b) *If the natural numbers $A$ and $B$ are congruent modulo $M$, then they have the same remainders when divided by $M$.*

PROOF. We first prove (a). Let the result of the division of $A$ by $M$ be $A = Q_1 M + R$, where $Q_1$ is the quotient and $R$ the remainder. Similarly, let the result of the division of $B$ by $M$ be $B = Q_2 M + R$, where $Q_2$ is the quotient and $R$ the remainder. (Note that we are permitted to use the same letter $R$ in both cases since, by hypothesis, the two numbers $A$ and $B$ have the same remainder when divided by $M$.) Then

$$A - B = (Q_1 M + R) - (Q_2 M + R) = Q_1 M - Q_2 M = M(Q_1 - Q_2),$$

so that $A - B$ is a multiple of $M$; that is, $A \equiv B$ (mod $M$). This establishes (a).

Now we prove (b). Since $A$ and $B$ are by hypothesis congruent modulo $M$, we can write

$$A - B = NM$$

where $N$ is some integer.

Let the result of the division of $A$ by $M$ be $A = Q_1 M + R_1$, where $Q_1$ is the quotient and $R_1$ the remainder. Similarly, let the result of the division of $B$ by $M$ be $B = Q_2 M + R_2$, where $Q_2$ is the quotient and $R_2$ the remainder. $R_1$ and $R_2$ are natural numbers, each at least 0 and at most $M - 1$; hence their difference $R_1 - R_2$ is at least $-(M - 1)$ and at most $M - 1$. But we also see that

$$R_1 - R_2 = (A - Q_1 M) - (B - Q_2 M) = A - B - M(Q_1 - Q_2)$$
$$= NM - M(Q_1 - Q_2).$$

Thus, since the right-hand side of this equation is divisible by $M$, the left-hand side, $R_1 - R_2$, is also divisible by $M$. It follows that since the difference $R_1 - R_2$ is on the one hand a multiple of $M$ and on the other hand an integer between $-(M - 1)$ and $M - 1$, it must equal 0, so that

$$R_1 - R_2 = 0 \quad \text{and} \quad R_1 = R_2.$$

This establishes (b).

The usefulness of this theorem will be shown by examples later in this chapter. For instance, we will easily answer the question "What is the remainder when $5^{110}$ is divided by 6?"

In the course of proving part (b) of Theorem 1 we showed that if two natural numbers less than $M$ are congruent modulo $M$, then they must be equal. This observation, which will be of use in the next chapter, we record in this

LEMMA. *Let $M$ be a positive natural number and $A$ and $B$ natural numbers less than $M$ such that $A \equiv B \pmod{M}$. Then $A = B$.*

Our next theorem will show that the symbol $\equiv$ behaves like the symbol $=$.

THEOREM 2. (a) $A \equiv A \pmod{M}$.
(b) *If $A \equiv B \pmod{M}$, then $B \equiv A \pmod{M}$.*
(c) *If $A \equiv B \pmod{M}$ and $B \equiv C \pmod{M}$, then $A \equiv C \pmod{M}$.*

PROOF. (a) To prove the statement $A \equiv A \pmod{M}$ we must show that $M$ divides $A - A$. But any natural number divides 0; hence $M$ divides $A - A$.

(b) To prove that $A \equiv B \pmod{M}$ implies that $B \equiv A \pmod{M}$ we proceed in this manner. We know that there is an integer $Q$ such that

$$A - B = QM.$$

Multiplying both sides by $-1$ gives

$$-A + B = (-Q)M,$$

and hence

$$B - A = (-Q)M.$$

Thus $M$ divides $B - A$. That is,

$$B \equiv A \pmod{M}.$$

(c) If we assume that $A \equiv B \pmod{M}$ and that $B \equiv C \pmod{M}$, then there are integers $Q_1$ and $Q_2$ such that

$$A - B = Q_1 M$$

and

$$B - C = Q_2 M.$$

Adding the left sides of these two equations together, we get

$$(A - B) + (B - C) = Q_1 M + Q_2 M.$$

Since the $B$'s cancel, and since we have

$$Q_1 M + Q_2 M = (Q_1 + Q_2)M,$$

we obtain

$$A - C = (Q_1 + Q_2)M;$$

and thus $A \equiv C \pmod{M}$. This ends the proof.

Furthermore, as the following theorems show, the symbol $\equiv$ behaves in the same way as does the symbol $=$ with respect to addition and multiplication.

THEOREM 3. *If $A \equiv B \ (mod \ M)$ and $a \equiv b \ (mod \ M)$, then*
$$a + A \equiv b + B \ (mod \ M).$$

PROOF. We again must go back to the definition of the symbol $\equiv$. We know that there are integers $Q_1$ and $Q_2$ such that
$$A - B = Q_1 M$$
and
$$a - b = Q_2 M.$$
Adding these equations, we get
$$(A - B) + (a - b) = Q_1 M + Q_2 M.$$
From this it follows that
$$(A + a) - (B + b) = (Q_1 + Q_2)M,$$
and consequently
$$A + a \equiv B + b \ (mod \ M).$$

THEOREM 4. *If $A \equiv B \ (mod \ M)$ and $a \equiv b \ (mod \ M)$, then $Aa \equiv Bb \ (mod \ M)$.*

Before proceeding to the proof, let us illustrate this theorem, which may be more surprising than Theorem 3. For example, it asserts that, since
$$17 \equiv 3 \ (mod \ 7)$$
and
$$5 \equiv 12 \ (mod \ 7),$$
we will have
$$17 \cdot 5 \equiv 3 \cdot 12 \ (mod \ 7),$$
or
$$85 \equiv 36 \ (mod \ 7),$$
a congruence that the reader may check.

Now we will present a

PROOF OF THEOREM 4. Since $A \equiv B \ (mod \ M)$ and $a \equiv b \ (mod \ M)$, there are integers $Q_1$ and $Q_2$ such that
$$A - B = Q_1 M$$
and                                                                                           (1)
$$a - b = Q_2 M.$$
Out of these two equations we hope to construct an integer $Q$ such that
$$Aa - Bb = QM.$$

Were we to proceed as in the proof of Theorem 3 (multiplying instead of adding the equations $A - B = Q_1M$ and $a - b = Q_2M$), we would get

$$(A - B) \times (a - b) = Q_1M \times Q_2M,$$

which, multiplied out, expands to give

$$Aa - Ba - Ab + Bb = Q_1Q_2M^2.$$

This would get us nowhere near a proof of Theorem 3, for there is too great a tangle of $A$'s, $a$'s, $b$'s, and $B$'s.

To avoid this difficulty let us start over, this time first separating $A$ from $B$ and $a$ from $b$ by replacing the opening equations (1) with

$$A = B + Q_1M$$

and

$$a = b + Q_2M.$$

If we multiply these together we obtain

$$Aa = (B + Q_1M) \times (b + Q_2M).$$

Multiplying out the right-hand side gives

$$Aa = Bb + BQ_2M + Q_1Mb + Q_1MQ_2M.$$

This can be rewritten as

$$Aa = Bb + (BQ_2 + Q_1b + Q_1MQ_2)M.$$

Thus

$$Aa - Bb = (BQ_2 + Q_1b + Q_1MQ_2)M.$$

Now, $BQ_2 + Q_1b + Q_1MQ_2$ is an integer, which we may call $Q$. We then have $Aa - Bb = QM$; and hence

$$Aa \equiv Bb \ (\mathrm{mod}\ M).$$

Theorem 4 says that we can multiply two congruences that involve the same modulus. It follows quickly that any number of congruences relative to the same modulus can be multiplied together to yield a valid congruence.

Now we can easily find the remainder when $5^{110}$ is divided by 6. Although it is tedious to multiply one hundred and ten 5's, it is a simple matter to multiply one hundred and ten $-1$'s. This will be the key to finding the remainder. We have $5 \equiv -1 \ (\mathrm{mod}\ 6)$, and we may write down the 110 congruences:

$$5 \equiv -1 \ (\mathrm{mod}\ 6), \quad 5 \equiv -1 \ (\mathrm{mod}\ 6), \ldots, \quad 5 \equiv -1 \ (\mathrm{mod}\ 6).$$

According to Theorem 4, we may multiply these congruences and obtain

$$5^{110} \equiv (-1)^{110} \ (\mathrm{mod}\ 6).$$

But $(-1)^{110} = 1$, since 110 is even. Thus

$$5^{110} \equiv 1 \ (\mathrm{mod}\ 6).$$

According to Theorem 1, both $5^{110}$ and 1 have the same remainder when divided by 6. But the remainder when 1 is divided by 6 is clearly 1. Hence the remainder when $5^{110}$ is divided by 6 is also 1.

Let us now apply congruences to explain an arithmetic device—**casting out 9's**—which the reader may have met in elementary school. This method is useful for finding out quickly whether a natural number (usually large) is divisible by 9. We illustrate the rule with the natural number 56,093,742. First we add the digits; $5 + 6 + 0 + 9 + 3 + 7 + 4 + 2 = 36$. The rule then says that since 9 divides 36, 9 divides 56,093,742.

To justify this, and even more, we will prove

THEOREM 5. *When a natural number N is divided by 9 the remainder is the same as that obtained when the sum of the digits of N is divided by 9.*

PROOF. Denote by $D_0$ the units digit of $N$. ($D_0$ is an integer between 0 and 9.) Denote by $D_1$ the tens digit of $N$. Similarly, define $D_2$, $D_3$, . . . until we stop at $D_d$, where $d$ is 1 less than the number of digits in $N$. (In the example above, $D_0 = 2$, $D_1 = 4$, $D_2 = 7$, . . . , $D_7 = D_d = 5$.) Thus

$$N = 10^d D_d + \cdots + 100 D_2 + 10 D_1 + D_0.$$

The sum, $S$, of the digits of $N$ is

$$S = D_d + \cdots + D_2 + D_1 + D_0.$$

To prove that $N$ and $S$ have the same remainder when divided by 9, in view of Theorem 1, we need only prove that $N \equiv S \pmod 9$.

Since $10^1 \equiv 1 \pmod 9$ and $D_1 \equiv D_1 \pmod 9$, we have by Theorem 4 that $10^1 D_1 \equiv 1 D_1 \pmod 9$; that is, $10 D_1 \equiv D_1 \pmod 9$.

Similarly, $10^2 \equiv 1 \pmod 9$, and $D_2 \equiv D_2 \pmod 9$; thus, by Theorem 4, $10^2 D_2 \equiv 1 D_2 \pmod 9$; that is, $10^2 D_2 \equiv D_2 \pmod 9$. Continuing in this fashion, we obtain $10^3 D_3 \equiv D_3 \pmod 9$, . . . , and finally, $10^d D_d \equiv D_d \pmod 9$.

With the aid of Theorem 3, which asserts that we can add congruences, we then have

$$10^d D_d + \cdots + 100 D_2 + 10 D_1 + D_0 \equiv D_d + \cdots + D_2 + D_1 + D_0 \pmod 9.$$

That is,

$$N \equiv S \pmod 9.$$

Our proof is complete.

Perhaps we should explain why the rule is called "casting out 9's." If we had erased the 9 before we computed the sum $5 + 6 + 0 + 9 + 3 + 7 + 4 + 2$ we would not have changed the remainder. Similarly, we could have crossed out the 7 and the 2 (which add up to 9) and the 3 and the 6, and the 5 and the 4. Thus the remainder of 56,093,742 when divided by 9 will be the same as the remainder of 0 when divided by 9.

As another example, let us find the remainder of 4,659,027 when divided by 9. Before adding the digits, we can "cast out" the 4 and the 5, the 9, and the 2 and the 7; thus, before adding the digits, we have 60. The sum of the digits left over is 6. The remainder of 6 when divided by 9 is obviously 6. The reader may check by division that the remainder of 4,659,027 when divided by 9 is indeed 6.

The next two theorems will be needed in Chapters 12 and 13. If the reader prefers, he may wait till then to examine them and skip now to the arithmetic of residue classes, p. 162. Both concern "solving" congruences, an idea that we will first illustrate by an example.

Is there an integer $X$ such that $5X \equiv 7 \,(\text{mod } 11)$? If we experiment with $X = 1, 2, 3, 4, 5, 6$, or 7, we see that none of them satisfies the demand $5X \equiv 7 \,(\text{mod } 11)$. But $5 \cdot 8 \equiv 7 \,(\text{mod } 11)$ for $5 \cdot 8 - 7 = 40 - 7 = 33$ is a multiple of 11. Thus 8 is a solution of the congruence $5X \equiv 7 \,(\text{mod } 11)$.

We cannot, however, always solve a congruence. For instance, we can show that there is no solution of the congruence $3X \equiv 1 \,(\text{mod } 6)$. For if $X$ is any integer, $3X$ is a multiple of 3: $\ldots, -3, 0, 3, 6, 9, \ldots$; thus $3X$ is congruent only to multiples of 3 modulo 6, and consequently can never be congruent to 1 modulo 6. Hence there is no integer $X$ such that $3X \equiv 1 \,(\text{mod } 6)$. In this example the modulus 6 and the number multiplying $X$, namely 3, have a common factor other than 1. This is the source of the trouble, as is suggested by

THEOREM 6. *If $A$ and $M$ are natural numbers such that $(A, M) = 1$, then for any integer $B$ we can find an integer $X$ such that*

$$AX \equiv B \,(\text{mod } M).$$

PROOF. Since $(A, M) = 1$, we know by Lemma 4 of Chapter 5 that there are integers $U$ and $V$ such that

$$1 = UA + VM.$$

(We call them $U$ and $V$ instead of $M$ and $N$ because the letter $M$ is already being used to name the modulus.)

Multiplying our equation by $B$ produces

$$B = BUA + BVM.$$

Thus $B - BUA = (BV)M$, and $B - BUA$ is a multiple of $M$; we can write $B \equiv (BU)A \,(\text{mod } M)$. Hence $BU$ serves as the desired $X$. This concludes the proof.

Once we have one solution to a congruence such as $AX \equiv B \pmod{M}$, we can easily manufacture more by adding (as often as we please) the modulus $M$ to the solution already found. For example, we saw above that 8 is a solution of $5X \equiv 7 \pmod{11}$. Suppose that $C$ is obtained from 8 by adding 11 to 8 one or more times. Then $C \equiv 8 \pmod{11}$ and hence, by Theorem 4, $5C \equiv 5 \cdot 8 \pmod{11}$. Theorem 2(c) then yields $5C \equiv 7 \pmod{11}$, thus $C$ is also a solution of $5X \equiv 7 \pmod{11}$. In general, then, if $S$ is a solution of the congruence $AX \equiv B \pmod{M}$, then all $C$ for which $C \equiv S \pmod{M}$, will also be solutions of the same congruence.

Conversely, if $S$ is any solution of $5X \equiv 7 \pmod{11}$, then, necessarily, $S \equiv 8 \pmod{11}$. This follows from

THEOREM 7 (Cancellation Theorem). *If $A$ and $M$ are natural numbers such that $(A, M) = 1$, and if $C$ and $D$ are integers satisfying*

$$AC \equiv AD \pmod{M},$$

then

$$C \equiv D \pmod{M}.$$

PROOF. We know that $M$ divides $AC - AD$, which is equal to $A(C - D)$. We wish to prove that $M$ divides $C - D$. This resembles the assertion that if a prime divides the product of two natural numbers, then it divides at least one of them (Theorem 3 of Chapter 5). Our argument will be almost the same as the one used in the proof of that theorem.

Since $(A, M) = 1$, we know by Lemma 4 of Chapter 5 that there are integers $U$ and $V$ such that

$$1 = UA + VM.$$

Multiplying both sides of this equation by $C - D$ gives

$$C - D = UA(C - D) + VM(C - D).$$

Since $M$ divides the right side of this equation, it divides $C - D$. Thus

$$C \equiv D \pmod{M}$$

and the proof is complete.

Theorem 7 justifies "cancellation" in congruences under certain circumstances. If we remove the restriction $(A, M) = 1$, then the conclusion of Theorem 7 need not hold. For example,

$$3 \cdot 1 \equiv 3 \cdot 5 \pmod{6},$$

yet it is *not* true that 1 is congruent to 5 modulo 6.

The arithmetic of congruences has its own addition and multiplication, which bear a close resemblance to the arithmetic of the rational numbers. Let us illustrate this with the modulus 4.

The set of integers congruent modulo 4 to a particular integer, $A$, we will denote $[A]$, *and we will call it a* **residue class** *modulo 4.* Thus $[5]$ consists of

$$\ldots, -15, -11, -7, -3, 1, 5, 9, 13, \ldots.$$

Note that $[-7] = [-3] = [1] = [5]$. Perhaps the best name for this particular residue class would be $[1]$, since this class consists of precisely those integers leaving a remainder of 1 when divided by 4.

(It would be more precise to use the symbol $[A]_4$ to indicate the residue class modulo 4 that contains $A$. It is usually simpler, however, just to keep the modulus in mind, and not use it as a subscript.)

Since there are four possible remainders (0, 1, 2, or 3) upon division by 4, there are four residue classes modulo 4: $[0]$, $[1]$, $[2]$, and $[3]$.

We can add two residue classes $[A]$ and $[B]$ by defining their sum to be the residue class containing $A + B$. This new addition we will denote by $\oplus$ (a symbol which, incidentally, is a letter in the Etruscan alphabet). For example,

$$[2] \oplus [3] = [5] = [1].$$

Similarly, we can define a multiplication $\otimes$ (another Etruscan letter) by

$$[A] \otimes [B] = [A \times B].$$

Thus, for example,

$$[2] \otimes [3] = [6] = [2].$$

The table for $\oplus$ is:

| $\oplus$ | [0] | [1] | [2] | [3] |
|---|---|---|---|---|
| [0] | [0] | [1] | [2] | [3] |
| [1] | [1] | [2] | [3] | [0] |
| [2] | [2] | [3] | [0] | [1] |
| [3] | [3] | [0] | [1] | [2] |

(mod 4)

It will be convenient to omit the brackets and write the table simply as:

| $\oplus$ | 0 | 1 | 2 | 3 |
|---|---|---|---|---|
| 0 | 0 | 1 | 2 | 3 |
| 1 | 1 | 2 | 3 | 0 |
| 2 | 2 | 3 | 0 | 1 |
| 3 | 3 | 0 | 1 | 2 |

(mod 4)

The multiplication table for ⊗, with the brackets omitted, is:

| ⊗ | 0 | 1 | 2 | 3 |
|---|---|---|---|---|
| 0 | 0 | 0 | 0 | 0 |
| 1 | 0 | 1 | 2 | 3 |
| 2 | 0 | 2 | 0 | 2 |
| 3 | 0 | 3 | 2 | 1 |

(mod 4)

These tables can be thought of as a record of how remainders of two natural numbers (when divided by 4) determine the remainder of their sum and product. For example, the entry, 1, for 2 ⊕ 3,

| ⊕ | | | 3 |
|---|---|---|---|
| | | | |
| 2 | | | 1 |
| | | | |

(mod 4)

records the following: if $A$ leaves a remainder of 2 when divided by 4, and $B$ leaves a remainder of 3 when divided by 4, then $A + B$ leaves a remainder of 1 when divided by 4.

We can make such tables not only for the modulus 4 but for any modulus. For the modulus 5 we have these two tables:

| ⊕ | 0 | 1 | 2 | 3 | 4 |
|---|---|---|---|---|---|
| 0 | 0 | 1 | 2 | 3 | 4 |
| 1 | 1 | 2 | 3 | 4 | 0 |
| 2 | 2 | 3 | 4 | 0 | 1 |
| 3 | 3 | 4 | 0 | 1 | 2 |
| 4 | 4 | 0 | 1 | 2 | 3 |

(mod 5)

| ⊗ | 0 | 1 | 2 | 3 | 4 |
|---|---|---|---|---|---|
| 0 | 0 | 0 | 0 | 0 | 0 |
| 1 | 0 | 1 | 2 | 3 | 4 |
| 2 | 0 | 2 | 4 | 1 | 3 |
| 3 | 0 | 3 | 1 | 4 | 2 |
| 4 | 0 | 4 | 3 | 2 | 1 |

(mod 5)

Let us do a little "mod 5" arithmetic to see how the ideas of ordinary arithmetic appear in this miniature, but complete, world that has only the five numbers 0, 1, 2, 3, and 4.

For instance, which number plays the role of $\frac{1}{3}$? To answer this question we must find a number $X$ in the miniature arithmetic such that

$$3 \otimes X = 1.$$

A glance at the $\otimes$ table shows that $3 \otimes 2 = 1$; hence $\frac{1}{3} = 2$.

Which number plays the role of $-1$? We must find a number $X$ in the miniature arithmetic such that

$$1 \oplus X = 0.$$

A glance at the addition table shows that $1 \oplus 4 = 0$. Hence $-1 = 4$.

What is $(-1) \otimes (-1)$? Since we already know that $-1 = 4$, we simply compute $4 \otimes 4$, which, as the table for $\otimes$ shows, is 1. Hence $(-1) \otimes (-1) = 1$, just as in ordinary arithmetic.

Does $-1$ have a square root in this miniature arithmetic? That is, is there a number $X$ such that

$$X \otimes X = -1?$$

Since $-1 = 4$, we are considering the equation

$$X \otimes X = 4.$$

Since $2 \otimes 2 = 4$ and $3 \otimes 3 = 4$, we see that $-1$ has two square roots, namely 2 and 3; this is quite a contrast with our customary arithmetic.

Another contrast between the "mod 5" arithmetic and the ordinary arithmetic for all integers is that there is no way to introduce the distinction "positive" versus "negative" in the "mod 5" arithmetic. For surely we would want 1 to be "positive" and the sum of "positive" numbers to be "positive." This would force each number in the "mod 5" arithmetic to be "positive," for $2 = 1 \oplus 1$, $3 = 1 \oplus 1 \oplus 1$, $4 = 1 \oplus 1 \oplus 1 \oplus 1$, and $0 = 1 \oplus 1 \oplus 1 \oplus 1 \oplus 1$. Even $-1$ would have to be "positive"! There would be no "negative" numbers.

In Chapters 12 and 13 we will study tables that, like the addition tables above, have no duplications in any row or column. Such tables are used in the design of experiments; furthermore, they will help us increase our understanding of our usual addition and multiplication. In those same chapters, we will find our arithmetic of congruences useful on several occasions.

## Exercises

1. Show that each of these statements is true: $4 \equiv 12 \pmod 8$, $4 \equiv 12 \pmod 1$, $4 \equiv 12 \pmod 2$, $4 \equiv 12 \pmod 4$.

2. Show that each of these statements is true: $5 \equiv 21 \pmod 4$, $5 \equiv 21 \pmod 8$, $5 \equiv 21 \pmod{16}$, $5 \equiv 21 \pmod 2$, $5 \equiv 21 \pmod 1$.

3. Show that each of these statements is true: $-3 \equiv 7 \pmod 5$, $57 \equiv 0 \pmod{19}$, $1 \equiv 1 \pmod{17}$.

4. Show that each of these statements is true: $8 \equiv -6$ (mod 7), $5 \equiv 5$ (mod 8), $117 \equiv 0$ (mod 39).

5. Translate these statements into the language of congruences:
   (a) The sum of two even integers is even.
   (b) The sum of an even integer and an odd integer is odd.
   (c) The sum of two odd integers is even.

6. Translate these statements into the language of congruences:
   (a) The product of an even integer and any integer is even.
   (b) The product of two odd integers is odd.

7. Translate these statements into the language of congruences.
   (a) 3 divides 12.
   (b) 4 divides 12.
   (c) 1 divides any natural number $N$.
   (d) Any integer $N$ divides itself.

8. Fill in each blank at least four different ways.
   (a) ___ $\equiv 1$ (mod 6),         (b) $11 \equiv 3$ (mod ___),
   (c) $6 \equiv -4$ (mod ___),         (d) $5 \equiv$ ___ (mod 3).

9. How can you tell at a glance whether two numbers, such as 419 and 627, given in the standard decimal notation, are congruent modulo 10?

10. Find the remainder when $5^{1001}$ is divided by 6.

11. Find the remainder when $3^{100}$ is divided by 8.

12. Find the remainder when $11 \cdot 9 \cdot 17 \cdot 6 \cdot 9^{100}$ is divided by 8.

13. Which of the following statements are true and which are false?
    (a) $7 \equiv -1$ (mod 2),         (b) $23 \equiv 4$ (mod 11),
    (c) $100 \equiv 1$ (mod 9),         (d) $25 \equiv 8$ (mod 3).

14. If $N \equiv -3$ (mod 7), what is the remainder when $N$ is divided by 7?

15. What single congruence is the basis of the theory behind "casting out 9's"?

16. That a natural number has the same remainder as the sum of its digits when divided by 9 can be proved without congruences. Notice that $10^N - 1$ is divisible by 9 since all its digits are 9. From this observation, write out a detailed proof, without using the language of congruences, for the rule of casting out 9's. (That is, prove Theorem 5 without using the symbol $\equiv$.)

17. (a) In Exercise 16 what important property of 9 is used?
    (b) Prove that remainders can be obtained for division by 3 in a manner similar to that for division by 9. (Use the method of E 16, not congruences.)

18. Using congruences, prove E 17(b).

19. Give three solutions in natural numbers to each of these congruences:
    (a) $X \equiv 1$ (mod 2),         (b) $3X \equiv 2$ (mod 5),         (c) $2X \equiv 3$ (mod 9).

    (A natural number, $N$, is a solution of a congruence if a true statement results when $X$ is replaced by $N$.)

20. Give three solutions in natural numbers to each of these congruences:
    (a) $X \equiv 7$ (mod 2),     (b) $2X \equiv 3$ (mod 5),     (c) $3X \equiv 1$ (mod 8).

21. Prove that there is no solution in integers to the congruence $2X \equiv 3$ (mod 8).

22. Prove that there is no solution in integers to the congruence $6X \equiv 8$ (mod 15).

23. The congruence $3X \equiv 2$ (mod 5) has an endless supply of solutions: ..., $-6, -1$, 4, 9, 14, ... [in fact, any integer congruent to 4 (mod 5)]. But the congruence $3X \equiv 2$ (mod 5) has only one solution if we demand that a solution be in the interval 0 through 4. Assuming that we place this same restriction on solutions, determine how many solutions there are of
    (a) $3X^2 \equiv 2$ (mod 5),     (b) $3X^2 \equiv 1$ (mod 5).

24. How many solutions are there from 0 through 7 of
    (a) $X^2 \equiv 1$ (mod 8),     (b) $X^3 \equiv 3$ (mod 8),     (c) $X^2 \equiv 0$ (mod 8)?

Exercises 25–32 develop simple divisibility tests for some small divisors.

25. Use the congruence $10 \equiv 0$ (mod 2) to obtain the following test for divisibility by 2: A number is divisible by 2 if and only if its units digit is divisible by 2.

26. Use the congruence $10 \equiv 1$ (mod 3) to obtain the following test for divisibility by 3: A number is divisible by 3 only when the sum of its digits is divisible by 3.

27. Use the congruence $10^2 \equiv 0$ (mod 4) to obtain the following test for divisibility by 4: A number is divisible by 4 if and only if the number formed by its last two digits is divisible by 4.

28. Use the congruence $10 \equiv 0$ (mod 5) to develop a test for divisibility by 5.

29. See E 25 and E 26. Prove that a number is divisible by 6 if and only if its last digit is even and the sum of its digits is divisible by 3.

We omit the test for divisibility by 7 since it is a bit cumbersome.

30. Use the congruence $10^3 \equiv 0$ (mod 8) to obtain the following test for divisibility by 8: A number is divisible by 8 if and only if the number formed by its last three digits is divisible by 8.

31. (a) Beginning with $10 \equiv -1$ (mod 11), prove that $100 \equiv 1$ (mod 11), $1000 \equiv -1$ (mod 11), $10^4 \equiv 1$ (mod 11), $10^5 \equiv -1$ (mod 11), and so on.
    (b) With the aid of (a), justify the following test for divisibility by 11: A natural number is divisible by 11 if and only if the units digit minus its tens digit plus its hundreds digit minus its thousands digit, and so on, is divisible by 11. We shall call this the "alternating sum" of the digits. (For example, 3564 is divisible by 11, since $4 - 6 + 5 - 3$ is divisible by 11.)

32. How would you test for divisibility by 12?

33. Is $8^{600} \equiv 6^{600}$ (mod 7)?

34. Find the remainder when (a) $2^{100}$ is divided by 5, (b) $2^{523}$ is divided by 5.

35. In E 31 it was proved, for example, that $10^2 \equiv 1$ (mod 11), $10^3 \equiv -1$ (mod 11), $10^4 \equiv 1$ (mod 11) and $10^5 \equiv -1$ (mod 11). Check each of these congruences by dividing 11 into the appropriate difference of two integers.

36. Theorem 7 implies that if $A$ is 1, 2, 3, or 4, then $AC \equiv AD$ (mod 5) implies $C \equiv D$ (mod 5). What does this mean about the multiplication table for the modulus 5?

37. Theorem 7 implies that if $A$ is 1, 2, 3, 4, 5, or 6, then $AC \equiv AD$ (mod 7) implies $C \equiv D$ (mod 7). What does this mean about the multiplication table for the modulus 7?

38. Make the addition and multiplication tables that correspond to these moduli: (a) 3, (b) 2, (c) 6. In which cases is the product of two numbers 0 only when at least one of the numbers is 0?

39. Find the remainder when $16 \cdot 15 \cdot 22 \cdot 29 \cdot 31$ is divided by 7 by using congruences and Theorem 4.

40. (a) What is the remainder of 9546 when divided by 9?
    (b) What is the remainder of 4965 when divided by 9?
    (c) The remainders in (a) and (b) are equal. Prove that whenever we rearrange the digits of a number, we do not change its remainder relative to division by 9.

41. Consider a number $N_1$. Rearrange the digits of $N_1$ and call the number thus obtained $N_2$. Prove that $N_1 - N_2$ is always divisible by 9.

●

42. With congruences it is easy to prove that 3 is special as follows. Assume that $A$ and $B$ are natural numbers, with $AB \equiv 0$ (mod 3). Then, to show that at least one of $A \equiv 0$ (mod 3) or $B \equiv 0$ (mod 3) holds, rule out one by one the following four cases:
    (a) $A \equiv 1$ (mod 3) and $B \equiv 1$ (mod 3),
    (b) $A \equiv 1$ (mod 3) and $B \equiv 2$ (mod 3),
    (c) $A \equiv 2$ (mod 3) and $B \equiv 1$ (mod 3),
    (d) $A \equiv 2$ (mod 3) and $B \equiv 2$ (mod 3).
    (*Hint:* Use Theorem 4.)

43. Using the method of E 42, prove that (a) 5 is special, (b) 7 is special.

44. Using the method of E 42, prove that 11 is special.

45. (a) Using Theorem 4, prove that the remainder of $AB$ when divided by 9 is congruent modulo 9 to the product (remainder of $A$ when divided by 9) $\times$ (the remainder of $B$ when divided by 9).
    (b) Use (a) to show that $73 \cdot 65$ is *not* equal to 4645.
    (c) Compute $73 \cdot 65$.
    (This exercise shows how we can use the rule of casting out 9's as a check on our multiplication.)

46. Using E 45(a), show that
    (a) $141 \cdot 625$ is not equal to 88025,
    (b) $58 \cdot 73$ is not equal to 4244.

47. (a) Show by computing the product correctly that $172 \cdot 251$ is not equal to 41372.
    (b) Does the rule of casting out 9's show that $172 \cdot 251$ is not 41372?

48. (a) Is the analogue of E 40 true for remainders relative to division by 2?
    (b) Is the analogue of E 40 true for remainders relative to division by 3?

49. Solve E 15(b) of Chapter 9 by using congruences.

50. Translate the technique of E 12 of Chapter 10 into the language of congruence modulo 2.

51. Deduce from Theorem 4 that if $A \equiv a$ (mod $M$), $B \equiv b$ (mod $M$), and $C \equiv c$ (mod $M$), then $ABC \equiv abc$ (mod $M$).

52. In ordinary arithmetic, if the product of two numbers is 0, then at least one of them is 0.
    (a) Is this true in the "mod 5" arithmetic?
    (b) In the "mod 4" arithmetic?
    (c) For which "mod" arithmetics is it true? Why?

53. In the "mod 5" arithmetic (a) find $\frac{2}{3}$, (b) find $\sqrt[3]{3}$, (c) show that 3 has no square root, (d) show that 1 has four 4th roots.

54. If the sum of the digits of $A$ is divisible by 9, what, if anything, can be said about the sum of the digits of $73A$? About the sum of the digits of $A + 73$? Why?

55. If the alternating sum of the digits of $A$ is 0, what, if anything, can be said about the alternating sum of the digits of $73A$? (See E 31.)

56. (a) In what ways does $\equiv$ behave like $=$?
    (b) In what ways does it not?

57. If $A \equiv B$ (mod 3) and $a \equiv b$ (mod 2), can we conclude that $Aa \equiv Bb$ (mod 6)? Prove your answer.

58. If $A \equiv B$ (mod 3) and $A \equiv B$ (mod 2), what further congruences modulo 6 can we deduce? Prove your answer.

59. We proved that Theorem 7 is true if $(A, M) = 1$. If $(A, M) = B$, where $B$ is greater than 1, the following theorem is true: If $A$ and $M$ are natural numbers such that $(A, M) = B$ and if $C$ and $D$ are integers for which

$$AC \equiv AD \text{ (mod } M),$$

then

$$C \equiv D \left( \text{mod } \frac{M}{B} \right).$$

Prove this theorem.

● ●

60. (a) For each natural number $A$ from 2 through 5, find the natural number $B$ from 2 through 5 such that $AB \equiv 1$ (mod 7). (For example, if $A$ is 2, then $B$ is 4.)
    (b) Using (a), show that $1 \cdot 2 \cdot 3 \cdot 4 \cdot 5 \cdot 6 \equiv 6$ (mod 7).

61. (a) For each natural number $A$ from 2 through 9 find the natural number $B$ between 1 and 10 such that $AB \equiv 1$ (mod 11). (For example, if $A$ is 3, then $B$ is 4.)
    (b) Using (a), show that

$$1 \cdot 2 \cdot 3 \cdot 4 \cdot 5 \cdot 6 \cdot 7 \cdot 8 \cdot 9 \cdot 10 \equiv 10 \text{ (mod 11)}.$$

62. In the proof of Theorem 3 of Chapter 4 we defined $N!$ as the product of all natural numbers from 1 through $N$. In E 48 of Chapter 4 it was shown that, for composite $N$ larger than 4,
$$(N - 1)! \equiv 0 \pmod{N}.$$
    (a) Show that E 60(b) proves that $6! \equiv -1 \pmod 7$.
    (b) Show that E 61(b) proves that $10! \equiv -1 \pmod{11}$.
    (c) Using the ideas of E 60 and E 61, prove that if $P$ is prime, then
$$(P - 1)! \equiv -1 \pmod{P}.$$

63. According to E 62, $4! \equiv -1 \pmod 5$. In fact, $4! \equiv -1 \pmod{5^2}$, as the reader may check. Show that $12! \equiv -1 \pmod{13^2}$. It is also known that $562! \equiv -1 \pmod{563^2}$. These are the only primes $P$ less than 200,000 such that $(P - 1)! \equiv -1 \pmod{P^2}$.

64. If $A \equiv B \pmod M$, is $(A, M) = (B, M)$? Prove your answer.

65. Fill in the blank with the largest modulus for which the statement is true: For any even number $A$, $A^2 \equiv 0 \pmod{\underline{\quad}}$. Prove your answer.

66. Fill in the blank with the largest modulus for which the statement is true: For any odd number $A$, $A^2 \equiv 1 \pmod{\underline{\quad}}$. Prove your answer.

67. In the "mod 7" arithmetic:
    (a) Find $-1$.
    (b) Compute $-1 \otimes -1$.
    (c) Does $-1$ have a square root?
    (d) Find $\frac{3}{5}$ and $\frac{5}{3}$.
    (e) Compute $\frac{3}{5} \otimes \frac{5}{3}$.
    (f) Compute $(\frac14)/(\frac23)$ and $(\frac14) \otimes (\frac23)$.
    (g) Is there anything that you want to say about (f)?

68. For odd $A$ prove that $A^3 \equiv A \pmod{24}$.

69. Use the "mod 11" arithmetic to answer the following:
    (a) Show that $\frac67 = \frac95$.
    (b) Compute $(\frac69) \otimes (\frac76)$.
    (c) Compute $\frac23 \otimes \frac57$.
    (d) Is $\frac12 = \frac48$?
    (e) Is $(-1) \otimes (-1) = 1$?

70. In Chapter 5 we obtained the Fundamental Theorem of Arithmetic from the (weighing) Lemma 4. We now show that Lemma 4 is a consequence of the Fundamental Theorem of Arithmetic.
    (a) Let $A$ and $B$ be natural numbers such that $(A, B) = 1$. Assuming the Fundamental Theorem of Arithmetic, prove that if $B$ divides $AC$ then $B$ divides $C$.
    (b) From (a) deduce that no two of the numbers $A \cdot 0, A \cdot 1, A \cdot 2, \ldots, A \cdot (B - 1)$ are congruent modulo $B$.
    (c) From (b) deduce that there is an integer $M$ such that $AM \equiv 1 \pmod B$.
    (d) From (c) deduce that there are integers $M$ and $N$ such that $MA + NB = 1$. This establishes Lemma 4 for the case $(A, B) = 1$.
    (e) Obtain Lemma 4 for the case when $(A, B)$ is not 1 by applying (d) to the numbers $A/(A, B)$ and $B/(A, B)$.

71. See E 70. Write a short essay on the theme, "The Fundamental Theorem of Arithmetic, the fact that every prime is special, and the (weighing) Lemma 4 of Chapter 5 are equivalent."

72. (a) Show that there is a positive integer $N$ such that $3^N$ has the remainder 1 when divided by 71.
    (b) Generalize.

73. (a) Arrange 25 dots in a five-by-five square. What is the largest number of them you find such that no three lie on a straight line?
    (b) Examine the same problem for dots arranged in a three-by-three and in a four-by-four square.
    (c) Explore this problem in other cases.
    We return to this exercise in E 91 of Chapter 16.

### References

1. R. Courant and H. Robbins, *What Is Mathematics?*, Oxford University Press, New York, 1960. (Congruences, pp. 31–40.)

2. I. Niven and H. S. Zuckerman, *An Introduction to the Theory of Numbers*, Wiley, New York, 1960. (Congruences, pp. 20–33.)

3. O. Ore, *Number Theory and Its History*, McGraw-Hill, New York, 1948. (Congruences, pp. 209–233.)

4. G. Birkhoff and S. MacLane, *A Survey of Modern Algebra*, Third edition, Macmillan, New York, 1965. (Congruences, pp. 22–28.)

*Chapter* **12**

# STRANGE ALGEBRAS

If we never travel, we may get into the habit of thinking that our way of life, with all its idiosyncrasies and peculiar customs, is the best and only reasonable way, and that all other peoples behave quite oddly. We may be like that child in Robert Louis Stevenson's *A Child's Garden of Verses*, who offers this sympathy to foreign children:

> You have curious things to eat,
> I am fed on proper meat;
> You must dwell beyond the foam,
> But I am safe and live at home.

In some lands the calendar presents the weeks vertically instead of horizontally; the American traveler must cock his head to read it. Even after years in such a land, he may still feel that the weeks should be horizontal. It is not easy to question the world we knew when we were children.

In this chapter we will visit algebras quite different from the one we met in grammar school. When we return to our own algebra, we will appreciate virtues in our system that we might never have noticed had we not made this journey. And, if we travel with an open mind, we might admit that the other systems have their own advantages.

Our journey begins in the countryside, with a problem suggested by agriculture. In these days of scientific farming, it is very important that a farmer get as great a return from his investment as possible. Thus a good number of experimental studies are made to find the best diet for each crop. How much water? How much nitrogen? How much potassium? In many small doses or in a few large ones?

Let us design an experiment for comparing the effect of four different doses of water on tomato plants. We might arrange four plants in a straight line and give each plant one of the four doses *A*, *B*, *C*, and *D*.

But scientists have observed that arranging plants in a line frequently introduces a bias that makes the experiment worthless. The statistician R. A. Fisher warns,

> In many fields there is found to occur either a gradient of fertility across a whole area, or parallel strips of land having a higher or lower fertility than the average. . . . Such soil variations may be due in part to the past history of the field, such as the lands in which it has been lain up for drainage producing variations in the depth and present condition of the soil, or to portions of it having been manured and cropped otherwise than the remainder.

As a precaution, we follow the advice and practice of the scientists, and instead of working with four plants in a straight line we devise without much effort a square arrangement of sixteen plants to which we distribute the four treatments like this:

| | | | |
|---|---|---|---|
| *A* | *C* | *D* | *B* |
| *D* | *B* | *A* | *C* |
| *B* | *D* | *C* | *A* |
| *C* | *A* | *B* | *D* |

*A horizontal line of plants we shall call a* **row**; *a vertical line of plants we shall call a* **column**. Each row contains each of the treatments once, as does each column.

We may think of the arrangement as a crossword puzzle in four-letter words, or perhaps we are reminded of the problem of forming memory words by making certain strings of 0's and 1's overlap in a special way. But what we should observe is the resemblance this agricultural experiment bears to the usual multiplication table.

The multiplication table that we all once memorized looks like this:

| | 1 | 2 | 3 | 4 | 5 | 6 | 7 | 8 | 9 |
|---|---|---|---|---|---|---|---|---|---|
| 1 | 1 | 2 | 3 | 4 | 5 | 6 | 7 | 8 | 9 |
| 2 | 2 | 4 | 6 | 8 | 10 | 12 | 14 | 16 | 18 |
| 3 | 3 | 6 | 9 | 12 | 15 | 18 | 21 | 24 | 27 |
| 4 | 4 | 8 | 12 | 16 | 20 | 24 | 28 | 32 | 36 |
| 5 | 5 | 10 | 15 | 20 | 25 | 30 | 35 | 40 | 45 |
| 6 | 6 | 12 | 18 | 24 | 30 | 36 | 42 | 48 | 54 |
| 7 | 7 | 14 | 21 | 28 | 35 | 42 | 49 | 56 | 63 |
| 8 | 8 | 16 | 24 | 32 | 40 | 48 | 56 | 64 | 72 |
| 9 | 9 | 18 | 27 | 36 | 45 | 54 | 63 | 72 | 81 |

Actually, the multiplication table does not stop at 9 by 9. It goes on to the right and downward without end. But with knowledge of the 9 by 9 table, together with the rules for carrying and adding, we can compute larger products, such as 193 × 64.

Both tables are square. The little table lacks a guide row and a guide column. These we can put in alphabetically and make of the tomato experiment a miniature and exotic algebra:

|   | A | B | C | D |
|---|---|---|---|---|
| A | A | C | D | B |
| B | D | B | A | C |
| C | B | D | C | A |
| D | C | A | B | D |

(*1*)

In such a miniature table, let us name each column by the guide letter above it. This column *C* is

$$D$$
$$A$$
$$C$$
$$B$$

Similarly, we shall name each row by the guide letter to its left. Thus row *B* is

$$D \; B \; A \; C$$

The letter that appears in the box where row *B* meets column *C* we shall denote by

$$B \circ C.$$

(We may read this "*B* circle *C*").

In our little table,

$$B \circ C = A.$$

In any such table, *what appears in the box where row X meets column Y, we shall denote by X ∘ Y.* Pictorially, it looks like this:

When we use our ordinary multiplication table, we write $X \times Y$ instead of $X \circ Y$. Similarly, in the usual addition table, we write $X + Y$ instead of $X \circ Y$.

Now let us leave our ordinary arithmetic and algebra far behind. Let us roam through various miniature arithmetics and, when we are done, compare them with our own.

A *table* will be a square arrangement of symbols without duplication in any row or in any column. (On occasion, the table might be of unlimited extent.) That is, for any $X$, $Y$, and $Z$, with $Y$ different from $Z$, we have

$$X \circ Y \text{ different from } X \circ Z$$

and

$$Y \circ X \text{ different from } Z \circ X.$$

The fact that row $X$ has no duplications is expressed by the requirement that $X \circ Y$ be different from $X \circ Z$:

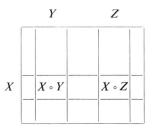

Another way to say this is

$$X \circ Y = X \circ Z \quad \text{forces} \quad Y = Z.$$

This "cancellation" point of view will be of use in constructing tables.

The fact that column $X$ has no duplications is expressed by the requirement that $Y \circ X$ be different from $Z \circ X$:

$X$

|   | $Y \circ X$ |   |
|---|---|---|
$Y$

$Z$ | $Z \circ X$ |

In other words, we have the "cancellation" rule

$$Y \circ X = Z \circ X \quad \text{forces} \quad Y = Z.$$

The letters used in the guide row will be the same as those used in the guide column; and they will be written in the same order in both. *Each box of the table will contain one of the letters that appear in the guide row.*

*The number of rows (or the number of columns) of a table we will call its* **order.** Looking at our little table of order 4, we observe that

$$A \circ A = A, \qquad B \circ B = B, \qquad C \circ C = C, \qquad D \circ D = D.$$

These four equations can be summarized in the brief assertion: for all $X$ we have $X \circ X = X$. The phrase "for all $X$" is short for "whenever $X$ is replaced by one of the letters of the table."

*A table satisfying the rule "$X \circ X = X$ for all $X$" we call* **idempotent** (from the Latin *idem*, self; *potens*, power). It is very easy to see whether a table is idempotent: the diagonal stretching from the upper left corner down to the lower right should read the same as the guide row (or guide column). *Henceforth let us call this diagonal the* **main diagonal.**

Neither our ordinary multiplication nor our addition is idempotent. For example, $5 \times 5$ is not 5, nor is $5 + 5$ equal to 5. We are certainly far from our usual daily routine.

But our ordinary multiplication and addition satisfies a rule that the little table of order 4 disobeys. In learning the multiplication table, once we had memorized a product, such as $6 \times 9 = 54$, we had the product $9 \times 6$ without further effort. We did not have to memorize $9 \times 6$ separately because ordinary multiplication satisfies the rule:

$$X \times Y = Y \times X \qquad \text{(for all } X \text{ and } Y\text{)}.$$

This is the so-called *commutative rule*. Ordinary addition and multiplication are both commutative. But the little table of order 4 is not; for example,

$$A \circ B = C,$$

whereas

$$B \circ A = D.$$

Daily life offers a simple example of an operation that is both commutative and idempotent—namely, let $X \circ Y$ denote the average of the numbers $X$ and $Y$,

$$X \circ Y = \frac{X + Y}{2},$$

or, equivalently,

$$2(X \circ Y) = X + Y.$$

For instance, $3 \circ 7 = (3 + 7)/2 = 5$, $4 \circ 4 = (4 + 4)/2 = 4$, $4 \circ 7 = (4 + 7)/2 = 11/2$, and $7 \circ 4 = (7 + 4)/2 = 11/2$.

It is almost as easy to check pictorially whether a table is commutative as it is to see whether it is idempotent. A table is commutative if it is symmetric with respect to the main diagonal. That is, the table can be made by first filling in the triangle above the main diagonal and then flipping it over that diagonal.

Here, for example, is a commutative table of order 4:

(2)

|   | A | B | C | D |
|---|---|---|---|---|
| A | A | B | C | D |
| B | B | A | D | C |
| C | C | D | A | B |
| D | D | C | B | A |

So far, we have visited miniature arithmetics only of order 4. But it is a simple matter to devise tables of other orders. Here are some tables of orders 1, 2, and 3:

|   | A |
|---|---|
| A | A |

(3)

|   | A | B |
|---|---|---|
| A | A | B |
| B | B | A |

(4)

|   | A | B | C |
|---|---|---|---|
| A | C | A | B |
| B | B | C | A |
| C | A | B | C |

(5)

The table of order 1 is the dullest arithmetic possible; nevertheless, we might acknowledge that it is idempotent and commutative. The table of order 2 is not idempotent, but it is commutative. Table (5), of order 3, is neither commutative nor idempotent.

With a little effort, such as we might use in solving a crossword puzzle, we can devise a commutative table of order 3:

(6)

|   | A | B | C |
|---|---|---|---|
| A | A | B | C |
| B | B | C | A |
| C | C | A | B |

Indeed, in the chapter on congruences we found a way of producing commutative tables of all orders. Let us illustrate the technique by producing a commutative table of order 5. Instead of the letters $A$, $B$, $C$, $D$, and $E$, we will use the numbers 0, 1, 2, 3, and 4. In the box where row $X$ meets column $Y$ we will place the remainder of $X + Y$ when divided by 5. For example, $3 \circ 4$ will be 2. Since ordinary addition is commutative, so will this table be commutative. Filling in the 25 boxes we obtain

|   | 0 | 1 | 2 | 3 | 4 |
|---|---|---|---|---|---|
| 0 | 0 | 1 | 2 | 3 | 4 |
| 1 | 1 | 2 | 3 | 4 | 0 |
| 2 | 2 | 3 | 4 | 0 | 1 |
| 3 | 3 | 4 | 0 | 1 | 2 |
| 4 | 4 | 0 | 1 | 2 | 3 |

(7)

Observe that since there are no duplications in any row or column, (7) is a table. We would translate it into letters by replacing 0 by *A*, 1 by *B*, 2 by *C*, 3 by *D*, and 4 by *E*. If we do so we obtain this table:

|   | *A* | *B* | *C* | *D* | *E* |
|---|-----|-----|-----|-----|-----|
| *A* | *A* | *B* | *C* | *D* | *E* |
| *B* | *B* | *C* | *D* | *E* | *A* |
| *C* | *C* | *D* | *E* | *A* | *B* |
| *D* | *D* | *E* | *A* | *B* | *C* |
| *E* | *E* | *A* | *B* | *C* | *D* |

(8)

Tables (7) and (8) really represent the same system. After all, if we had obtained a new table from (7) by painting all the numbers red, we would still think of it as the same table in disguise. The basic thing is that Tables (7) and (8) have the same *form* (just as, in Chapter 8, it was discovered that certain electrical networks and tilings of a rectangle by rectangles have the same underlying form). *Two tables that are obtainable from each other by simply renaming all the letters* (or all the numbers, as the case may be) and possibly interchanging certain rows (or columns) we will call isomorphic (from the Greek *iso*, same; *morph*, form).

The same technique that gave us Table (7) can be used for any order. Thus we have

THEOREM 1. *There are commutative tables of all orders.*

Having discovered that there are commutative tables of all orders, let us extend our journey by finding out whether there are tables of all orders that are both commutative and idempotent.

We have already noticed that the table of order 1 is both commutative and idempotent. Let us try to build a table of order 2 that is both commutative and idempotent. Since it is to be idempotent, we can begin by filling in the main diagonal:

(9)

|     | A   | B   |
| --- | --- | --- |
| A   | A   |     |
| B   |     | B   |

But what can we put in the box where row A meets column B? That is, what shall we have for the value of A ∘ B? We see that A ∘ B cannot be A, for then row A would have a duplication. Nor can A ∘ B be B, for then column B would have a duplication. This information we record in

THEOREM 2. *There is no commutative and idempotent table of order 2.* (*In fact, there is no idempotent table of order 2.*)

Theorem 2, asserting that we cannot do something, is an example of an "impossibility" theorem. We shall meet a much more surprising "impossibility" theorem in Chapter 17.

Let us attack order 3. Starting with the main diagonal we must have

(10)

|     | A   | B   | C   |
| --- | --- | --- | --- |
| A   | A   |     |     |
| B   |     | B   |     |
| C   |     |     | C   |

What can we define A ∘ B to be? Not A, for row A already has an A. Not B, for column B already has a B. We have no choice. We must set A ∘ B = C. By commutativity, we must then have B ∘ A = C. Continuing in this fashion, we obtain this idempotent and commutative table of order 3:

(11)

|     | A   | B   | C   |
| --- | --- | --- | --- |
| A   | A   | C   | B   |
| B   | C   | B   | A   |
| C   | B   | A   | C   |

Let us attack order 4. Is there an idempotent and commutative table of order 4? The main diagonal must be:

(12)

|     | A   | B   | C   | D   |
| --- | --- | --- | --- | --- |
| A   | A   |     |     |     |
| B   |     | B   |     |     |
| C   |     |     | C   |     |
| D   |     |     |     | D   |

We have two choices for $A \circ B$: either $C$ or $D$. Let us choose $C$ first and follow up the consequences. If the table is to be commutative we must also have $B \circ A = C$. Thus, with the choice of $A \circ B = C$, we would have:

(13)

|     | A   | B   | C   | D   |
| --- | --- | --- | --- | --- |
| A   | A   | C   |     |     |
| B   | C   | B   |     |     |
| C   |     |     | C   |     |
| D   |     |     |     | D   |

What will $A \circ C$ be? Since $A$ and $C$ already appear in row $A$, $A \circ C$ cannot be $A$ or $C$. If we made $A \circ C = B$, then we would be forced to have $A \circ D = D$ (since $D$ has to appear in row $A$). But $D$ is already present in column $D$. So the only possibility for $A \circ C$ is $D$. Then, by commutativity, $C \circ A = D$. Since row $A$ must have $B$ somewhere, we must then have $A \circ D = B$. By commutativity, $D \circ A = B$. Let us record all the boxes already filled in (with the only free choice being $A \circ B = C$).

(14)

|     | A   | B   | C   | D   |
| --- | --- | --- | --- | --- |
| A   | A   | C   | D   | B   |
| B   | C   | B   |     |     |
| C   | D   |     | C   |     |
| D   | B   |     |     | D   |

Where can $D$ appear in row $B$? Since $D$ already appears in column $C$, we cannot have $B \circ C = D$, and since it already appears in column $D$, we cannot have $B \circ D = D$. We are blocked. And, as the reader may easily check, we would be blocked eventually if we had made the other choice at the beginning; namely, $A \circ B = D$. Thus there is no idempotent commutative table of order 4.

For orders 1 and 3 there are idempotent commutative tables; for orders 2 and 4 there are none at all. This suggests that we are once again involved with the duality of the odd and even natural numbers. The next two theorems show that this suspicion is justified.

THEOREM 3. *There are idempotent commutative tables of all odd orders.*

PROOF. Let $N$ be an odd natural number. We will construct a table of order $N$ that is both idempotent and commutative. It will be convenient to build it on the numbers 0, 1, 2, 3, 4, . . . , $N - 1$, just as Table (7) was built on numbers rather than letters. Our method will be a variation of the one we used to make Table (7).

Recalling that the averaging of two numbers is commutative and idempotent in ordinary arithmetic, we will define $X \circ Y$ as that natural number less than $N$ such that

$$2(X \circ Y) \equiv X + Y \,(\text{mod } N).$$

That $X \circ Y$ exists and is unique is *guaranteed* by Theorems 6 and 7 (the Cancellation Theorem) of Chapter 11, together with the fact that $(2, N) = 1$.

For instance, take the case $N = 9$. Let us compute $3 \circ 8$. We have

$$2(3 \circ 8) \equiv 3 + 8 \,(\text{mod } 9).$$

Since $3 + 8 = 11$ and $2 \cdot 1 \equiv 11 \,(\text{mod } 9)$, we see that $3 \circ 8 = 1$. The reader is urged to fill out the complete table for $N = 5$ before proceeding further into the proof.

Thus we have specified how we will fill in all the boxes. Now we must check that this construction provides an idempotent and commutative table of order $N$.

Is $X \circ X$ always $X$? By the definition of $\circ$, $2(X \circ X) \equiv X + X \,(\text{mod } N)$. Since $2X = X + X$, we also have $2X \equiv X + X \,(\text{mod } N)$ and therefore, since the congruence has a unique solution, $X$ and $X \circ X$ coincide. The table is idempotent.

Is $X \circ Y$ always equal to $Y \circ X$? What we do know is that

$$2(X \circ Y) \equiv X + Y \,(\text{mod } N)$$

and that

$$2(Y \circ X) \equiv Y + X \,(\text{mod } N).$$

Since ordinary addition is commutative, $X + Y$ is equal to $Y + X$. Thus we see that

$$2(X \circ Y) \equiv 2(Y \circ X) \,(\text{mod } N).$$

Again, using the Cancellation Theorem (Theorem 7 of Chapter 11), we conclude that

$$X \circ Y \equiv Y \circ X \,(\mathrm{mod}\ N).$$

But $X \circ Y$ and $Y \circ X$ are both less than $N$. Hence (by the lemma of p. 156) $X \circ Y = Y \circ X$. Therefore our system is commutative.

Finally, is it a table? We must show that there are no duplications in any row or column. Let us look at a typical row—say, row $X$. We must prove that if

$$X \circ Y = X \circ Z, \qquad \text{then} \qquad Y = Z.$$

By definition,

$$2(X \circ Y) \equiv X + Y \,(\mathrm{mod}\ N)$$

and

$$2(X \circ Z) \equiv X + Z \,(\mathrm{mod}\ N).$$

Since we are assuming that $X \circ Y = X \circ Z$, we have

$$X + Y \equiv X + Z \,(\mathrm{mod}\ N),$$

hence

$$Y \equiv Z \,(\mathrm{mod}\ N).$$

But $Y$ and $Z$ are in the range from 0 to $N - 1$ and, being congruent, they must coincide. (Again the Lemma of p. 156.) Since $Y = Z$, we have no duplications in any row.

Similarly, there are no duplications in any column. (In a commutative table, row $X$ and column $X$ look alike.) Thus our way of filling in the boxes indeed produces a table. This concludes the proof of Theorem 3.

We shall next show that there is no idempotent commutative table of any even order. Our proof will be in the same spirit as the one proving that it is impossible to find two natural numbers whose quotient is $\sqrt{2}$ or to find solutions to certain Fifteen puzzles. We will prove that if there were such a table, then a certain natural number would have to be both odd and even. Theorem 4 will be another "impossibility" result.

THEOREM 4. *There is no idempotent commutative table of even order.*

PROOF. We will prove that there is no idempotent commutative table of order $N$ when $N$ is an even natural number larger than 0. Suppose, on the contrary, that there does exist such a table. We will examine the boxes of this table that are occupied by the letter $A$ and obtain a contradiction.

Since the table is commutative, the number of boxes occupied by $A$ and not on the main diagonal is even. Since the table is idempotent, there is exactly one box on the main diagonal occupied by the letter $A$. Thus the number of boxes occupied by $A$ is one more than an even number. That is, the number of boxes occupied by $A$ is *odd*.

On the other hand, since $A$ appears exactly once in each of the $N$ rows, $A$ must occupy an *even* number of boxes. This contradiction proves Theorem 4.

There are algebras stranger than those that are idempotent and commutative. Indeed, there are algebras, as yet unnamed, that satisfy the rule

$$X \circ (X \circ Y) = Y \circ X \qquad \text{(for all } X \text{ and } Y).$$

Table (*1*), which was used to arrange the sixteen tomato plants, is an example of such a weird algebra. To show that Table (*1*) does satisfy the rule $X \circ (X \circ Y) = Y \circ X$, we must make 16 checks, since there are four choices for $Y$ and, for each choice of $Y$, four choices of $X$. Let us carry out just two of these 16 checks.

Does $A \circ (A \circ C) = C \circ A$? To answer this we must compute both sides of the equation in question and see whether they are equal. Before we can compute $A \circ (A \circ C)$ we have to find $A \circ C$. Looking at Table (*1*) we see that $A \circ C = D$. Thus $A \circ (A \circ C) = A \circ D$. But we have still not finished the computation of $A \circ (A \circ C)$; we must compute $A \circ D$. Looking again at Table (*1*) we see that $A \circ D = B$. Thus $A \circ (A \circ C) = B$. Next, we compute the other side of the equation we are checking; namely, the expression $C \circ A$. Looking at Table (*1*), we see that $C \circ A = B$. Since both $A \circ (A \circ C)$ and $C \circ A$ are $B$, we see that $A \circ (A \circ C) = C \circ A$. This completes only one of the 16 checks that are to be made.

Another of the 16 checks would be, for example, to show that the rule $X \circ (X \circ Y) = Y \circ X$ is satisfied when $X$ and $Y$ are both replaced by $D$; that is, to show that

$$D \circ (D \circ D) = D \circ D.$$

Glancing at Table (*1*) we see that $D \circ D = D$. Hence both sides of this equation in question are equal to $D$. Thus the equation $X \circ (X \circ Y) = Y \circ X$ is valid in this case too. The reader is invited to make a few more checks for himself.

Most of what is known about tables that satisfy the rule $X \circ (X \circ Y) = Y \circ X$ is reported in the following two theorems:

THEOREM 5. *Any table that satisfies the rule* $X \circ (X \circ Y) = Y \circ X$ *for all $X$ and $Y$ is idempotent.*

PROOF. Since $X \circ (X \circ Y) = Y \circ X$ for all $X$ and $Y$, we have, when $Y$ and $X$ coincide,

$$X \circ (X \circ X) = X \circ X.$$

Since row $X$ has no duplications, $X \circ X$ must be the same as $X$.

THEOREM 6. *Any table that satisfies the rule* $X \circ (X \circ Y) = Y \circ X$ *has the property that distinct letters do not commute. That is, if X is different from Y, then* $X \circ Y$ *is different from* $Y \circ X$.

PROOF. Let $U$ and $V$ be any two distinct letters in a table satisfying the rule $X \circ (X \circ Y) = Y \circ X$. We will prove that $U \circ V$ must be different from $V \circ U$.

Assume for the moment that $U \circ V = V \circ U$. We will obtain a contradiction. Indeed, we have $U \circ (U \circ V) = V \circ U$, and if $U \circ V = V \circ U$, it follows that

$$U \circ (U \circ V) = U \circ V.$$

Since there can be no duplication in row $U$,

$$U \circ V = V.$$

But, by Theorem 5,

$$V \circ V = V.$$

Thus

$$U \circ V = V \circ V,$$

and, since there can be no duplication in column $V$, it follows that

$$U = V.$$

This contradiction completes the proof of the theorem.

Theorems 5 and 6 help us a little in deciding which orders are possible for tables satisfying the rule $X \circ (X \circ Y) = Y \circ X$. First, there can be none of order 2, since there is no idempotent table of order 2. Secondly, there can be none of order 3, since any idempotent table of order 3 must be commutative, as the reader may easily check. But no one knows in general for which orders there are tables satisfying the rule $X \circ (X \circ Y) = Y \circ X$. This is quite a contrast to the complete solution we found for the rules $X \circ X = X$ and $X \circ Y = Y \circ X$ (Theorems 3 and 4).

Though there are many other exotic tables, which satisfy other strange rules, we must now end our journey and come back to our own arithmetic and look at it with the fresh viewpoint of the returning traveler.

First of all, we are glad that our addition and multiplication are commutative, for once we memorize $X \times Y$ and $X + Y$ we also know $Y \times X$ and $Y + X$.

But there is a deeper virtue in our addition and multiplication. Observe that when we compute $3 \times 6 \times 7$ we can proceed in two ways. We can either consider it as $(3 \times 6) \times 7$, in which case $3 \times 6$ is computed first and the result then multiplied by 7; or we can consider it as $3 \times (6 \times 7)$, in which case $6 \times 7$ is computed first and the result then multiplied by 3. Both methods give the same final result, 126. A similar remark can be made about addition.

Consider someone who must work with the arithmetic of Table (*1*). If he were to write down $A \circ B \circ C$ he would be in trouble. If he interprets it as $(A \circ B) \circ C$ he would first compute $A \circ B$, which is $C$, and then $C \circ C$, which is $C$. Therefore $(A \circ B) \circ C$ is $C$. On the other hand, as the reader may check, $A \circ (B \circ C)$ is $A$. Parentheses are important in computing with Table (*1*), since their omission would cause confusion.

We can omit parentheses when working with multiplication because multiplication satisfies the rule

$$X \times (Y \times Z) = (X \times Y) \times Z \qquad \text{(for all } X, Y, \text{ and } Z).$$

*Any table satisfying this rule is called* **associative**—it does not matter whether $Y$ is "associated" with $X$ or with $Z$. Since $X + (Y + Z) = (X + Y) + Z$ (for all $X, Y,$ and $Z$), addition is also associative.

But let us not get the impression that we have the only associative arithmetics, the only systems in which parentheses can be dropped. Table (*2*) is associative, as the reader may check (a complete check would require testing 64 equations, since there are four choices for each of $X, Y,$ and $Z$). Table (*7*) is also associative.

*An associative table is called a* **group**. Groups have been studied by mathematicians for over a century, and what is now known about them fills thousands of pages. Among the many results we will mention only two: Any group whose order is a prime or the square of a prime is commutative. Any group has exactly one row that reads the same as the guide row and exactly one column that reads the same as the guide column; moreover, the row and column referred to have the same name.

It is not easy to tell at a glance whether a table is associative. There is no nice pictorial interpretation of associativity. Associativity is best thought of in terms of "dispensing with parentheses." If a system is associative, we do not have to use parentheses, even when we have four or more letters (or numbers) to work with.

For example, there are five ways of placing parentheses in $A \circ B \circ C \circ D$ meaningfully:

$$A \circ (B \circ (C \circ D)), \qquad A \circ ((B \circ C) \circ D), \qquad (A \circ B) \circ (C \circ D),$$
$$(A \circ (B \circ C)) \circ D, \qquad ((A \circ B) \circ C) \circ D.$$

But *if the table is a group, then all five of these computations lead to the same value.*

To see why, observe that since $B \circ (C \circ D) = (B \circ C) \circ D$, the first two of the five have the same value. Similarly, the last two have the same value. Consider next the third expression, $(A \circ B) \circ (C \circ D)$, and the fifth, $((A \circ B) \circ C) \circ D$. Thinking of $A \circ B$ as $X$, $C$ as $Y$, and $D$ as $Z$, we have, using the associative property,

$$(A \circ B) \circ (C \circ D) = X \circ (Y \circ Z) = (X \circ Y) \circ Z = ((A \circ B) \circ C) \circ D.$$

In a similar manner, by associating $B$ with $C \circ D$ and then with $A$, we can prove

$$A \circ (B \circ (C \circ D)) = (A \circ B) \circ (C \circ D);$$

thus the first and the third expressions have the same value.

Combining these observations, we obtain

THEOREM 7. *In a group, no matter how we parenthesize* $A \circ B \circ C \circ D$ *to describe a computation, the result will be the same.*

The reader is invited to show that the fourteen ways of parenthesizing $A \circ B \circ C \circ D \circ E$ all lead to the same value if the table is a group. It can be proved that Theorem 7 holds for any number of letters, not just four or five. Thus in a group no ambiguity is caused by omitting the parentheses.

In ordinary arithmetic we need parentheses when working with expressions that involve both addition and multiplication. For example, were we to omit parentheses from $3 \times (6 + 7)$, we would not know whether to compute $(3 \times 6) + 7$ or $3 \times (6 + 7)$. As the reader may check, the two results are quite different.

As a matter of fact, there is a general agreement that when parentheses are omitted in an expression involving both multiplication and addition, we should multiply first, then add. Omission of the multiplication sign helps us carry out this agreement. Thus $ab + 1$ is short for $(a \times b) + 1$. We abided by this agreement, for example, in Chapter 5, where we wrote $MA + NB$.

Now we see that our algebra is one of many possible algebras, just as our planet is one of several planets. We are glad that our addition and multiplication are commutative and associative. But perhaps we envy the hypothetical beings who work with small tables and fascinating rules. And if, as in the next chapter, one of these exotic algebras should turn out to be of use to us (in designing experiments more complicated than the one described at the beginning of this chapter), we will not hesitate to exploit it. We will not cock our heads and complain, "this is not proper meat."

### Exercises

1. Prove that every idempotent table of order 3 is commutative.
2. Prove that every table of order 2 is commutative.
3. (a) Make a table of order 2.
   (b) Make the eight checks necessary to prove that it satisfies the rule $X \circ (Y \circ Z) = (X \circ Y) \circ Z$ for all $X$, $Y$, and $Z$ (and thus is a group).
4. Use the method we used to make Table (7) to produce a commutative table of order 6.

5. Make four of the 27 checks necessary to prove that Table (6) is a group.

6. Use the method we used to make Table (7) to produce a commutative table of order 3.

7. Is the table you made in E 6 isomorphic to Table (6)?

8. Is Table (5) isomorphic to Table (6)?

9. We showed that Table (12) could not be filled in so as to be idempotent and commutative if we made the choice $A \circ B = C$. Prove that the choice $A \circ B = D$ also fails to produce such a table.

10. In Table (1) does (a) $B \circ C = C \circ B$? (b) $B \circ C = C \circ D$?

11. In Table (2) (a) show that $B \circ D = C \circ A$, (b) show that $A \circ A = C \circ C$, (c) compute $((A \circ A) \circ A) \circ A$, (d) compute $((B \circ B) \circ B) \circ B$.

12. Using the construction employed in the proof of Theorem 3, fill in five boxes of the table for order (a) 11, (b) 21.

13. In a commutative table of order 200, at most how many different symbols appear on the main diagonal? Prove your answer.

14. Make five of the 16 checks necessary to prove that Table (1) satisfies the rule $X \circ (X \circ Y) = Y \circ X$.

15. Show that Table (7) does *not* satisfy the rule $X \circ (X \circ Y) = Y \circ X$.

16. Show that Table (6) does *not* satisfy the rule $X \circ (X \circ Y) = Y \circ X$.

17. Make four of the 64 checks necessary to prove that Table (2) is associative (that is, is a group).

18. Make four of the 125 checks necessary to prove that Table (7) is associative (that is, is a group).

19. (a) In Table (5), what are the two possible interpretations of $A^3$?
    (b) What is it about ordinary multiplication that assures us that $5^3$ has only one possible value?
    (c) What is it about ordinary multiplication that assures us that $5^4$ has only one possible value?

20. (a) Why can we omit parentheses when we are dealing only with addition ($+$)?
    (b) Why can we omit parentheses when we are dealing only with multiplication ($\times$)?
    (c) Why must we be careful about parentheses in expressions that involve both addition and multiplication? Illustrate your statement by an example.

21. In a table, for a given $A$ and $B$, is there always an $X$ such that $A \circ X = B$?

22. Prove that, in a group, $(A \circ B) \circ ((C \circ D) \circ E) = (A \circ (B \circ C)) \circ (D \circ E)$.

23. (a) Prove that, in a group, $(A \circ (B \circ C)) \circ (D \circ E) = A \circ ((B \circ (C \circ D)) \circ E)$.
    (b) Prove that, in a group, $((A \circ B) \circ (C \circ D)) \circ E = A \circ ((B \circ C) \circ (D \circ E))$.

24. List the 14 different ways there are of parenthesizing $A \circ B \circ C \circ D \circ E$ so as to describe a sequence of computations.

25. (a) Prove that, in a commutative group, $B \circ (C \circ A) = (C \circ B) \circ A$.
    (b) Prove that, in a commutative group, $(A \circ B) \circ C = (C \circ A) \circ B$.
    (c) Make a conjecture.

26. (a) Prove that, in a commutative group, $(A \circ (C \circ B)) \circ D = B \circ (D \circ (C \circ A))$.
    (b) Prove that, in a commutative group, $(A \circ B) \circ (C \circ D) = (A \circ C) \circ (B \circ D)$.
    (c) Make a conjecture.

27. Let the symbol $\circ$ denote the operation of ordinary subtraction. For example, $4 \circ 5 = -1$, and $7 \circ 2 = 5$. Prove that the operation $\circ$
    (a) is not commutative.
    (b) is not associative.
    (c) is not idempotent.
    (d) does satisfy the rule $X \circ X = Y \circ Y$.
    (e) does satisfy the rule $X \circ (X \circ Y) = Y$.
    (f) does satisfy the rule $(X \circ Y) \circ (Z \circ W) = (X \circ Z) \circ (Y \circ W)$.

28. Let the symbol $\circ$ denote ordinary division. For example, $6 \circ 2 = 3$, and $3 \circ 4 = \frac{3}{4}$. Prove that statements (a) through (f) of E 27 hold for this operation.

29. Prove that Table ($I$) satisfies the rule $(X \circ Y) \circ (Y \circ X) = X$ for all $X$ and $Y$.

30. Prove that Table ($I$) satisfies the rule $X \circ (Y \circ X) = (X \circ Y) \circ X$ for all $X$ and $Y$.

●

31. Prove that every letter appears exactly once on the main diagonal of a commutative table of odd order.

32. Using the letters $A$, $B$, $C$, $D$, and $E$, construct a table of order 5 that satisfies the rule $X \circ (X \circ Y) = Y \circ X$. (To simplify your bookkeeping as you gradually fill in the 25 boxes, be sure to use Theorems 5 and 6.)

33. Build a table of order 5 on the numbers 0, 1, 2, 3, and 4 by defining $X \circ Y$ to be the remainder when $4X + 2Y$ is divided by 5.
    (a) Fill in the 25 boxes.
    (b) Check that no row or column has duplications.
    (c) Check that the table satisfies the rule $X \circ (X \circ Y) = Y \circ X$ by using the definition of the table and the congruences (not by checking the 25 cases).

34. Show that the tables you constructed in E 32 and E 33 are isomorphic.

35. Prove that
$$X + Y = X - ((X - X) - Y).$$
This equation shows that addition can be expressed in terms of a combination of several subtractions. (Of course, $X + Y$ is also equal to $X - (-Y)$, but the second minus in $X - (-Y)$ has a different meaning from the first one; these two meanings are defined in Appendix C.)

36. Prove that subtraction cannot be expressed in terms of a combination of several additions (compare with E 35).

37. Comparing E 35 and E 36 why might we think that "subtraction is more fundamental than addition"?

38. Prove that $X \times Y = X/((X/X)/Y)$. This equation shows that multiplication can be expressed in terms of division (compare with E 35).

39. Prove that division cannot be expressed in terms of a combination of multiplications. In view of E 41, why might we think that "division is more fundamental than multiplication"? (Compare with E 37.)

40. Prove that any table satisfying the two rules

$$X \circ (X \circ Y) = Y \quad \text{and} \quad (Y \circ X) \circ X = Y \quad \text{(for all } X \text{ and } Y\text{)}$$

must be commutative.

● ●

41. Prove that, if a square is filled in to satisfy the rule $X \circ (X \circ Y) = Y \circ X$ and if no column has a duplication, then no row has a duplication.

42. Prove that an idempotent group must be of order one.

43. Prove that in a table satisfying the rule $(X \circ Y) \circ (Y \circ X) = X$ for all $X$ and $Y$, distinct letters do not commute.

44. Prove that any table satisfying the rule $X \circ (X \circ Y) = Y \circ X$ for all $X$ and $Y$ must also satisfy the rule

$$X \circ (Y \circ X) = (X \circ Y) \circ X.$$

45. Prove that Table (7) is associative, not by making the 125 checks, but by making use of the original definition of the table:
    (a) First prove that $X \circ (Y \circ Z) \equiv X + (Y + Z) \pmod 5$.
    (b) Next prove that $(X \circ Y) \circ Z \equiv (X + Y) + Z \pmod 5$.
    (c) Using the fact that $+$ is associative, deduce that $X \circ (Y \circ Z) \equiv (X \circ Y) \circ Z \pmod 5$.
    (d) Finally, prove that $X \circ (Y \circ Z) = (X \circ Y) \circ Z$.

46. Using the method of E 45, prove that there are groups of all orders, 1, 2, 3, 4, . . . .

47. Define a table of order 11 on the numbers 0, 1, 2, 3, 4, 5, 6, 7, 8, 9, and 10 by letting $X \circ Y$ be the remainder when $9X + 3Y$ is divided by 11.
    (a) Compute $3 \circ 3$.
    (b) Compute $4 \circ 8$.
    (c) Without filling in the 121 boxes of the table, prove by using congruences that no row or column has duplications.
    (d) Without filling in the table, prove that it satisfies the rule $X \circ (X \circ Y) = Y \circ X$.

48. Let $A$ be one of the letters in a table for a group. Since $A$ appears somewhere in column $A$, there is a letter, $T$, such that $T \circ A = A$.
    (a) Prove that $T \circ X = X$ for any letter $X$ in the group.
    (b) How does row $T$ compare with the guide row?

49. Using the method of E 48, prove that any group has a column reading the same as the guide column.

50. (a) Prove that the row exhibited in E 48 and the column exhibited in E 49 have the same name, $T$.
    (b) For ordinary addition of integers, what is $T$?
    (c) For ordinary multiplication of positive rationals, what is $T$?

51. Prove that the tables constructed in Theorem 3 satisfy the rule

$$(X \circ Y) \circ (Z \circ W) = (X \circ Z) \circ (Y \circ W) \qquad \text{(for all } X, Y, Z, \text{ and } W).$$

52. If $A$ and $B$ are natural numbers other than 0, denote their greatest common divisor $A \circ B$ instead of $(A, B)$. For example, $8 \circ 12 = 4$.
    (a) Is the operation denoted by $\circ$ idempotent, commutative, and associative?
    (b) Give an example of $A$, $B$, and $C$ such that $B$ is different from $C$, yet $A \circ B = A \circ C$.

53. (a) Show that the tables

| $\circ$ | $A$ | $B$ |
|---|---|---|
| $A$ | $A$ | $B$ |
| $B$ | $B$ | $A$ |

and

| $*$ | $B$ | $A$ |
|---|---|---|
| $B$ | $A$ | $B$ |
| $A$ | $B$ | $A$ |

   are isomorphic.
   (b) Show that they cannot be obtained from each other by relabeling.
   (c) Show that they record the same "multiplication facts."

54. Show that these two tables are isomorphic:

| $\oplus$ | 0 | 1 | 2 | 3 |
|---|---|---|---|---|
| 0 | 0 | 1 | 2 | 3 |
| 1 | 1 | 2 | 3 | 0 |
| 2 | 2 | 3 | 0 | 1 |
| 3 | 3 | 0 | 1 | 2 |

(mod 4)

| $\otimes$ | 1 | 2 | 3 | 4 |
|---|---|---|---|---|
| 1 | 1 | 2 | 3 | 4 |
| 2 | 2 | 4 | 1 | 3 |
| 3 | 3 | 1 | 4 | 2 |
| 4 | 4 | 3 | 2 | 1 |

(mod 5)

## Reference

1. R. A. Fisher, *The Design of Experiments*, Hafner, New York, 1951. (The warning is quoted from p. 70.)

# ORTHOGONAL
# TABLES

"A very curious question, which has exercised for some time the ingenuity of many people," wrote Euler in 1779 (in a paper called "On a New Type of Magic Square") "has involved me in the following studies, which seem to open a new field of analysis, in particular in the study of combinations. The question revolves around arranging 36 officers to be drawn from 6 different ranks and at the same time from 6 different regiments so that they are also ranged in a square so that in each line (both horizontal and vertical) there are 6 officers of different ranks and different regiments."

If the "very curious question" had asked instead for an arrangement of 9 officers in a 3 by 3 square, such that in each row and in each column the 3 officers were from different ranks and different regiments, then it would be (as Euler himself noted) a simple matter to give a solution. If we call the three regiments $a$, $b$, and $c$, and the three ranks $A$, $B$, and $C$, then one possible arrangement would be

(1)

| $aA$ | $bC$ | $cB$ |
|------|------|------|
| $bB$ | $cA$ | $aC$ |
| $cC$ | $aB$ | $bA$ |

Let us look for a moment at this arrangement of 9 officers. The regiments $a$, $b$, and $c$ themselves form the table

(2)

| $a$ | $b$ | $c$ |
|-----|-----|-----|
| $b$ | $c$ | $a$ |
| $c$ | $a$ | $b$ |

This is of the type met in the last chapter—in each row and in each column

each letter appears exactly once. Looking at the arrangement of the ranks
*A*, *B*, and *C*, we have the table

(3)

| A | C | B |
|---|---|---|
| B | A | C |
| C | B | A |

again of the type met in the last chapter.

Thus the problem of selecting 9 officers is the same as that of making two
tables like (2) and (3) such that, after placing one on top of the other, we may
look down on them and see each of the 9 pairs *aA*, *aB*, *aC*, *bA*, *bB*, *bC*, *cA*, *cB*,
and *cC* exactly once. In perspective we can think of them placed like two trays
of muffins in an oven:

(4)

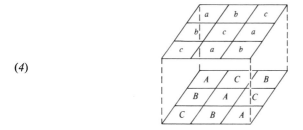

Another way of thinking of the relation between Tables (2) and (3) is this:
If one looks at the three boxes of Table (3) that are occupied by officers of a
fixed rank, such as *A*, then in the corresponding boxes of Table (2) one finds
each of the three regiments *a*, *b*, and *c*.

*We will say that two tables, of order N, one of which uses small letters and the
other capital letters, are* **orthogonal** *if when we put one on top of the other we
see all N² combinations of small and capital letters.* The problem of the 36
officers now reads: devise a pair of orthogonal tables of order 6. All that we
have shown is that there is a pair of orthogonal tables of order 3.

Before considering order 6, let us treat another simpler problem. Is there an
orthogonal pair of order 2? Recall that entries on the main diagonal of any
table of order 2 are the same; that is, the two shaded boxes must always have
the same letter:

From this fact we can see that two tables of order 2 are never orthogonal. If we place one table of order 2 above another and look only at the shaded boxes, we will already see pairs duplicated. Thus we have proved

THEOREM 1. *There are no orthogonal tables of order 2. There are orthogonal tables of order 3.*

It will sometimes be convenient to have numbers instead of letters in the boxes. If we start with the tables

(5)

| a | b | c |
|---|---|---|
| b | c | a |
| c | a | b |

and

| A | C | B |
|---|---|---|
| B | A | C |
| C | B | A |

and, throughout, make the replacements 1 for $a$, 2 for $b$, 3 for $c$ and 1 for $A$, 2 for $B$, 3 for $C$, we obtain these two tables:

(6)

| 1 | 2 | 3 |
|---|---|---|
| 2 | 3 | 1 |
| 3 | 1 | 2 |

and

| 1 | 3 | 2 |
|---|---|---|
| 2 | 1 | 3 |
| 3 | 2 | 1 |

If we put one of these new tables above the other we see that they too ought to be called orthogonal; as we look down we see each of the 9 possible pairings of the numbers 1, 2, 3, with 1, 2, 3: the pairs (1, 1), (1, 2), (1, 3), (2, 1), (2, 2), (2, 3), (3, 1), (3, 2), and (3, 3). For example, the pair (1, 1) appears in the boxes of row 1, column 1 and only there.

We could have replaced $A$ by antelope, $B$ by bear, and $C$ by crocodile without losing information. We would still find ourselves with a pair of orthogonal tables. Or we could have replaced $a$ by 1, $b$ by 2, $c$ by 3, and $A$ by 1, $B$ by 3, $C$ by 2, and found ourselves with this orthogonal pair:

(7)

| 1 | 2 | 3 |
|---|---|---|
| 2 | 3 | 1 |
| 3 | 1 | 2 |

and

| 1 | 2 | 3 |
|---|---|---|
| 3 | 1 | 2 |
| 2 | 3 | 1 |

Notice that the top row in each of these tables consists simply of the numbers 1, 2, and 3 in order.

The ideas displayed by this example of order 3 we summarize in

THEOREM 2. *From any two orthogonal tables of order N we can get two orthogonal tables of order N whose boxes are filled with the numbers 1, 2, . . . , N. Moreover, the latter tables can be chosen in such a way that the first row of each consists of the numbers 1, 2, . . . , N in their usual order.*

We shall use Theorem 2 soon.

Before we examine orthogonal tables in greater detail, however, we ought to have a definition of them that does not involve such geometric ideas as "above" and "looking down." To find a better definition, let us look at the orthogonal Tables (7). If we furnish each of them with a guide row and guide column, they become

|   | 1 | 2 | 3 |
|---|---|---|---|
| 1 | 1 | 2 | 3 |
| 2 | 2 | 3 | 1 |
| 3 | 3 | 1 | 2 |

(8)

and

|   | 1 | 2 | 3 |
|---|---|---|---|
| 1 | 1 | 2 | 3 |
| 2 | 3 | 1 | 2 |
| 3 | 2 | 3 | 1 |

(9)

In Table (8) we denote the entry located where row $X$ meets column $Y$ by $X \circ Y$. For example, $2 \circ 3 = 1$. In Table (9) to the entry located where row $X$ meets column $Y$ we must give a different name; say, $X * Y$ (which we can read as "$X$ star $Y$"). Thus $2 * 3 = 2$.

The fact that Tables (8) and (9) are orthogonal can be expressed as follows: For each of the 9 possible pairs $(M, N)$ formed from the numbers 1, 2, 3, we can find an $X$ and a $Y$ such that the following two equations both hold:

$$(10) \qquad X \circ Y = M; \qquad X * Y = N.$$

Equations (10) state that if $(M, N)$ is any one of the 9 pairs formed from the numbers 1, 2, and 3, then $M$ and $N$ appear in a pair of corresponding boxes in Tables (8) and (9). For example, if $M$ is 3 and $N$ is 2, the reader may check that $X = 3$ and $Y = 1$ are solutions (and the only solutions) of equations (10); that is,

$$3 \circ 1 = 3,$$

and

$$3 * 1 = 2.$$

Clearly Equations (10) provide an algebraic criterion for testing whether two tables, furnished with guide row and guide column, are orthogonal.

Let us now try to make a pair of orthogonal tables of order 4. As one table of the pair let us try one of the most orderly tables of order 4:

(11)

| a | b | c | d |
|---|---|---|---|
| b | c | d | a |
| c | d | a | b |
| d | a | b | c |

The challenge is to find a table of order 4 (say, with the letters $A$, $B$, $C$, and $D$) that is orthogonal to the first table.

Let us, for a moment, think only of the four boxes that the letter $A$ will occupy, and disregard the other twelve boxes. These four boxes will come from four different rows and four different columns. And the four boxes of (11) below them will each contain a different one of the letters, $a$, $b$, $c$, and $d$.

Let us attack the problem by looking for four boxes in (11), no two in the same column, no two in the same row, and no two containing the same one of the letters $a$, $b$, $c$, and $d$. *These four boxes may be thought of as a* **sampling** *quartet, in the sense that they simultaneously sample all the rows, all the columns, and all the letters $a$, $b$, $c$, and $d$.* The four boxes in which we hope to place $B$ will also form a sampling quartet. Similarly for $C$ and $D$. Building a table orthogonal to (11) is thus the same as grouping the sixteen boxes into four sampling quartets.

If we try to find even a single sampling quartet for this particular table, we will have no luck. To see this, note that Table (11) is simply the table

(12)

| 0 | 1 | 2 | 3 |
|---|---|---|---|
| 1 | 2 | 3 | 0 |
| 2 | 3 | 0 | 1 |
| 3 | 0 | 1 | 2 |

in disguise. Put in a guide row and a guide column:

(13)

|   | 0 | 1 | 2 | 3 |
|---|---|---|---|---|
| 0 | 0 | 1 | 2 | 3 |
| 1 | 1 | 2 | 3 | 0 |
| 2 | 2 | 3 | 0 | 1 |
| 3 | 3 | 0 | 1 | 2 |

We see that $X \circ Y$ is just the remainder when the usual $X + Y$ is divided by 4. For example, the remainder is 1 when $2 + 3$ is divided by 4, and $2 \circ 3 = 1$ in the table. The reader may check the other boxes without trouble. Note that $X \circ Y \equiv X + Y \pmod 4$ since any integer is congruent modulo 4 to its remainder when divided by 4.

Let us suppose, now, that we can find a sampling quartet. Then there is some arrangement of the column guide numbers (say, $P$, $Q$, $R$, $S$) such that the four numbers $0 \circ P$, $1 \circ Q$, $2 \circ R$, and $3 \circ S$ are, in turn, some arrangement of the numbers 0, 1, 2, and 3. Then, clearly,

$$(0 \circ P) + (1 \circ Q) + (2 \circ R) + (3 \circ S) = 6.$$

Now

$$0 \circ P \equiv 0 + P \pmod 4,$$
$$1 \circ Q \equiv 1 + Q \pmod 4,$$
$$2 \circ R \equiv 2 + R \pmod 4,$$
$$3 \circ S \equiv 3 + S \pmod 4.$$

Adding these four congruences we obtain

$$(0 \circ P) + (1 \circ Q) + (2 \circ R) + (3 \circ S)$$
$$\equiv (0 + P) + (1 + Q) + (2 + R) + (3 + S) \pmod 4,$$

and hence

$$6 \equiv 0 + 1 + 2 + 3 + P + Q + R + S \pmod 4.$$

But $P, Q, R, S$ is some arrangement of 0, 1, 2, 3; therefore $P + Q + R + S = 6$. Thus

$$6 \equiv 6 + 6 \pmod 4,$$

which is clearly not true.

We see, then, that there is no sampling quartet for Table (*11*) and that there is no chance of finding a table orthogonal to it. We have proved

THEOREM 3. *There is no table orthogonal to Table (11).*

We are not the first to run into difficulty in building orthogonal tables of order 4. On March 12, 1842, Schumacher, in a letter to Gauss, wrote about orthogonal tables of order $N$:

> . . . For $N = 2$ this is impossible, for $N = 3$, easy, and I believed to have understood earlier from you, that it is also impossible for $N = 4 \cdots$ but I must have been wrong since Clausen brought me this solution . . . :

| a | b | c | d |
|---|---|---|---|
| c | d | a | b |
| d | c | b | a |
| b | a | d | c |

| D | A | B | C |
|---|---|---|---|
| A | D | C | B |
| B | C | D | A |
| C | B | A | D |

(14)

May I ask, as long as the investigation costs you no effort, for which values of $N$ is it impossible?

Three weeks later, April 2, 1842, Gauss replied:

I could devote only a few minutes to the question of your last letter. That I at any time should have asserted that the problem in question was impossible for $N = 4$, I can hardly believe since one sees on first glance that Clausen has not even imposed the most stringent demands. One can even add this demand: that in both diagonals all of the $a$, $b$, $c$, $d$ and all of the $A$, $B$, $C$, $D$ occur; the diagrams you sent me do not satisfy this condition. But here is an example:

| a | b | c | d |
|---|---|---|---|
| c | d | a | b |
| d | c | b | a |
| b | a | d | c |

| D | A | B | C |
|---|---|---|---|
| C | B | A | D |
| A | D | C | B |
| B | C | D | A |

(15)

(The question: "for which values of $N$ is it impossible?" Gauss did not answer. We will soon consider it.)

Evidently neither Schumacher nor Gauss was familiar with Euler's work of 1779, for he had already found these three tables, any two of which are orthogonal:

| a | b | c | d |
|---|---|---|---|
| b | a | d | c |
| c | d | a | b |
| d | c | b | a |

| A | C | D | B |
|---|---|---|---|
| B | D | C | A |
| C | A | B | D |
| D | B | A | C |

| A | D | B | C |
|---|---|---|---|
| B | C | A | D |
| C | B | D | A |
| D | A | C | B |

(16)

This suggests new questions, but let us concentrate first on the problem of seeking just two orthogonal tables.

For the next order, 5, we can obtain an orthogonal pair by flipping the table

(17)

| A | C | D | E | B |
|---|---|---|---|---|
| D | B | E | C | A |
| E | A | C | B | D |
| B | E | A | D | C |
| C | D | B | A | E |

upside down along the axis of its main diagonal, which gives

(18)

| A | D | E | B | C |
|---|---|---|---|---|
| C | B | A | E | D |
| D | E | C | A | B |
| E | C | B | D | A |
| B | A | D | C | E |

Tables (*17*) and (*18*) are orthogonal, as the reader may verify.

So far, then, we have managed to find orthogonal tables of orders 3, 4, and 5, and have shown that order 2 is impossible. Before we discuss the next order, 6, let us read what Euler, writing in 1779, had to say about the problem it presents:

> I have examined a very great number of tables . . . and I do not hesitate to conclude that one cannot produce an orthogonal pair of order 6 and that the same impossibility extends to the orders 10, 14, . . . and in general all the orders which are twice an odd number.

In the year 1900, Gaston Tarry, assisted by his brother Herbert Tarry, by a careful and laborious listing of all tables of order 6, proved that no two of them are orthogonal. This method was criticized for failing to show *why* the problem could not be solved; what was the mysterious property of 6? For the order 2 (which also is twice an odd number) we had an insight; but for the order 6 there was just a long list. Anyway, by the year 1900 it was certain that there are no orthogonal tables of order 2 or 6. Orders 10, 14, 18, . . . (doubles of odd numbers) remained a mystery.

Although mathematicians failed to find a solution to the problem, they nevertheless raised a more general question. They asked: How long a list of tables of order $N$ can be found such that any two tables in the list will be orthogonal?

We shall call such a list simply *a list of orthogonal tables*. For order 4 we have, for example, Euler's list of three orthogonal tables. For orders 2 and 6, since there are no orthogonal pairs whatsoever, we will say that the list must stop with just one table. (We use "one" rather than "zero" to record the fact that there is a table of order 6.) Euler conjectured that any list for orders twice an odd number must stop at one.

Some restraint is put on the length of such lists, for a fixed order $N$, by

THEOREM 4. *For any given order N, it is impossible to make a list of N orthogonal tables.*

PROOF. For simplicity let us take as a specific example the order 5. We will show that it is impossible to find a list of 5 orthogonal tables of order 5. Our reasoning will work not only for 5, but for any order.

Picture the 5 tables as 5 muffin trays in an oven, stacked one above the other. Using Theorem 2, we will first, if need be, relabel the muffins on these trays such that the muffins in the first row of each tray will be numbered 1, 2, 3, 4, 5, in order.

Our oven looks like this:

(19)

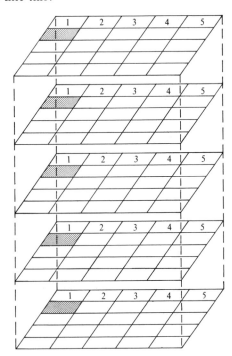

Although the remaining twenty muffins on each tray are numbered in some way or another, let us consider just the five muffins found in the first box of the second row in each tray [shaded in (*19*)]. Since in each tray there is already a muffin labeled 1 in column 1, none of the five muffins under consideration can be labeled 1, which means that we have five muffins and only four labels: 2, 3, 4, and 5. By the "pigeonhole principle" (discussed in the introduction to E 44), at least two of these five muffins must have the same number. But then the two trays carrying these two muffins with the same label could not be orthogonal; for, looking down on the first row of these two trays, we already see the five pairs (1, 1), (2, 2), (3, 3), (4, 4), and (5, 5), in which the two numbers are the same. Thus there is no list having five orthogonal tables of order 5. Since similar reasoning applies to any order $N$, the theorem is proved.

Theorem 4 says, for example, that we can never find four tables of order 4 of which each pair is orthogonal. But we already have an example of a list of three orthogonal tables of order 4. Thus the longest list for order 4 has three tables.

Theorem 4 says also that we can never find six tables of order 6 of which each two are orthogonal. As mentioned above, the Tarry brothers proved that we cannot find even two orthogonal tables of order 6.

Theorem 4 says, further, that we cannot find five tables of order 5, of which each two are orthogonal. But, using the machinery of Chapter 12, we can easily build four tables of order 5 of which each two are orthogonal. Here is how:

Number the rows and columns 0, 1, 2, 3, and 4. One table is made by placing in the box formed by row $X$ and column $Y$ the remainder when $X + Y$ is divided by 5. Calling this entry $X \circ Y$, we have, for example, $3 \circ 4 = 2$. Filling in all 25 boxes, we have the first of our four tables:

|   | 0 | 1 | 2 | 3 | 4 |   |
|---|---|---|---|---|---|---|
| 0 | 0 | 1 | 2 | 3 | 4 |   |
| 1 | 1 | 2 | 3 | 4 | 0 |   |
| 2 | 2 | 3 | 4 | 0 | 1 | $X \circ Y$ |
| 3 | 3 | 4 | 0 | 1 | 2 |   |
| 4 | 4 | 0 | 1 | 2 | 3 |   |

To get the second table, we will *distort* our formula somewhat. In the box of row $X$, column $Y$, let us *not* put $X \circ Y$, but rather $X \circ 2Y$; that is, the remainder when $X + 2Y$ is divided by 5. For example, in the box of row 3 and column 4 we place $3 \circ (2 \cdot 4) = 3 \circ 8 = 1$ (the remainder when 11 is divided by 5). Here is our second table:

|     | 0 | 1 | 2 | 3 | 4 |
|-----|---|---|---|---|---|
| 0   | 0 | 2 | 4 | 1 | 3 |
| 1   | 1 | 3 | 0 | 2 | 4 |
| 2   | 2 | 4 | 1 | 3 | 0 |
| 3   | 3 | 0 | 2 | 4 | 1 |
| 4   | 4 | 1 | 3 | 0 | 2 |

$X \circ 2Y$

In the box where row $X$ meets column $Y$, the third table will have $X \circ 3Y$; the fourth table $X \circ (4Y)$. If we stack all four of these muffin pans in the oven, we can check that the four trays are orthogonal:

(20)

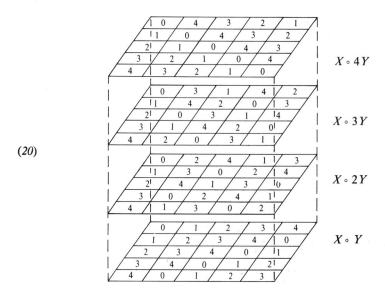

To be on the safe side, let us prove that the tables corresponding to $X \circ 3Y$ and $X \circ 2Y$ are orthogonal. In view of the remark leading up to (10) on page 193, we will show that for each pair, $M$ and $N$, of integers chosen from 0 through 4, there is a pair, $X$ and $Y$, in the same range, such that both these demands are met:

$$X \circ 3Y = M \quad \text{and} \quad X \circ 2Y = N.$$

Recalling the definition of $\circ$, we must show that there is a pair, $X$ and $Y$, such that

$$X + 3Y \equiv M \text{ (mod 5)} \quad \text{and} \quad X + 2Y \equiv N \text{ (mod 5)}.$$

(Thus we wish to solve a pair of congruences similar to the "simultaneous" equations of Chapter 8.) Subtraction of the second congruence from the first shows that

$$Y \equiv M - N \text{ (mod 5)}.$$

This determines a unique $Y$ in the interval from 0 through 4. To find $X$, we use the first congruence, $X + 3Y \equiv M$ (mod 5), obtaining

$$X \equiv M - 3Y \equiv M - 3(M - N) \equiv 3N - 2M \text{ (mod 5)}.$$

This determines a unique $X$, and we already know that $X$ and $Y$ satisfy the first congruence. Do they also satisfy the second congruence, $X + 2Y \equiv N$ (mod 5)? We have

$$X + 2Y \equiv (3N - 2M) + 2(M - N) \equiv N \text{ (mod 5)},$$

as we desired. Thus the two tables considered are orthogonal. The other pairs of trays can be checked similarly.

The same method can be used to prove

THEOREM 5. *If $N$ is a prime, then there is a list of $N - 1$ orthogonal tables of order $N$.*

If $N$ is not a prime this method fails to produce a list of $N - 1$ orthogonal tables of order $N$. Indeed, some of the square arrangements that it produces for such an $N$ will have duplications in some rows. (As an example, the reader might take $N$ to be 6, fill in the arrangement given by the formula $X \circ 2Y$, and see why duplications occur.)

In order to summarize what we have learned about lists of orthogonal tables, let us denote by $L(N)$ the largest number of tables of order $N$ that can be orthogonal to each other. This is what we have seen about $L(N)$:

$$L(2) = 1, \quad L(3) = 2, \quad L(4) = 3,$$
$$L(5) = 4, \quad L(6) = 1, \quad L(N) = N - 1$$

for any prime $N$; and finally, Euler's conjecture of 1779, $L(N) = 1$ if $N$ is twice an odd integer.

In 1923, H. F. MacNeish wrote a paper in which he did three things. He proved that whenever $N$ is a power of a prime, $L(N) = N - 1$ [for example, $L(27) = 26$. Theorem 4 tells us only that $L(27)$ is less than 27]. Next, he gave an incorrect proof of Euler's conjecture. Finally, he made this more general conjecture: $L(N)$ can be computed by (1) writing $N$ as the product of powers of distinct primes, and then (2) subtracting 1 from the smallest of these powers.

As an example, let us use his conjecture to compute $L(36)$. We begin by writing 36 as the product of powers of distinct primes: $36 = 4 \times 9$. (By the Fundamental Theorem of Arithmetic, this can be done in only one way.) The smaller of these powers is 4. The conjecture implies that $L(36) = 4 - 1$; that is, $L(36) = 3$. Similarly, the conjecture implies, as the reader may compute, that $L(9) = 8$ (in agreement with MacNeish's theorem), that $L(6) = 1$, that $L(10) = 1$ (in agreement with Euler's conjecture), and that $L(21) = 2$. The reader may easily show that Euler's conjecture is just a special case of Mac-Neish's.

In a sense, MacNeish's conjecture, though unproved, did suggest that Euler's conjecture was right and that the latter was, indeed, describing only part of the truth. More evidence in favor of Euler's conjecture came in 1944, when H. B. Mann proved that there is no table orthogonal to a group (that is, a table satisfying the rule $X \circ (Y \circ Z) = (X \circ Y) \circ Z$), whose order is twice an odd integer.

Many mathematicians became interested in Euler's conjecture after R. A. Fisher, in the 1920's, showed that orthogonal tables are useful in the design of certain kinds of experiments. For example, we might want to test the effect on tomato plants of various doses of water combined with various doses of nitrogen fertilizer. The orthogonal tables (2) and (3) show how to apply the doses $a$, $b$, and $c$ of water and $A$, $B$, and $C$ of nitrogen on nine plants arranged in a square. Each of the nine possible combinations is represented once. As a result of this new use for orthogonal tables, many research papers devoted to them appeared in the mathematical and statistical journals in the following decades.

In 1958, to the surprise of the mathematical world, MacNeish's conjecture was shown to be false. Using group theory and finite geometries, E. T. Parker constructed a list of three orthogonal tables of order 21, contradicting Mac-Neish's conjecture that $L(21) = 2$. But Euler's conjecture still remained.

A few months later, the statisticians Bose and Shrikhande, putting a new twist in Parker's technique, constructed two orthogonal tables of order 22, which is twice an odd number. Euler's conjecture, which had stood for 180 years, was proved false. After a furious correspondence, the three mathematicians together proved that the only orders for which Euler's conjecture is true are 2 and 6. In particular, their methods produced the two orthogonal tables of order 10 that appear at the top of the next page.

In 1960, Sade noticed that a technique he had invented in 1950 for constructing a table of order $a + bc$ out of tables of orders $a$, $b$, $a + b$, and $c$ behaves well with respect to orthogonality. In particular, it yields orthogonal tables of order 22 ($= 1 + 3 \cdot 7$) and thus, again disproves Euler's conjecture.

Now that the two conjectures have been shown to be false, mathematicians face the more difficult question: what is $L(N)$ for each $N$? The first unsolved

| 0 | 4 | 1 | 7 | 2 | 9 | 8 | 3 | 6 | 5 |
|---|---|---|---|---|---|---|---|---|---|
| 8 | 1 | 5 | 2 | 7 | 3 | 9 | 4 | 0 | 6 |
| 9 | 8 | 2 | 6 | 3 | 7 | 4 | 5 | 1 | 0 |
| 5 | 9 | 8 | 3 | 0 | 4 | 7 | 6 | 2 | 1 |
| 7 | 6 | 9 | 8 | 4 | 1 | 5 | 0 | 3 | 2 |
| 6 | 7 | 0 | 9 | 8 | 5 | 2 | 1 | 4 | 3 |
| 3 | 0 | 7 | 1 | 9 | 8 | 6 | 2 | 5 | 4 |
| 1 | 2 | 3 | 4 | 5 | 6 | 0 | 7 | 8 | 9 |
| 2 | 3 | 4 | 5 | 6 | 0 | 1 | 8 | 9 | 7 |
| 4 | 5 | 6 | 0 | 1 | 2 | 3 | 9 | 7 | 8 |

| 0 | 7 | 8 | 6 | 9 | 3 | 5 | 4 | 1 | 2 |
|---|---|---|---|---|---|---|---|---|---|
| 6 | 1 | 7 | 8 | 0 | 9 | 4 | 5 | 2 | 3 |
| 5 | 0 | 2 | 7 | 8 | 1 | 9 | 6 | 3 | 4 |
| 9 | 6 | 1 | 3 | 7 | 8 | 2 | 0 | 4 | 5 |
| 3 | 9 | 0 | 2 | 4 | 7 | 8 | 1 | 5 | 6 |
| 8 | 4 | 9 | 1 | 3 | 5 | 7 | 2 | 6 | 0 |
| 7 | 8 | 5 | 9 | 2 | 4 | 6 | 3 | 0 | 1 |
| 4 | 5 | 6 | 0 | 1 | 2 | 3 | 7 | 8 | 9 |
| 1 | 2 | 3 | 4 | 5 | 6 | 0 | 9 | 7 | 8 |
| 2 | 3 | 4 | 5 | 6 | 0 | 1 | 8 | 9 | 7 |

case is $L(10)$, which is at least 2 and at most 9. Work is being done on this problem. Also, using techniques of number theory, Chowla, Erdös, and Strauss proved that when $N$ gets large, $L(N)$ is larger than $\sqrt[91]{N}$ and hence becomes (and remains) large. If there is a formula for computing $L(N)$, it is probably much more complicated than the one MacNeish had conjectured.

One question may still be troubling the reader: why did Euler title his paper, "On a New Type of Magic Square"? Let us answer that question. By a "magic square of order $N$" we mean an arrangement of the $N^2$ natural numbers from 1 to $N^2$ in a square array of $N^2$ boxes such that the sums of the entries in each row, each column, and each of the two diagonals are the same.

Euler showed how to construct magic squares with the aid of two orthogonal tables if these tables have no duplications in any one of their four diagonals. For example, let us construct a magic square of order 4 starting with the orthogonal tables (*15*) given by Gauss.

Replace *a*, *b*, *c*, *d* by 1, 2, 3, 4 and *A*, *B*, *C*, *D* by 0, 4, 8, 12. Then form the sums of entries in corresponding boxes:

(*21*)

| 1 + 12 | 2 + 0 | 3 + 4 | 4 + 8 |
|--------|-------|-------|-------|
| 3 + 8 | 4 + 4 | 1 + 0 | 2 + 12 |
| 4 + 0 | 3 + 12 | 2 + 8 | 1 + 4 |
| 2 + 4 | 1 + 8 | 4 + 12 | 3 + 0 |

This is clearly a magic square, since each row, column, and diagonal adds up to $1 + 2 + 3 + 4 + 0 + 4 + 8 + 12$. Written out in a manner that conceals its origin, it is

(22)

| 13 | 2  | 7  | 12 |
|----|----|----|----|
| 11 | 8  | 1  | 14 |
| 4  | 15 | 10 | 5  |
| 6  | 9  | 16 | 3  |

(If the pair were of order 5 instead of order 4, we would replace $a$, $b$, $c$, $d$, $e$ by 1, 2, 3, 4, 5 and $A$, $B$, $C$, $D$, $E$ by 0, 5, 10, 15, 20.)

Just as Euler did not foresee that his work on the Koenigsberg bridge problem would some day be applied to telegraphy and electronic computers, he did not foresee that his work on the problem of the 36 officers would be applied to the design of statistical experiments.

The assertion he made at the end of his memoir is thus partly right and partly wrong:

> . . . Leaving it to geometers to see whether there are ways of achieving an enumeration of all possible cases—that which appears to provide a vast field for new and interesting research—I place an end here to my thoughts on a question which, although itself of little use, has led us to important observations both in the theory of combinations and the general theory of magic squares.

**Exercises**

1. Find a sampling quartet for this table

| a | b | c | d |
|---|---|---|---|
| c | d | a | b |
| d | c | b | a |
| b | a | d | c |

2. Find as many sampling quartets as you can for the table of E 1.

3. The second table of (14) provides 4 sampling quartets for the table of E 1. What are they? Are they on the list you made for E 2?

4. Find the 3 sampling triplets for the table

| | | |
|---|---|---|
| *a* | *b* | *c* |
| *b* | *c* | *a* |
| *c* | *a* | *b* |

determined by each of the letters *A*, *B*, and *C* of the following table orthogonal to it:

| | | |
|---|---|---|
| *A* | *C* | *B* |
| *B* | *A* | *C* |
| *C* | *B* | *A* |

Are there any more sampling triplets for the first table?

5. Each of the letters *a*, *b*, and *c* in E 4 determines a sampling triplet for the second table of E 4. Find these triplets. Are there any more sampling triplets for this second table?

6. Prove that this table has no sampling sextet.

| | | | | | |
|---|---|---|---|---|---|
| 0 | 1 | 2 | 3 | 4 | 5 |
| 1 | 2 | 3 | 4 | 5 | 0 |
| 2 | 3 | 4 | 5 | 0 | 1 |
| 3 | 4 | 5 | 0 | 1 | 2 |
| 4 | 5 | 0 | 1 | 2 | 3 |
| 5 | 0 | 1 | 2 | 3 | 4 |

7. Replace Tables (*14*) with tables of the same form, but built on the integers 1, 2, 3, and 4 instead of letters, and in such a way that the top row of both tables reads 1, 2, 3, 4 in the usual order.

8. The right pair of tables in (*16*) are both orthogonal to the left table of (*16*). Indicate the 8 sampling quartets that the right pair determines in the left table.

9. (a) Replace all the letters in the three tables of (*16*) by 1, 2, 3, 4 in order that the top row of each table will read 1, 2, 3, 4 in the usual order.
   (b) Check that each two of the three tables constructed in (a) are orthogonal.

10. How many sampling quartets does the left table of (*16*) have?

11. (a) The bottom three tables of (20) provide 15 sampling quintets for the top table of (20). Find and list them.
    (b) Does the top table of (20) have any sampling quintets not listed in (a)?

12. Table (17) satisfies the rule $X \circ (X \circ Y) = Y \circ X$ for all $X$ and $Y$. Check this for five choices of $X$ and $Y$.

13. Can two commutative tables be orthogonal?

14. Of which of the tables of (15) and (16) can it be said that the table is orthogonal to the table obtained by rotating it about its main diagonal?

15. If a table is orthogonal to the table obtained by rotating it about its main diagonal, what must be true about its main diagonal?

16. Prove that the trays in (20) that correspond to $X \circ Y$ and $X \circ 4Y$ are orthogonal.

17. We proved Theorem 4 for order 5.
    (a) Prove it for order 12.
    (b) Prove it for any order.

18. Show that table $X \circ 3Y$ of (20) is orthogonal to table $X \circ 2Y$ of (20) by finding the 5 sampling quintets that the second table determines for the first.

19. (a) Give an important application of orthogonal tables. Explain.
    (b) Why did Euler study orthogonal tables? How did he apply them?

20. Using congruences, prove that each of the tables in (20) has no duplication in any row or in any column. What property of the number 5 is important here?

21. Construct a list of 6 orthogonal tables of order 7.

22. A square of order 9 is filled in with the numbers 0 through 8 by placing in the box of row $X$ and column $Y$, $X \circ 4Y$, the remainder when $X + 4Y$ is divided by 9. Prove that there are no duplications in any row or column.

23. A square of order 9 is filled in as in E 22, except that we use $X \circ 3Y$, the remainder when $X + 3Y$ is divided by 9. Do any rows have duplications? Do any columns have duplications?

24. Using the method of E 22, construct two tables of order 9 by using $X \circ 2Y$ and $X \circ 5Y$. Prove that neither of these two tables of order 9 has duplications in any row or column. Are they orthogonal?

25. Can a scientist figure out a way of applying 4 doses of water, 4 doses of fertilizer, and 4 doses of sunlight to 16 tomato plants arranged in a square such that
    (a) each dose of each factor is applied to 4 plants, of which no two come from the same row or column;
    (b) each of the 16 combinations of water and fertilizer is represented;
    (c) each of the 16 combinations of water and sunlight is represented; and
    (d) each of the 16 combinations of fertilizer and sunlight is represented?
    If so, show how it can be done.

26. Show how the four tables in (20) can be used for testing the effect of combinations of water, sunlight, potassium, and nitrogen on tomato plants.

27. Prove that if MacNeish's conjecture had been correct, then Euler's conjecture would also have been correct.

28. Is it possible to design an experiment that will test the effects of 6 doses of water in combination with 6 doses of nitrogen on 36 plants arranged in a square?

29. We constructed the magic square (*22*) from (*15*) by replacing letters by numbers in a certain way. What magic square results if we replace *a*, *b*, *c*, and *d* by 3, 1, 2, and 4, respectively; and *A*, *B*, *C*, and *D* by 4, 0, 12, and 8, respectively?

30. See E 29. Construct still another magic square from (*15*) by making a replacement with a different order of 1, 2, 3, 4 and 0, 4, 8, 12.

31. Use the middle two squares of (*20*) to produce a magic square of order 5. (Note that neither contains duplications in the diagonals.)

32. *Let us call an arrangement of the $N^2$ natural numbers from 1 through $N^2$ an* **almost-magic square** *if all the rows and columns have the same sum (we disregard the diagonals).* Construct an almost-magic square of order 3 by using the orthogonal tables (*2*) and (*3*).

33. Construct an almost-magic square (see E 32) of order 4 by using the orthogonal tables (*14*).

34. Construct an almost-magic square (see E 32) of order 4 by using the left two orthogonal tables of (*16*).

35. Show that this almost-magic square (see E 32) comes from a pair of orthogonal tables of order 4. (*Hint:* Write each entry as the sum of one of 1, 2, 3, 4 and one of 0, 4, 8, 12).

| 13 | 2 | 7 | 12 |
|----|----|----|----|
| 3 | 16 | 9 | 6 |
| 8 | 11 | 14 | 1 |
| 10 | 5 | 4 | 15 |

36. Show that this magic square comes from a pair of orthogonal tables.

| 16 | 3 | 2 | 13 |
|----|----|----|----|
| 5 | 10 | 11 | 8 |
| 9 | 6 | 7 | 12 |
| 4 | 15 | 14 | 1 |

37. (a) What did Euler conjecture?
    (b) How much of his conjecture turned out to be true?
    (c) What did MacNeish conjecture?
    (d) What did he prove?

●

38. We describe another way of recording and thinking of tables. Consider the left table in (*6*) and insert a guide row and guide column. Then we have, for instance, $2 \circ 3 = 1$. Let us record this by writing 2 3 1 vertically:

$$\begin{array}{ll} 2 & \text{row} \\ 3 & \text{column} \\ 1 & \text{entry} \end{array}$$

Each of the 9 boxes will be recorded by one such vertical triplet. These 9 triplets are (their order being irrelevant):

$$\begin{array}{llllllllll} \text{row} & 1 & 1 & 1 & 2 & 2 & 2 & 3 & 3 & 3 \\ \text{column} & 1 & 2 & 3 & 1 & 2 & 3 & 1 & 2 & 3 \\ \text{entry} & 1 & 2 & 3 & 2 & 3 & 1 & 3 & 1 & 2 \end{array}$$

(a) In the 3 by 9 rectangle just formed, cover the numbers that record row. Why are the 9 pairs that show all different?

(b) Cover the numbers that record column. Why are the 9 pairs that show all different?

(c) Cover the numbers that record entry. Why are the 9 pairs that show all different?

39. Record the table in (*13*) by triplets in the manner described in E 38.

40. The bookkeeping technique introduced in E 38 can easily be extended to record a pair of tables. For instance, we can record the two tables in (*6*) by 9 vertical quadruplets that record row, column, entry in left table, and entry in right table. One of the quadruplets would be

$$\begin{array}{ll} 2 & \text{row} \\ 3 & \text{column} \\ 1 & \text{entry in left table} \\ 3 & \text{entry in right table} \end{array}$$

(a) Record the 9 quadruplets for the tables in (*6*) as a 4 by 9 rectangle.

(b) Cover the numbers in (a) that record row and column. Why are the 9 pairs remaining all different?

(c) Why is it that whenever you cover two horizontal lines in (a), the 9 pairs remaining are all different?

(d) How can you describe the concept "two orthogonal tables of order $N$" by an arrangement of the numbers $1, 2, \ldots, N$ in a single 4 by $N^2$ rectangle?

41. Can a 4 by 36 array of boxes be filled with the numbers 1, 2, 3, 4, 5, 6 in such a way that whenever you erase two of the horizontal lines there is no repetition among the 36 pairs that still remain? (The long side of the rectangle is horizontal.)

42. See E 40.

(a) Record the three tables in (*16*) by a single 5 by 16 rectangular arrangement.

(b) Why is it that whenever you cover 3 horizontal lines in the arrangement in (a), the 16 pairs showing are all different?

43. Show that a list of $k$ orthogonal tables of order $N$ is equivalent to a certain arrangement of the numbers $1, 2, \ldots, N$ in a $(k + 2)$ by $N^2$ rectangle.

INTRODUCTION TO E 44 THROUGH E 56

In the proof of Theorem 4 we used the fact that if we have 5 muffins, each labeled with 1, 2, 3, or 4, then at least two of the muffins must have the same label. This is an application of the "pigeonhole principle": if one has more letters than pigeonholes, then when one places the letters in the pigeonholes, at least one pigeonhole will contain more than one letter. The following exercises can be solved easily if the pigeonhole principle is employed; these exercises are otherwise unrelated to the chapter.

44. See the preceding introduction. Prove that in a crowd of 400 people at least two people celebrate their birthday the same day.

45. See the introduction preceding E 44.
    (a) Prove that if 12 natural numbers are given, there must be among them a pair with the same remainder when divided by 11.
    (b) Using (a), prove that there are two different natural numbers $M$ and $N$ such that
    $$2^M \equiv 2^N \pmod{11}.$$
    (c) Using (b), prove that there is a natural number $A$, bigger than 0, such that
    $$2^A \equiv 1 \pmod{11}.$$
    (d) Find an example of such an $A$.

46. (a) Using the ideas of E 45 prove the following

    THEOREM. *If $P$ is a prime (other than 2), then there is a natural number $A$, bigger than 1, such that*
    $$2^A \equiv 1 \pmod{P}.$$
    (b) Find an example of such an $A$ for $P = 5$, for $P = 7$, and for $P = 13$.

47. (a) Must the modulus in E 46 be prime?
    (b) Must the number being raised to a power be 2?
    (c) Generalize the theorem of E 46 as far as possible.

48. Prove that if five pins are stuck into a piece of cardboard cut in the form of an equilateral triangle of side 2, then at least two of the pins must be within a distance 1

of each other. (*Hint:* Break up the triangle into four smaller congruent triangles, and apply the pigeonhole principle.)

49. See E 48. Prove that if 17 pins are stuck into a piece of cardboard cut in the form of an equilateral triangle of side 2, then at least two of the pins are within a distance $\frac{1}{2}$ of each other.

50. (a) See E 48. Prove that if five pins are stuck into a piece of cardboard cut in the form of a 2 by 2 square, then at least two of the pins are within a distance of $\sqrt{2}$ of each other.

(b) Show that nine pins can be stuck into the square described in (a) without any two of the pins being closer than 1 to each other.

51. (a) See E 50. How close must some pair of pins be if 17 pins are stuck into a rectangle of dimensions 4 by 4?
    (b) Show that 25 pins can be stuck into the square described in (a) without any two of the pins being closer than 1 to each other.

52. (a) Exhibit 100 natural numbers, in the range from 1 to 200, none of which is a divisor of any of the other 99.
    (b) Prove that any list of 101 natural numbers, in the range from 1 to 200, must contain at least two natural numbers such that one of them divides the other. (*Hint:* Write every number in the list as an odd number times a power of 2, and apply the pigeonhole principle to these odd numbers.)

53. In Chapter 1 we studied arrangements of natural numbers and backward pairs. Let us define backward triplets, backward quadruplets, . . . and forward triplets, forward quadruplets, . . . . For example, a backward triplet (or 3-tuplet) is an arrangement $ABC$ of three different natural numbers with $A$ bigger than $B$ and $B$ bigger than $C$. Within the arrangement

$$3, 8, 4, 1, 5, 6, 2, 7, 9$$

we have, for example, the backward triplet 8 5 2, the forward 4-tuplet 3 4 5 6, the forward 6-tuplet 3 4 5 6 7 9, and so on. A natural number by itself is regarded as both a backward and a forward 1-tuplet.

Prove that any nonrepetitious arrangement of the first $N$ natural numbers that has no backward couplet must be a forward $N$-tuplet (pigeonhole principle not needed).

54. Exercise 53 shows that an arrangement that is not at all backward is completely forward. This is a special case of a more general theorem, which we now state.

THEOREM. *Let $B$ and $F$ be natural numbers bigger than 1, and let $N$ be a natural number bigger than the product $(B - 1) \times (F - 1)$. Then any nonrepetitious arrangement of the numbers $1, 2, 3, \ldots, N - 1, N$ that has no backward $B$-tuplet must have a forward $F$-tuplet.*

We outline a proof:

(a) Assume that there are no backward $B$-tuplets and no forward $F$-tuplets.
(b) For each natural number $m$ in the arrangement let $b_m$ be the number of natural numbers in a longest backward tuplet whose right member is $m$; let $f_m$ be the number of natural numbers in a longest forward tuplet whose left member is $m$.
(c) Show for each $m$ that $b_m$ is no more than $B - 1$ and no less than 1, and that $f_m$ is no more than $F - 1$ and no less than 1.
(d) Let us call $b_m$ the "given name of $m$" and $f_m$ the "family name of $m$." Prove that no two $m$'s have both the same given name and family name.
(e) Compare (c) and (d), and show that the pigeonhole principle has been violated. Why does this prove the theorem?

55. (a) Using the result of E 54, prove that any arrangement of 10 (or more) natural numbers must have a backward 4-tuplet or a forward 4-tuplet.
    (b) Devise an arrangement of the natural numbers 1, 2, 3, 4, 5, 6, 7, 8, and 9 that has no backward quadruplet or forward quadruplet.

56. Show that E 53 is a special case of E 54.

57. Prove that if a table satisfies the rule $X \circ (X \circ Y) = Y \circ X$, then it is orthogonal to the table obtained by rotating it about the axis of its main diagonal. Compare this with the relation between (*17*) and (*18*).

### References

1. *New York Times*, April 24, 1959, p. 1. (News story on the work of Bose, Parker, and Shrikhande.)

2. *A. Seidenberg, A simple proof of a theorem of Erdös and Szekeres, *Journal of the London Mathematical Society*, vol. 34, 1959, p. 352. (This is the basis of E 46 and E 47.)

3. M. Gardner, Mathematical games, *Scientific American*, November, 1959, pp. 181–188. (An article on the disproof of Euler's conjecture.)

4. W. W. Rouse Ball, *Mathematical Recreations and Essays*, Macmillan, New York, 1947. (Chapter 7 describes several other ways of constructing magic squares.)

5. R. A. Fisher, *The Design of Experiments*, Hafner, New York, 1951. (See p. 86, where orthogonal tables are used to find a defective bobbin in a cotton mill.)

6. *R. C. Bose, S. S. Shrikhande, and E. T. Parker, Further results in the construction of mutually orthogonal Latin squares and the falsity of Euler's conjecture, *Canadian Journal of Mathematics*, vol. 12, 1960, pp. 189–203.

7. *S. Chowla, P. Erdös, and E. G. Strauss, On the maximal number of pairwise orthogonal Latin squares of a given order, *ibid.*, pp. 204–208.

8. *M. Hall, Jr., Block designs, Chapter 13, pp. 369–405, *in* E. F. Beckenbach (ed.), *Applied Combinatorial Problems*, Wiley, New York, 1964.

## Chapter 14

# MAP COLORING

The squares of a checkerboard can be colored with only two colors in such a way that any two squares sharing an edge will be of different colors. As the reader may easily check, we can color this map of sixteen countries in two colors such that countries sharing an edge have different colors.

*(1)*

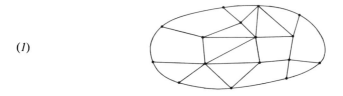

But we cannot do so with this map:

*(2)*

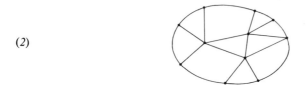

The reader may make some experiments of his own and discover for himself a simple rule for checking whether a map can be colored in two colors.

For the sake of brevity, we will say that "a map can be colored in two colors" or that "a map has a two-coloring" if we can color the countries with only two colors in such a way that countries sharing at least one edge have different colors. Similarly, later in the chapter, we will speak of "coloring a map in three colors" or of "a five-coloring." We will always mean, whatever the coloring, that countries sharing an edge have different colors.

Let us think of (*1*) and (*2*) as maps of islands divided into countries. *All vertices not on the shore of the island we shall call* **dry vertices.** A map of an island that has a dry vertex of odd degree cannot be colored in two colors, since the colors must alternate as we run through the countries meeting at a dry vertex. (Recall that the degree of a vertex is the number of edges that meet at that vertex.) We thus have

THEOREM 1. *If the countries of an island can be colored in two colors, then each dry vertex is of even degree.*

We would hope that the opposite (or converse) would also be true; namely, that if each dry vertex were of even degree, then the countries of the island could be colored in two colors. As an example, consider this map, in which each dry vertex is even:

(3)

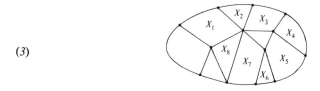

If we color one of the countries red (say, country $X_1$), one of its neighbors black (say, country $X_2$), and continue to alternate colors along a typical journey (say $X_1$, $X_2$, $X_3$, $X_4$, $X_5$, $X_6$, $X_7$, $X_8$) until we return to a neighbor of $X_1$, we find that $X_1$ and $X_8$ fortunately have different colors. Though we worried only about coloring each country differently from its predecessor at each step, we encountered no trouble when we returned to the first country. That we never will run into this trouble is guaranteed by

THEOREM 2. *If each dry vertex is of even degree, then the countries of an island can be colored in two colors.*

PROOF. Stick a tack into each country. Select one of the tacks, and call it Rome. Consider the (typical) tack $T$ indicated below:

(4)

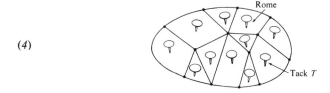

To simplify our reasoning it will be convenient to place Rome such that no straight line from Rome to any other tack crosses a vertex.

The reader, traveling on the island from tack $T$ to Rome, and passing over no vertex, will notice that each such trip crosses an even number of edges.

(Of course, for another tack the number of crossings might be odd.) For example, this trip has ten crossings,

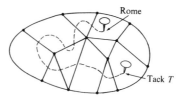

whereas the most direct trip has two crossings:

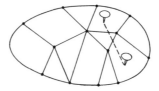

To see why each trip from $T$ to Rome must have an even number of crossings, let us consider such a trip and trace out its course with a taut elastic, taping the latter to the map along the way (we suppose that one end of the elastic is tied to $T$; and the other end, to Rome). Now we remove the tape from the elastic; when we do this, the elastic contracts until it forms a straight line joining $T$ and Rome.

Watch what happens to the number of crossings as the shrinking elastic passes a vertex:

 (5)

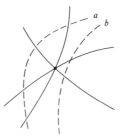

In particular, consider the number of crossings in the vicinity of the vertex shown in (5) as the elastic shrinks from position (a) to position (b). In position (a) there are four crossings; in position (b), after the elastic passes the vertex, there are two crossings (a change by an even number). Why will this change always be even at each vertex? The answer is simple: the number of crossings near the vertex in position (a) plus the number of crossings near the vertex in position (b) equals the degree of the vertex, which is even. (If two integers have an even sum, then their difference must also be even, as the reader may easily check.)

Thus, as the contracting elastic passes a vertex, the number of crossings changes by an even number. Furthermore, as the contracting elastic frees itself of a border in this manner,

the change is again even.

Finally, the elastic forms the direct straight line trip from *T* to Rome. Since this direct trip has an even number of crossings, so must the original trip, which we had turned into rubber.

Thus any trip from *T* to Rome has an even number of crossings. *We will call any tack with this property* **even.** In particular, Rome itself is even (since we can travel from Rome to Rome on a path that crosses no edge at all). On the other hand, consider a tack *T'* whose direct trip to Rome has an odd number of crossings. Then, in view of the above discussion, it is clear that any trip from *T'* to Rome must also have an odd number of crossings. *We will call any tack with this property* **odd.**

Now observe that if two tacks are adjacent in the sense that we can travel from one to the other with just one crossing, then, clearly, one of them is odd and the other even. It follows that we can obtain a two-coloring of the island by coloring the countries with even tacks red and the countries with odd tacks blue. Theorem 2 is proved.

In order to avoid the necessity of distinguishing dry vertices from shoreline vertices we shall from now on insist that the ocean also be colored. That is, we will treat the ocean as just another country. We can think of the ocean, together with all the countries, as covering the surface of the earth.

Figure (*1*), with the ocean included, cannot be colored in two colors (since some of the dots on the shore are of odd degree). Figure (*6*), however, can be colored in two colors:

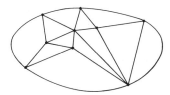

The same technique that we used to prove Theorem 2 provides a proof of

THEOREM 3. *If each vertex is of even degree, then the ocean and the countries on an island can be colored in two colors.*

Theorem 3 is more elegant than Theorem 2 because it does not have to distinguish the dry vertices from the wet. However, it has seemingly complicated matters by distinguishing between the ocean and the countries on the island. But this distinction is quite artificial. After all, if we were to draw our map on a ball instead of on a flat paper, the ocean would look just like another country. For the sake of objectivity and clarity let us remove these geographical overtones and refer to either the countries or the ocean as **regions**. With this agreement we may rephrase Theorem 3 as

THEOREM 4. *If the surface of a sphere is divided into regions, and if each vertex is of even degree, then these regions can be colored in two colors.*

We have completely solved the two-color problem. That is to say, we have discovered a simple test for determining when a map can be colored in only two colors: simply check that all the vertices are of even degree.

From Theorem 4 we can obtain (for later use) this

LEMMA. *If the surface of a sphere is divided into regions such that each region has an even number of edges, then the vertices can be colored red or blue in such a way that the two vertices of each edge have different colors.*

PROOF. Consider, for example, such a map as this one, which represents eleven regions covering the sphere (on the sphere the borders would, of course, be curved):

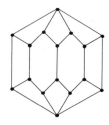

Each region (including the one outside the figure) has an even number of edges. We obtain from this map a new map on the sphere by placing a dot in each of the eleven regions and joining two dots if their regions share an edge (if two regions share $k$ edges, we join the corresponding dots by $k$ lines):

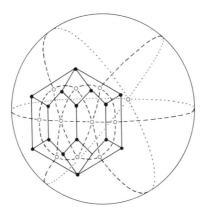

(For the sake of clarity the second map is drawn in perspective as it appears on a sphere.)

The number of vertices in the new map formed of broken lines is the same as the number of regions in the original map. Each vertex of this new map has even degree because each region of the old map has an even number of edges.

Theorem 4 then tells us that the new map can be colored in two colors. This indirectly provides a coloring of the vertices of the original map, of the type promised in the lemma. A similar argument proves the lemma in general.

Let us turn next to the problem of determining which maps can be colored in three colors. The results in this direction are quite incomplete. In 1941, R. L. Brooks proved

THEOREM 5. *If the surface of a sphere is divided into at least five regions, each of which shares its borders with three neighboring regions, then the regions can be colored in three colors or less.*

Though we will not prove Theorem 5, we suggest that the reader check it on some examples, including the following:

(7)

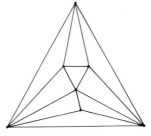

(Be sure to color the outside region, too.)

Another three-color result, found by Kempe in 1880, is

THEOREM 6. *If the surface of a sphere is divided into regions, each of which has an even number of edges, and if three regions meet at each vertex, then this map can be colored with exactly three colors.*

An illustration of Theorem 6 is Figure (*8*), which, as the reader may check, can be colored with three colors.

(8)

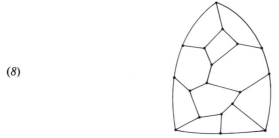

Note that the outside region also has an even number of edges. It is to be colored along with all the other regions.

If we start to color (*8*) by coloring one of the regions, then we must alternate the two other colors as we run through the regions that border the original region. Since there are an even number of regions bordering that region, we have no trouble. If we were to color a region with the color $A$, then we would color the regions adjacent to this region $B$ and $C$, as in this diagram:

(9)

$$\begin{array}{ccc} C & | & B \\ B & A & C \\ C & | & B \end{array}$$

But being able to color in the vicinity of each region does not, of course, assure us that we can color the whole map with three colors.

Though we will not prove Theorem 6 in detail, we will describe one technique for coloring such maps. The first step consists in placing at each vertex either a

clockwise circle ↻ or a counterclockwise circle ↺ in such a way that

the two vertices at the ends of each edge are furnished with one clockwise and one counterclockwise circle. (That this can be done is guaranteed by the lemma proved after Theorem 4.) Doing this for figure (*8*), we obtain

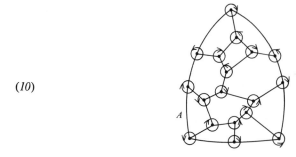

(10)

Next, using the colors *A*, *B*, and *C*, we shall color the edges in the following manner. First we choose an edge and color it *A* [as in Figure (*10*)]. The two edges that meet this edge at one of its ends we color *B* and *C* in such a way that the direction of turning (starting with the edge colored *A*, then the edge colored *B*, then the edge colored *C*) will be the same as the direction given by the little circle where the three edges meet. For example if we have

then the edges are to be colored

If we have

then the three edges are to be colored

As we travel along the edges we continue coloring uncolored edges such that the turning direction *A*, then *B*, and then *C* always agrees with the little circle. We get this coloring for the edges of (*8*):

(11)

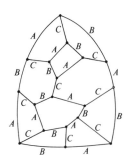

The reader may check that at each vertex the direction given by *A*, *B*, *C* is the same as that given by the arrowed circle of (*10*). Note that the border of each region (even the outside one) has only two colors. Now we color each region with the color that does not appear on its border. Neighboring regions will, of course, receive different colors. Thus the map can be colored in three colors.

Theorems 5 and 6 cover most of what is known about the three-color problem. No one has discovered a general rule for determining which maps can be colored in three colors.

The problem of coloring maps with four colors goes back to October 23, 1852, as K. O. May has established. On that day Francis Guthrie posed it to his teacher, De Morgan, who wrote W. R. Hamilton,

> A student of mine asked me today to give him a reason for a fact which I did not know was a fact, and do not yet. He says that if a figure be anyhow divided, and the compartments differently coloured, so that figures with any portion of common boundary line are differently coloured—four colors may be wanted, but no more. . . .
>
> What do you say? And has it, if true, been noticed? My pupil says he guessed it in colouring a map of England. The more I think of it, the more evident it seems.

In 1879 Kempe published an erroneous proof in a paper that contained these remarks:

> Some inkling of the nature of the difficulty of the question, unless its weak point be discovered and attacked, may be derived from the fact that a very small alteration in one part of a map may render it necessary to recolor it throughout. After a somewhat arduous search, I have succeeded, suddenly, as might be expected, in hitting upon the weak point, which proved an easy one to attack. The result is, that the experience of the map makers has not deceived them, the maps they had to deal with, viz: those drawn on a [sphere] can in every case be painted with four colors.

The flaw in Kempe's proof was exposed in 1890 in a paper by P. J. Heawood, which began:

> The Descriptive-Geometry Theorem that any map whatsoever can have its divisions properly distinguished by the use of but four colors, from its generality and intangibility, seems to have aroused a good deal of interest a few years ago when the rigorous proof of it appeared to be difficult if not impossible, though no case of failure could be found. The present article does not profess to give a proof of this original Theorem; in fact its aims are so far rather destructive than constructive, for it will be shown that there is a defect in the now apparently recognized proof. . . .

In the same paper Heawood showed that Kempe's technique could be used to prove that every map on a sphere can be colored with five (or fewer) colors. We shall divide this proof into several lemmas.

Before we begin, let us point out three assumptions that have so far been taken for granted. The first assumption is that the border of a region can be thought of as a single piece of string in the form of a loop that does not meet or cross itself. Thus a region looks like this:

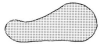

Note that the following figures are *not* regions:

(*12*)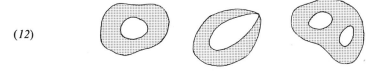

(Nor is the whole surface of the sphere itself a region.) Without this assumption, Lemmas 2 and 3, which follow, would not always hold. We will also assume that an edge in a map has two different ends (that is, does not close up to form a loop). Thus the border of a region is divided into at least two edges, and hence has at least two vertices. Furthermore, we will assume that any map in question has a finite number of regions.

As we did for the proof of the Fundamental Theorem of Arithmetic, we now supply a diagram showing how the six lemmas leading up to the five-color theorem fit together. All these lemmas refer to maps covering a sphere.

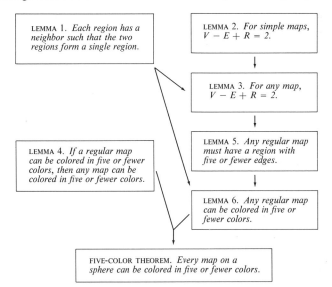

In any map on a sphere, let $V$ be the number of vertices, $E$ the number of edges, and $R$ the number of regions. A list of $V$, $E$, and $R$ for some of the maps we have met in this chapter is given in this table (which the reader is invited to check):

|  | MAP | $V$ | $E$ | $R$ |
|---|---|---|---|---|
|  | (1) | 16 | 31 | 17 |
| (13) | (2) | 11 | 20 | 11 |
|  | (3) | 13 | 23 | 12 |
|  | (6) | 9 | 20 | 13 |

Inspection of this table shows that, in each case,

$$V - E + R = 2.$$

That is, the sum of the number of vertices and the number of regions is two more than the number of edges. That $V - E + R = 2$ is true for any map covering a sphere was proved first by Descartes in 1640 and was later proved independently by Euler in 1752. Before we can prove Descartes' theorem, we will need

LEMMA 1. *In any map (with more than two regions) that covers a sphere, each region has a neighbor such that their common border consists of a single arc. (In other words, each region has a neighbor such that the two regions together form a region.)*

PROOF. Observe that if two of the regions, say $X$ and $Y$, meet in more than just a single arc, then $X$ and $Y$ must surround at least one other region. This is illustrated by the following examples:

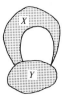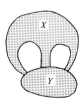

With this observation in mind, let $X_1$ be any one of the regions of the map. Suppose that, contrary to the assertion of the lemma, $X_1$ has no neighbor $Y$ such that their common border consists of a single arc.

Let $X_2$ be a region that meets $X_1$ along at least one arc. Then, applying the observations made immediately above, we see that $X_2$ meets $X_1$ along more than just one arc, and, consequently, $X_1$ and $X_2$ surround at least one other region. Among such regions there must be at least one that meets $X_1$ along at least one arc; pick one such region, and call it $X_3$. Arguing just as before, we can show that $X_1$ and $X_3$ surround at least one region that meets $X_1$ along at least one arc; pick one such region, and call it $X_4$. If we continue in this manner, we will, of course, obtain an unending list

$$X_1, X_2, X_3, X_4, \ldots$$

of distinct regions. But since there is only a limited supply of regions on the map, we have arrived at a contradiction, which proves Lemma 1.

As indicated by the diagram that shows the order of the lemmas, we will need Lemma 1 twice.

LEMMA 2. *If a map has just two regions, then* $V - E + R = 2$.

PROOF. We may think of the two regions as two hemispheres meeting on the equator. On the equator, there appear exactly as many vertices (all of degree 2) as edges ($V$ spokes divide a rim into $V$ edges), thus $V = E$. Since $R = 2$, we have $V - E + R = 2$.

Now it will be easy to prove

LEMMA 3. *For any map covering the sphere,* $V - E + R = 2$.

PROOF. Choose any map on the sphere, and let $K$ denote the number $V - E + R$. We wish to prove that $K = 2$. To do so we are going to change the given map into a map with only two regions by repeatedly

(1) deleting any vertex of degree 2,
(2) deleting the edge shared by two regions that, together, form a region.

If we delete a vertex of degree 2, we obtain a map having the same regions, but with two short edges replaced by one longer edge. (The reader should provide a picture for this.) Let $V'$, $E'$, and $R'$ be the number of vertices, edges, and regions in this new map. We have

$$V' = V - 1,$$

and (since two edges merge into one when a dot is removed)

$$E' = E - 1.$$

Since there is no change in the number of regions,

$$R' = R.$$

Thus

$$V' - E' + R' = (V - 1) - (E - 1) + R.$$

But

$$(V - 1) - (E - 1) + R = V - E + R = K,$$

hence

$$V' - E' + R' = K.$$

Now let us see what happens if, instead, we delete an edge (but not its two ends). In the new map, $V' = V$, $E' = E - 1$, and $R' = R - 1$, as the reader may check. Thus, we again have

$$V' - E' + R' = K.$$

By repeated applications of the procedures (1) and (2), we can go from the original map to a map with just two regions. The preceding remarks show that, after each step,

*(14)* (number of vertices) − (number of edges) + (number of regions) = $K$.

In particular, *(14)* holds for the final map, which has just two regions. Thus by Lemma 2, $K = 2$, and Lemma 3 is proved.

We will need Lemma 3 in the proof of Lemma 5.

LEMMA 4. *If we can color in five or fewer colors every map on the sphere whose vertices are all of degree 3, then we can color any map on the sphere in five or fewer colors.*

PROOF. Place a small region (or "patch") over each vertex whose degree is more than 3. This maneuver replaces a vertex such as

by several vertices of degree 3:

Since each vertex in the new map is of degree 3, then by our assumption the map can be colored in five or fewer colors. If we remove the patches and color the little "pieces of pie" thus exposed the same as the regions to which they belong, we will obtain a coloring of the original map in five or fewer colors.

Maps having every vertex of degree 3 will play an important role in the following lemmas. *For convenience we name them* **regular maps.**

LEMMA 5. *A regular map on the sphere must have at least one region with five or fewer edges.*

PROOF. We shall prove this by supposing that each region has six or more edges and obtaining a contradiction.

Let us consider any vertex; we place a little mark close to it on each of the three edges that meet at that vertex. We do this for each vertex. All told, we make $3V$ marks on the edges. But each edge will have two such marks, one near each of its two ends. Thus $3V = 2E$. For later convenience we write this in the equivalent form

$$(15) \qquad\qquad 6V = 4E.$$

Next, let us consider the regions. Within each region we place a pebble near each of its edges. Since each region has six or more edges, there are six or more pebbles inside each region. Now, we have $R$ regions, each with six or more pebbles inside it; hence

$(16)$ the total number of pebbles is at least $6R$.

On the other hand, there are two pebbles near each of the $E$ edges (one on each side); hence

$(17)$ the total number of pebbles $= 2E$.

Taken together, $(16)$ and $(17)$ tell us that

$(18)$ $2E$ is at least $6R$.

Now recall that, by Lemma 3, $V - E + R = 2$, or equivalently,

$$(19) \qquad\qquad 6V - 6E + 6R = 12.$$

By equation $(15)$, we may substitute $4E$ for $6V$ in $(19)$, obtaining

$$(20) \qquad\qquad 4E - 6E + 6R = 12,$$

or

$$(21) \qquad\qquad 6R - 2E = 12.$$

Since, from $(18)$, we have that $2E$ is at least $6R$, the left side of $(21)$ is either 0 or a negative number, and therefore certainly cannot be as large as 12. Thus we have a contradiction.

With the aid of Lemma 5 we can easily prove

LEMMA 6. *Any regular map covering a sphere can be colored in five (or fewer) colors.*

PROOF. Clearly, if a map has at most five regions, then it can be colored with five or fewer colors. Now, let $R$ be a natural number, and let us assume that any regular map with at most $R$ regions can be colored with five colors.

On the basis of this assumption we shall show that any regular map with $R + 1$ regions can be colored with five colors.

Let us consider a regular map with $R + 1$ regions. By Lemma 5 there is at least one region present with five or fewer edges. We shall divide the argument into two cases:

*Case 1: There is a region X having four or fewer edges.* By Lemma 1, there is a region $Y$ bordering $X$ such that $X$ and $Y$ together form a region. If we erase the border between $X$ and $Y$, together with its end vertices, we obtain a new regular map with only $R$ regions. (The reader should draw a picture of this.) By our assumption, this map with only $R$ regions can be colored in five colors. If we now replace the border we find that $X$ and $Y$ have the same color. But we can erase the color of $X$, and give $X$ a color that differs from any of its neighbors' colors. (Since $X$ has at most four neighbors, and we have five colors available, this is possible.)

*Case 2: There is a region X having exactly five edges.* This time, we do not have enough colors to use the same argument as in Case 1. But as the reader may check, $X$ has two neighboring regions $Y$ and $Z$ such that $Y$ and $Z$ do not meet and such that $X$, $Y$, and $Z$ together form a region. For example:

If we delete the border between $X$ and $Y$, the border between $X$ and $Z$, and also the end vertices of these borders, we obtain a regular map with $R - 1$ regions (two less than the original $R + 1$ regions). By assumption, this map with $R - 1$ regions is colorable with at most five colors (since $R - 1$ is less than $R$).

Next we put the deleted borders (and vertices) back. When we do so, we see that the regions $X$, $Y$, and $Z$ all have the same color. Since $Y$ and $Z$ do not meet, we do not need to change their color. But the color of $X$ must be changed. Since $Y$ and $Z$ have the same color, no more than four different colors will appear in the regions bordering $X$. But since we have five colors available, we can select one to give $X$ a color different from those of its neighbors. This completes the argument for Case 2.

Since we can color any map that has five or fewer regions with five or fewer colors, the argument just given (for $R = 5$) tells us that we can color any regular map that has six regions. Applying the argument again (with $R = 6$), we see that we can color any regular map that has seven regions with five or fewer colors. Finally, using this argument as many times as necessary, we see that any regular map can be colored with five or fewer colors, which proves Lemma 6.

Combining Lemmas 4 and 6, we obtain

THE FIVE-COLOR THEOREM: *Every map on a sphere can be colored with five or fewer colors.*

In 1940 Winn proved that any map with 35 or fewer regions can be colored with four or fewer colors; in 1968 Ore and Stempel announced that they had raised the number from 35 to 39. No one knows whether there is a map on a sphere that requires five colors.

If we consider maps on a doughnut or inner tube, instead of maps on a sphere, then the coloring problem is completely solved. Heawood proved that any map on an inner tube can be colored with seven or fewer colors. Moreover, there are maps on an inner tube that require seven colors (see E 41). That the coloring problem on an inner tube is easier than that on a sphere seemed odd to Heawood, who remarks in his introduction (which we interrupted in order to prove the five-color theorem):

> . . . [my] main object is to contrast with that simple proposition [the four-color conjecture] some remarkable generalizations of it, of which strangely the rigorous proof is much easier.

In spite of the work of many mathematicians since 1852, the original question: "Can every map on a sphere be colored in four colors?" still remains unanswered.

## Exercises

1. Give an example of a map of an island that requires four colors.

2. (a) Draw a map of an island such that each dry vertex has even degree and each wet one has odd degree.
   (b) Color the countries on your island in two colors.
   (c) Check that there are an even number of edges along the coast line.
   (d) Using Theorem 2, prove that the number of edges in (c) must be even.

3. What is the similarity between the proof of Lemma 1 of the present chapter and the proof of Theorem 8 of Chapter 7?

4. If a map on a sphere consists only of triangles, then, when it is drawn on a flat piece of paper, exactly three countries have a coastline.

    (a) Draw a map in which each country shares its border with three neighbors (we take the ocean as a neighbor) and in which the ocean borders three countries.

    (b) Color it in three colors.

    (c) Does this map agree with Theorem 5?

5. See E 4. Draw a map of your own that illustrates Theorem 5.

6. In the map below, each region (including the ocean) has an even number of edges.

    (a) Prove that this map cannot be colored in three colors.

    (b) Does this map contradict Theorem 6?

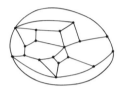

7. (a) Draw a map in which every vertex is of degree 3 and in which each region (including the ocean) has an even number of edges.

    (b) Place at each vertex a clockwise or counterclockwise circle, alternating the circles as in the informal proof of Theorem 6.

    (c) Using these circles, color the edges in three colors.

    (d) From the coloring of the edges, obtain a coloring of the map in three colors.

8. Do E 7(a), (b), (c), and (d) for still another map of your choice.

9. Draw three different maps, and check for each that $V - E + R = 2$. (*Hints:* Be sure to count the ocean as a region; to help avoid errors, number each vertex, edge, or region as you count it.)

10. In the proof of Lemma 3 we indicated that any map on a sphere can be gradually reduced to a map with just two regions by using steps (1) and (2).

    (a) Draw a map with seven regions (on a flat piece of paper the ocean represents a region).

    (b) Reduce this map to one with just two regions by repeatedly applying steps (1) and (2).

11. The same as E 10 for a map with ten regions (again count the ocean as a region).

12. Consider a regular map covering a sphere.

    (a) What happens to $V$, $E$, and $R$ when you remove from this map the border between two regions together with both vertices of degree three at the ends of this border?

    (b) What happens to $V - E + R$ in that case?

13. Does Lemma 4 remain true if "five or fewer" is replaced by "four or fewer"? Prove your answer.

14. A map covering a sphere has 200 regions, each with five edges. What can we say about the number of vertices?

15. At what point does the type of argument that proved the five-color theorem break down in proving the four-color theorem?

16. Can a sphere be wrapped in a net in such a way that each hole is a hexagon and three hexagons meet at each knot?

17. On a sphere are placed 100 dots. There are many maps, all of whose regions have three edges and whose vertices are the given 100 dots. Prove that any two such maps with 100 vertices have the same number of regions.

18. Explain why all the maps in E 21 of Chapter 2 have the same number of triangles. (See E 17.) (*Hint:* Take two (elastic) copies of the big triangle, and glue these copies together along their borders. Stretch the resulting figure until it becomes a sphere, and compare with E 17.)

19. Why did we introduce the concept of a regular map in the proof of the five-color theorem?

20. Consider a regular map covering a sphere. Let $R_2$ be the number of regions with two edges, $R_3$ the number of regions with three edges, and so on.
    (a) Prove that $R_2 + R_3 + R_4 + \cdots = R$.
    (b) Prove that $2R_2 + 3R_3 + 4R_4 + \cdots = 2E$.
    (c) Combining (a), (b), and the equations $3V = 2E$ and $V - E + R = 2$, prove that
    $$4R_2 + 3R_3 + 2R_4 + R_5 = 12 + R_7 + 2R_8 + 3R_9 + \cdots.$$
    (d) From (c) obtain another proof of Lemma 5.

21. See E 20(c). If a regular map covering a sphere has no regions with less than five edges, what can we say about the number of five-edged regions in it?

22. See E 20(c).
    (a) Fill in the blank with the largest number for which the following statement is always true: Any regular map covering a sphere has at least ____ regions that have five or fewer edges.
    (b) Give an example of a map that has as few such regions as claimed in (a).

23. See E 20(c). A regular map covering a sphere has two regions with seven edges and three regions with eight edges but no regions with more than eight edges. This map must have at least how many regions with five or fewer edges?

24. See E 20(c). A regular map has four regions with four edges and two regions with three edges, but no regions with two or five edges.
    (a) How many regions in this map have nine edges?
    (b) What can you say about the number of regions that have seven or eight edges?

25. See E 20(c). Assume that the four-color conjecture is false. Consider a regular map covering a sphere with the smallest number of regions such that the map cannot be colored in four colors.
    (a) Why are $R_2$, $R_3$, and $R_4$ all 0 for this map?
    (b) What can be said about the number of five-edged regions in this map?

26. Twelve wires are put together to form the edges of a cube. A light bulb is put inside this cube, and the cube is put inside a sphere. The shadows of the wires can be thought of as edges of a map on the sphere. Each of the six faces of the cube

corresponds to a region. Thus any theorem we have proved about coloring maps on the surface of the sphere will apply to coloring each of the faces of the cube such that faces meeting along a wire have different colors.

   (a) Color the faces of the cube in as few colors as you can (such that adjacent faces have different colors).

   (b) Show that the cube illustrates Theorem 6.

27. See E 26. A tetrahedron is a pyramid having four triangular faces:

   (a) Color the faces of the tetrahedron with as few colors as possible (such that adjacent faces have different colors).

   (b) Show that the tetrahedron illustrates the need for the assumption in Theorem 5 that there are at least five regions.

28. See E 26. An octahedron has eight triangular faces with four meeting at each vertex:

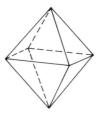

   (a) Color the faces of the octahedron with as few colors as possible.

   (b) Show that the octahedron illustrates Theorem 3.

29. Cut out of paper twelve circles all of the same diameter (at least three inches). Inscribe in each a regular pentagon like this:

Fold up the five flaps on each circle, and glue the twelve pieces of paper along the flaps in such a way that they form a surface in which three pentagons meet at each vertex (the flaps can show, or can be glued inside, as you prefer). This surface is called a dodecahedron.

   (a) Prove that the dodecahedron cannot be colored in two colors.

   (b) Prove that it cannot be colored in three colors.

   (c) Color it in four colors.

30. The instructions are similar to those of E 29. Cut out twenty circles all of the same diameter, and inscribe in each an equilateral triangle. Then fold up the flaps, and glue the twenty pieces of paper together such that five triangles meet at each vertex. The resulting twenty-sided figure is called an icosahedron.
    (a) Prove that the twenty triangles cannot be colored in two colors.
    (b) What theorem assures us that we can color them in three colors?
    (c) Color them in three colors.

31. Consider a map covering the sphere such that all vertices have the same degree, $b$, and all the regions have the same number of edges, $a$.
    (a) Show that $bV = 2E = aR$.
    (b) From (a) and the equation, $V - E + R = 2$, deduce that

$$\left(\frac{2}{a} + \frac{2}{b} - 1\right) E = 2.$$

32. See E 31(b). Let $a$ and $b$ be positive integers, each not less than 3. What possibilities are there of such $a$ and $b$ if

$$\frac{1}{a} + \frac{1}{b} \text{ is larger than } \frac{1}{2}?$$

33. The surfaces described in E 26–30 are *regular polyhedra*. A *regular polyhedron* is a polyhedron whose faces all have the same number of edges (at least three) and whose vertices all have the same degree (at least three). Prove that there are only five regular polyhedra, a fact known to the geometers of Euclid's time. (*Hint:* See E 31 and E 32.)

34. Given a polyhedron, we may obtain from it another polyhedron by using the centers of the faces of the given one as the vertices of the new polyhedron. What is the new polyhedron if the given one is: (a) A cube? (b) An octahedron? (c) A dodecahedron (12 faces)? (d) An icosahedron (20 faces)? (e) A tetrahedron? (f) A regular polyhedron?

35. Draw a map consisting of several regions, one of which, contrary to our agreement, has a hole in it. Color this map with as few colors as possible.

36. See E 35.
    (a) Using the five-color theorem, prove that even if we allow one region to have holes (such that it surrounds islands made up of other regions), we can still color any such map on the sphere with five colors. (*Hint:* Show that a region with a hole creates two "simpler" coloring problems: one for the regions in the hole, and one for the regions outside.)
    (b) The same as (a), but allow two regions to have holes.

●

37. Prove Theorem 6, the three-color theorem, modeling your proof on that of the five-color theorem.

38. On the border of a disk select an even number, $e$, of points. Draw $e/2$ nonoverlapping curves in the disk whose ends are the $e$ dots. For instance, in the case $e = 10$ we may have

The curves cut the disk into $e + 1$ regions. Prove that they can be colored in two colors.

39. Consider a regular map covering a sphere. The edges and vertices form a highway system of the type discussed in Chapter 9. Show that if in this system a salesman has a route that takes him once through each vertex and then back to his starting position, then the map can be colored in four colors. (*Hint:* Note that his route cuts the surface into two pieces, each of which can be distorted into a disk of the type shown in E 38.)

40. The dressmaker's pattern for an inner tube is

If the edges marked $c$ are sewn together such that $A$ and $D$ coincide and $B$ and $C$ coincide, we obtain a pipe. Sewing the edges marked $d$ together joins the ends of the pipe to form an inner tube. This pattern enables us to represent an inner tube on a flat piece of paper.

Check that this diagram, drawn on the dressmaker's pattern for an inner tube, represents seven regions on the tube, each of which shares a border with the other six regions. (This map is due to P. Ungar, 1953.)

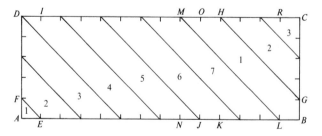

For example, region 1 borders region 2 on $EF$ and $GH$, 3 on $LB = RC$, 4 on $AE = DI$, 5 on $MO = NJ$, 6 on $JK = OH$, and 7 on $LM$. If this map is drawn on an inner tube, then each region will twist like this:

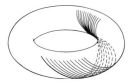

41. (a) Why does the map of E 40 require seven colors?
    (b) Draw the map of E 40 on an inflated inner tube. First draw a circle on the inner tube, as indicated below, and then mark off on this circle fourteen regularly spaced divisions to serve as a guide.

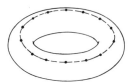

42. Show that in the map of E 40, $V - E + R = 0$.

43. Assume that for any map covering the inner tube we have $V - E + R = 0$. Following our reasoning for Lemma 5 as closely as possible, prove that any map covering the inner tube and having each vertex of degree 3 must have a region with six or fewer edges.

44. See E 43. Prove that any map covering the inner tube and having each vertex of degree 3 must have a region with at least six edges.

45. Study this question: If each vertex of a map covering the surface of an inner tube has even degree, does the map have a two-coloring?

46. If a hole is made in a (very elastic) inner tube, then it can be turned inside out:

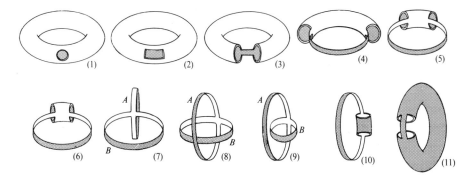

From (1) to (7) we shrink the tube. To go from (7) to (8) we turn band *A* inside out without moving any part of band *B* (this is easy to do if you make a paper model, or simply use a belt, to represent *A*). From (8) to (9) we shrink band *B*. From (9) to (11) we stretch the tube. Comparison of (11) and (3) shows that we have turned the inner tube inside out. (This exercise is not related to coloring problems.)

Make (7) with paper and glue, and use it to demonstrate the crucial step from (7) to (8).

47. Show that if a map covers the surface of a sphere and has no vertices of degree 2, then it must have at least one region with five or fewer edges. (See E 16.)

48. Let us assume that all the edges of the maps satisfying the hypothesis of Theorem 3 are straight.
   (a) Place a lighthouse in each country and in the ocean in such a way that no lighthouse lies on the line determined by any edge.
   (b) Assume that the rotating beam from each lighthouse always has one ray of light.
   (c) Prove that, for a given lighthouse, the rotating beam always crosses an even number of edges or else is always directly above an odd number of edges. Disregard times when the beam passes over a vertex.
   (d) How can such lighthouses be used to give another proof of Theorem 3 (at least when the edges are straight)?

49. As in Chapter 2, dots are drawn on the surface of a sphere and are then used to divide the surface into triangles (with curved sides, of course). Prove this

   THEOREM. *If each dot is of even degree, then the dots can be labeled with the letters A, B, and C such that each of the triangles has all three of the letters.*

● ●

50. The regions of a regular map that covers a sphere are painted, at random, red, blue, or green. In such a painting is it possible that there is exactly one vertex at which all three colors appear?

51. In the year 1880 Tait offered what he called a "very simple solution of the problem of map-coloring with four colors." We describe Tait's technique for coloring regular maps in (a) and (b) below:
   (a) Label the edges 1, 2, and 3 in such a way that the three edges meeting at each vertex have all three numbers.
   (b) First color one of the regions with the color $A$, and then color the remaining regions with colors $A$, $B$, $C$, and $D$ in accordance with the following rules: Any edge labeled 1 is to separate regions colored either $A$ and $B$ or $C$ and $D$; any edge labeled 2 is to separate regions colored either $A$ and $C$ or $B$ and $D$; and any edge labeled 3 is to separate regions colored either $A$ and $D$ or $B$ and $C$.
   (c) Draw a map in which each vertex is of degree 3. Does Tait's technique work for your map (including the ocean)? (See E 52.)
   (d) Prove that if Tait's technique always works, then any map on the sphere (regular or not) can be colored in four or fewer colors.

52. See E 51. Regrettably, Tait's proof that the edges can always be numbered as he required in E 51(a) was false. Prove that if a regular map can be colored with the colors $A$, $B$, $C$, and $D$, then the edges can be numbered 1, 2, and 3 in such a way

that at each vertex all three numbers are represented. (*Hint:* Use the rule of E 51(b) in the reverse direction.) Comparing E 51 and E 52, show that the problem of coloring the edges of a regular map, as in E 51(a), is equivalent to the four-color problem.

53. A flat island has one hundred towns. Some pairs of towns are joined directly by roads; we call such towns "adjacent." Two or more roads may meet at a town, but none cross outside of towns. Prove that it is possible for each town to offer exactly one of these five services: hospital, library, radio station, newspaper, and school, in such a way that no two adjacent towns offer the same service.

54. Kempe, in the same 1879 paper that contains his imperfect proof of the four-color theorem, observed:

> If we lay a sheet of tracing paper over a map and mark a point on it over each district and connect the points corresponding to districts which have a common boundary, we have on the tracing paper a diagram of a "linkage," and we have as the exact analogue of the question we have been considering, that of lettering the points in the linkage with as few letters as possible, so that no two directly connected points shall be lettered with the same letter.

In our terminology, a linkage is a system of towns and highways. Let us even allow a linkage to have bridges, in order that we can connect as many towns as we wish directly to each other. (Kempe's linkages have no bridges.) Draw a linkage of eight towns, of which any two are adjacent. Clearly this linkage could not be "colored" (that is, lettered) in fewer than eight colors.

55. See E 54. An "$n$-cage" in a linkage consists of $n$ towns, named $A_1, A_2, \ldots, A_n$, and paths in the linkage between each two of the $n$ towns such that no two paths meet except at the $n$ towns. (That the paths might pass through other towns in the linkage does not concern us.) Thus (a) is a 2-cage (with 1 path), (b) is a 3-cage (with 2 paths), (c) is a 4-cage (with 6 paths), and (d) is a 5-cage (with 10 paths):

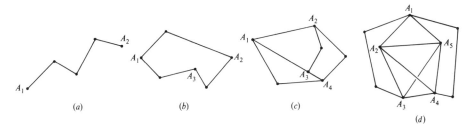

(Note that in (d), because of the bridge, the path from $A_3$ to $A_5$ does not meet the path from $A_2$ to $A_4$.) Prove that if there is no 2-cage in a linkage, then the towns of the linkage can be colored with just one color. (*Hint:* Show that there are no highways at all.)

56. See E 55. Prove that if a linkage contains no 3-cage, then it can be colored in two colors. (*Hint:* First show that the linkage is a tree.)

57. See E 55 and E 56. Dirac proved in 1952 that if a linkage contains no 4-cage, then it can be colored in three colors. Draw a linkage containing a 3-cage but no 4-cage, and check Dirac's theorem for your linkage.

58. See E 55, E 56, and E 57.
    (a) Show by a few experiments that any linkage that can be drawn on a flat piece of paper without the use of bridges has no 5-cage.
    (b) What theorem (like those of E 55, E 56, and E 57) would you like to be true?
    (c) Why would such a theorem prove the four-color theorem for countries on an island or on a sphere?

59. (a) Draw a map in which each vertex is of degree 3.
    (b) Place the number 1 or the number $-1$ at each vertex in such a way that the sum of the labels you meet as you go around each region is a multiple of 3.
    (c) Replace each 1 by a clockwise circle and each $-1$ by a counterclockwise circle.
    (d) Using the technique employed in the informal proof of Theorem 6, obtain a labeling of the edges with the numbers 1, 2, and 3 such that the three edges meeting at each vertex carry all three numbers.
    (e) Using the technique of E 51, color the map in four colors.
    No one has been able to prove that step (b) can always be made.

60. Show that if a map on a sphere cannot be colored in four colors, then such a map with the fewest regions must be regular. (*Warning:* The technique used in the proof of Lemma 4 increases the number of regions.)

61. By a "town" we shall mean one of the arrangements of the natural numbers 1, 2, 3, 4, 5, 6 of the type studied in Chapter 1. Two towns will be "adjacent" if one can be obtained from the other by a single switch. Is it possible to label each town $A$ or $B$ in such a way that adjacent towns have different labels?

62. See E 54 for the definition of "linkage." Is there a linkage that has ten towns such that no two of the towns have the same degree?

63. We outline a proof that a "5-cage" (see E 55) cannot be drawn on the surface of a sphere. Imagine that it is drawn, and cuts the surface of the sphere into $R$ regions with $E$ edges and $V$ vertices.
    (a) Show that $V = 5$, $E = 10$, and thus $R = 7$.
    (b) Show that each region has at least three edges.
    (c) From (b) and $R = 7$, deduce that $E$ is at least $\frac{21}{2}$.
    (d) Obtain a contradiction from (a) and (c).

### References

1. H. S. M. Coxeter, Map coloring problems, *Scripta Mathematica*, vol. 23, 1957, pp. 11–25.

2. W. W. Rouse Ball, *Mathematical Recreations and Essays*, Macmillan, 1947. (Especially Chapters 5 and 8.)

3. P. Franklin, The four-color problem, *Scripta Mathematica* Library, Number 5.

4. H. Weyl, *Symmetry*, Princeton University Press, 1952. (In particular, pp. 75–76 discuss the regular polyhedra. Part of this book is reprinted in J. R. Newman, *The World of Mathematics*, Simon and Schuster, vol. 1, 1956. The illustrations on p. 721 show the 8-, 12-, and 20-faced polyhedra as they appear in nature.)

5. *R. L. Brooks, On colouring the nodes of a network, *Proceedings of the Cambridge Philosophical Society*, vol. 37, 1941, pp. 194–197.

6. *G. A. Dirac, A property of 4-chromatic graphs and some remarks on critical graphs, *Journal of the London Mathematical Society*, vol. 27, 1952, pp. 85–92.

7. D. Hilbert and S. Cohn-Vossen, *Geometry and the Imagination*, Chelsea, New York, 1952. (Regular polyhedra, pp. 89–93, 289–293; map-coloring problems on various surfaces, pp. 333–340.)

8. M. Gardner, *The Second Scientific American Book of Mathematical Puzzles and Diversions*, Simon and Schuster, New York, 1961. (Regular polyhedra, pp. 13–23.)

9. P. Franklin, The four-color problem, *Scripta Mathematica*, vol. 6, 1939, pp. 149–156 and 197–210.

10. K. O. May, The origin of the four-color problem, *Isis*, vol. 56, 1965, pp. 346–348.

11. O. Ore, The *Four-color Problem*, Academic, New York, 1967.

12. A. Beck, M. N. Bleicher, D. W. Crowe, *Excursions into Mathematics*, Worth, New York. (The chapter on coloring maps includes problems equivalent to the four-color conjecture, and a treatment of coloring maps whose regions consist of two separate pieces.)

## Chapter 15

# THE
# REPRESENTATION
# OF NUMBERS

The basketball scorekeeper, making a mark for each point scored, is using the oldest and most cumbersome method of writing the natural numbers. The number 13, which we write with just two symbols, he would write as

*///////////////*

or, to make the marks easier to count, perhaps as

*₦ ₦ ///*.

Even with this device, his method is still a far from ideal notation for the natural numbers. For a larger number, such as 500, which we can write with just three symbols, he requires 500 marks.

Our system of representing the natural numbers has virtues other than brevity. For example:

1. We can see easily whether one number is larger than another by looking at their representations.

2. If we know the representations of the natural numbers $A$ and $B$, we can easily compute the representation of $A + B$.

3. If we know the representation of the natural numbers $A$ and $B$, we can easily compute the representation of $AB$.

But let us not conclude that ours is necessarily the best of all possible systems.

What is the decimal system? To refresh our understanding, let us examine the meaning behind the brief notation 3,102. Checking off numbers from right to left, we see that the 2 records the number of ones; the 0, tens; the 1, hundreds; the 3, thousands. Thus 3,102 is short for the sum

$$3 \text{ thousands} + 1 \text{ hundred} + 0 \text{ tens} + 2 \text{ ones.}$$

One, ten, hundred, and thousand are powers of ten:

$$1 = 10^0, \quad \text{ten} = 10^1, \quad \text{hundred} = 10^2, \quad \text{thousand} = 10^3.$$

The decimal system is based on powers of the number ten. Each natural number is represented as a sum of powers of ten, no single power used more than 9 times. We say that the "coefficients" or "digits" may not exceed 9.

In the decimal notation

$$1,000 = 10^3 \quad \text{and} \quad 10,000 = 10^4,$$

the number of 0's records the power of ten. It is not hard to check that

$$10^3 \cdot 10^4 = 10^7.$$

Hence

$$\underbrace{1,000}_{10^3} \cdot \underbrace{10,000}_{10^4} = \underbrace{10,000,000,}_{10^7}$$

a multiplication fact that says simply, "If you multiply the product $10 \cdot 10 \cdot 10$ by the product $10 \cdot 10 \cdot 10 \cdot 10$, you obtain $10 \cdot 10 \cdot 10 \cdot 10 \cdot 10 \cdot 10 \cdot 10$." This shows why 0's are tallied up in multiplying two numbers whose representations end in a string of 0's.

But there is nothing sacred about the number ten and its powers: one, ten, hundred, thousand, and so on. Let us see what happens if we use the number two and its powers

$$1 = 2^0, \quad 2 = 2^1, \quad 4 = 2^2, \quad 8 = 2^3, \quad 16 = 2^4,$$
$$32 = 2^5, \quad 64 = 2^6, \quad 128 = 2^7, \quad 256 = 2^8, \quad 512 = 2^9,$$

and so on. As an example, let us see if we can write 153 as the sum of powers of 2. Since 153 is between $2^7 = 128$ and $2^8 = 256$, $2^7$ is the largest power of 2 less than 153. We have

$$153 = 2^7 + 25.$$

Now let us work on 25 in a similar manner. Twenty-five is trapped between $2^4 = 16$ and $2^5 = 32$. We have $25 = 2^4 + 9$, and hence

$$153 = 2^7 + 2^4 + 9.$$

Finally, let us work on 9. Nine is trapped between $2^3$ and $2^4$; in fact, $9 = 2^3 + 1$, which means we can write

$$153 = 2^7 + 2^4 + 2^3 + 1.$$

Thus we have expressed 153 as a sum of powers of 2.

If we wish to omit the plus signs, as we do in our usual representation of natural numbers with powers of 10, we proceed as follows. Instead of $2^7 + 2^4 + 2^3 + 1$, we first write

$$1 \cdot 2^7 + 0 \cdot 2^6 + 0 \cdot 2^5 + 1 \cdot 2^4 + 1 \cdot 2^3 + 0 \cdot 2^2 + 0 \cdot 2^1 + 1 \cdot 2^0,$$

and then record just the coefficients:

$$1\ 0\ 0\ 1\ 1\ 0\ 0\ 1$$

(pronounced "one-zero-zero-one-one-zero-zero-one"). In order to keep in mind that we are recording powers of two, not powers of ten, we shall use the subscript 2 (and the subscript "ten" for the decimal system)

$$153_{ten} = 10011001_2.$$

The scorekeeper's /N/ /N/ /N/ /N/ // has the representation $22_{ten}$, which records the arrangement

2 ten's and 2 one's

and the representation $10110_2$ ($2^4 + 2^2 + 2^1$), which records the arrangement

1 sixteen, 0 eight's,
1 four, 1 two, 0 one's

The decimal system, depending on the powers of ten and the digits 0, 1, 2, 3, 4, 5, 6, 7, 8, 9, is also called the *base-ten* system. The system depending on the powers of two and using only the digits 0 and 1 is called the *binary* or *base-two* system.

We can just as well use 3 as a base and build up numbers as the sums of the powers of 3,

$$1 = 3^0, \quad 3 = 3^1, \quad 9 = 3^2, \quad 27 = 3^3, \quad 81 = 3^4, \quad 243 = 3^5, \ldots .$$

For instance, let us see how to express $64_{ten}$ in base 3. First of all, 64 lies between $3^3 = 27$ and $3^4 = 81$. Or, to put it another way

$$64 \text{ lies between } 1 \cdot 3^3 \text{ and } 3 \cdot 3^3.$$

We show this on a number line, on which we also include $2 \cdot 3^3$ ($= 54$):

$$\overset{\longleftarrow 3^3 \longrightarrow\!\!\!\times\!\!\!\longleftarrow 3^3 \longrightarrow\!\!\!\times\!\!\!\longleftarrow 3^3 \longrightarrow}{\underset{\underset{54 = 2 \cdot 3^3}{0 \qquad 27 = 1 \cdot 3^3 \qquad 64 \qquad 81 = 3 \cdot 3^3}}{\rule{0.4\textwidth}{0.4pt}}}$$

In fact, 64 is trapped between $2 \cdot 3^3$ and $3 \cdot 3^3$, two numbers that differ by $3^3$. Thus

$$64 = 2 \cdot 3^3 + \text{some number less than } 3^3.$$

Since $64 - 2 \cdot 3^3 = 64 - 54 = 10$, we may say more precisely:

$$64 = 2 \cdot 3^3 + 10.$$

We have started to build up 64 as a sum of powers of 3. We already know that we need two of the $3^3$'s.

To complete the representation, we express 10 as a sum of powers of 3. Since $10 = 9 + 1 = 3^2 + 1$, we see that

$$64 = 2 \cdot 3^3 + 1 \cdot 3^2 + 1.$$

To turn this into the base-three shorthand, we first record that no $3^1$'s appear,

$$64 = 2 \cdot 3^3 + 1 \cdot 3^2 + 0 \cdot 3^1 + 1$$

and then simply write the digits in order:

$$2101_3.$$

In base 5 we would use the powers of 5, such as 1 (penny), 5 (nickel), 25 (quarter), $125 = 5^3$, and so on; the digits are 0, 1, 2, 3, 4. In base twelve we would use the powers of 12, such as 1, 12 (a dozen), 144 (a gross), and so on; the digits are 0, 1, 2, 3, 4, 5, 6, 7, 8, 9, $t$, $e$, where $t$ and $e$ are introduced as symbols to denote ten and eleven.

The reader should pause and check the entries in this table, which shows how a few sample numbers are denoted in the bases mentioned.

| *Scorekeeper* | *Base ten* | *Base two* | *Base three* | *Base five* | *Base twelve* |
|---|---|---|---|---|---|
| / | 1 | 1 | 1 | 1 | 1 |
| Ⅳ | 5 | 101 | 12 | 10 | 5 |
| Ⅳ // | 7 | 111 | 21 | 12 | 7 |
| Ⅳ Ⅳ | 10 | 1010 | 101 | 20 | $t$ |
| Ⅳ Ⅳ // | 12 | 1100 | 110 | 22 | 10 |
| Ⅳ Ⅳ Ⅳ Ⅳ //// | 24 | 11000 | 220 | 44 | 20 |

Notice how quickly the number of digits grows in the representation of large numbers by small bases, such as 2 or 3. Notice also that in each system the base is always represented by 10.

So far we have treated only the representation of the natural numbers. What are the methods for representing (positive) numbers that are not natural numbers? Since any positive number, rational or irrational, can be written as a natural number plus a number less than 1, we can concentrate on the representation of numbers between 0 and 1.

In the decimal system, when we write $\frac{5}{8}$ as 0.625, we are saying that

$$\frac{5}{8} = \frac{6}{10} + \frac{2}{10^2} + \frac{5}{10^3} = \frac{6}{10} + \frac{2}{100} + \frac{5}{1000}.$$

In the decimal system we are recording how a number less than 1 can be expressed as a sum of 1/10's, 1/10²'s, 1/10³'s, and so on, if we allow none of these expressions to appear more than 9 times in the sum. The representation of $\frac{5}{8}$, for example, tells us that $\frac{5}{8}$ is the sum of six 1/10's, two $1/10^2$'s, and five $1/10^3$'s.

A very simple number, less than 1, can have a terrible representation in the decimal system. For example, recall that

$$\frac{1}{3} = 0.333333\ldots,$$

where the ellipsis ($\cdots$) indicates that the 3's go on and on. But what exactly does the equation

$$\frac{1}{3} = 0.333333\ldots$$

mean? We might translate it into

$$\frac{1}{3} = \frac{3}{10} + \frac{3}{10^2} + \frac{3}{10^3} + \frac{3}{10^4} + \frac{3}{10^5} + \cdots,$$

but what does this mean? In particular, what does "$\cdots$" mean? Surely it cannot mean "the sum of all rationals of the form 3 divided by a power of 10," since not even the fastest electronic computer can ever add an endless list of numbers.

The expression "$\frac{1}{3} = 0.333333\cdots$" is a very terse summary of an observation that goes like this: the sequence of numbers

$$\frac{3}{10}, \qquad \frac{3}{10} + \frac{3}{10^2}, \qquad \frac{3}{10} + \frac{3}{10^2} + \frac{3}{10^3}, \qquad \cdots$$

*approaches* $\frac{1}{3}$. (Note that at no point do we add an endless supply of numbers; we simply make a statement about the way a certain list of numbers behaves. This is discussed in Appendix E.)

The decimal representation of a number less than 1 describes how that number can be written as a sum of rationals of the form $A/10^N$ where $A$ is a natural number at most 9. Once again, there is nothing sacred about the number 10. Let us show that we can just as well use the base 2.

Take, for example, $\frac{5}{8}$. Since $\frac{5}{8}$ is larger than $\frac{1}{2}$, when we subtract $\frac{1}{2}$ from $\frac{5}{8}$ we obtain a positive result. Thus

$$\frac{5}{8} - \frac{1}{2} = \frac{1}{8},$$

$$\frac{5}{8} = \frac{1}{2} + \frac{1}{8}.$$

This gives a representation of $\frac{5}{8}$ as a sum of rationals of the form $1/2^N$. We can say that

$$\frac{5}{8} = \frac{1}{2} + \frac{0}{4} + \frac{1}{8}.$$

and record this by

$$\frac{5}{8} = 0.101_2.$$

Thus $17\frac{5}{8} = 17.625_{ten} = 10001.101_2$.

A ruler, which is divided into halves, quarters, eighths, sixteenths, and so on, of an inch, is partially on base 2. The walls of many garages and shops have charts that show the decimal equivalents of these awkward measures, such as $\frac{11}{16} = 0.6875$. Surely, if England can finally put its monetary system into base ten, we should be able to do the same for the measurement of length, and switch to the metric system, where a meter is cut into 100 centimeters, and a centimeter is cut into 10 millimeters. The mechanic would require no conversion chart on the wall. But those who enjoy computing one-third of 7 feet $7\frac{7}{16}$ inches will no doubt delay this humanitarian change.

Let us see how $\frac{1}{3}$ looks when written in base 2. The rationals of the form $1/2^N$ are $\frac{1}{2}, \frac{1}{4}, \frac{1}{8}, \frac{1}{16}, \frac{1}{32}, \ldots$ . The largest of these that we can subtract from $\frac{1}{3}$ and obtain a nonnegative result is $\frac{1}{4}$. We have $\frac{1}{3} - \frac{1}{4} = \frac{1}{12}$. Thus $\frac{1}{3} = \frac{1}{4} + \frac{1}{12}$. Next let us work on $\frac{1}{12}$. The largest fraction of the form $1/2^N$ that is not larger than $\frac{1}{12}$ is $\frac{1}{16}$. We have $\frac{1}{12} - \frac{1}{16} = \frac{4}{192} = \frac{1}{48}$. Thus $\frac{1}{12} = \frac{1}{16} + \frac{1}{48}$. So far we have

$$\frac{1}{3} = \frac{1}{4} + \frac{1}{16} + \frac{1}{48}.$$

We can continue by subtracting $\frac{1}{64}$ from $\frac{1}{48}$, and so on, at each step obtaining a closer approximation of $\frac{1}{3}$ by a sum of distinct fractions of the type $1/2^N$. Our first approximation of $\frac{1}{3}$ was $\frac{1}{4}$, with an error of $\frac{1}{12}$; our second approximation of $\frac{1}{3}$ was $\frac{1}{4} + \frac{1}{16}$, with a smaller error, $\frac{1}{48}$; the reader may check that the approximation of $\frac{1}{3}$ by $\frac{1}{4} + \frac{1}{16} + \frac{1}{64}$ has an error of only $\frac{1}{192}$.

Just as the decimal representation of $\frac{1}{3}$ does not stop, the binary representation of $\frac{1}{3}$ does not stop. For if it did stop, we would have $\frac{1}{3}$ expressed as a sum of rationals of the type $1/2^N$. This sum could be written as one rational of the form $A/2^M$, where $2^M$ is the largest power of 2 present in the sum and where $A$ is a natural number. We would thus have

$$\frac{1}{3} = \frac{A}{2^M},$$

and thus

$$2^M = 3A.$$

Hence 3 would divide $2^M$, in contradiction with Theorem 3 in Chapter 5, which asserts that if a prime divides the product of several natural numbers, it must divide at least one of them. It turns out that the binary representation of $\frac{1}{3}$ is $0.01010101 \ldots$ . (See E 40.)

We could also use bases other than 2 or 10. The Babylonians used the base 60 (but required coefficients as large as 59). With the base 3, every number less than 1 can be expressed as a sum of rationals of the form $1/3^N$, in which sum no such fraction appears more than twice. To see this, divide the line segment from 0 to 1 into 3 equal pieces, and each of these pieces into 3 equal pieces, and so on.

As an example, let us consider $\frac{7}{8}$. It appears in the segment from $\frac{2}{3}$ to 1. Thus $\frac{7}{8}$ is between $\frac{2}{3}$ and 1. The segment from $\frac{2}{3}$ to 1 is divided into three equal pieces:

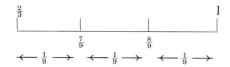

It can be checked that $\frac{7}{8}$ appears in the little segment from $\frac{7}{9}$ to $\frac{8}{9}$. We have already shown that $\frac{7}{8}$ is between

$$\frac{2}{3} + \frac{1}{9} \quad \text{and} \quad \frac{2}{3} + \frac{2}{9}.$$

This process can be continued as long as we please. It contrasts with our procedure in the decimal system in that we cut each segment into three pieces, not ten.

In the base 3, the rational $\frac{1}{3}$ has the simple representation $\frac{1}{3} = 0.10000 \ldots$. But the rational $\frac{1}{2}$, which has a simple representation in the decimal and binary systems, has an endless representation in the base 3 system. It can be shown that this representation is $0.11111 \ldots$, an endless string of 1's. Thus, to represent $\frac{1}{3}$ conveniently, we sacrificed our simple representation of $\frac{1}{2}$. Clearly, then, no one base is best when it comes to representing numbers less than 1.

Now let us see how addition and multiplication look in other bases, say in the base three. Let us pretend that we live in a base-three world and have *never heard* of the decimal system. First of all, in kindergarten we learn how to count and write 0, 1, 2, for these will be the digits.

In first grade, perhaps, we make the addition and multiplication tables for these three numbers. We find, by counting pebbles or working with such teaching aids as Cuisenaire rods, that $2 + 2 = 11$, since two pebbles, together with two more pebbles, can be arranged into one group of three with one left over:

rearranged as

The entire addition table is simply

| + | 0 | 1 | 2 |
|---|---|---|---|
| 0 | 0 | 1 | 2 |
| 1 | 1 | 2 | 10 |
| 2 | 2 | 10 | 11 |

(base-three addition table)

The multiplication table for the base-three digits is also much simpler than the corresponding table for base ten:

| × | 0 | 1 | 2 |
|---|---|---|---|
| 0 | 0 | 0 | 0 |
| 1 | 0 | 1 | 2 |
| 2 | 0 | 2 | 11 |

(base-three multiplication table)

The only entry that requires more than a moment's computation is $2 \cdot 2 = 11$. Paradoxically, the multiplication table is easier to construct than the addition table.

In this base-three world, let us compute

$$\begin{array}{r} 102 \\ + \ 21 \end{array} \quad \text{(base three)}$$

Starting from the right we meet $2 + 1$, which we have memorized in our addition table: $2 + 1 = 10$. So we may say that the sum is equal to the sum

$$\begin{array}{r} 100 \\ 20 \\ + \ 10 \end{array} \quad \text{(base three)}$$

Again recalling that $2 + 1 = 10$, we see that $10 + 20 = 100$, hence our sum is the same as

$$\begin{array}{r} 100 \\ + \ 100 \end{array} \quad \text{(base three)}$$

which is simply 200. Once we have gone through these steps in "slow motion," we may use this short bookkeeping, which records above the columns some additions "done in our head":

$$\begin{array}{r} ^{(1)(1)}102 \\ + \ \ 21 \\ \hline 200 \end{array} \quad \text{(base three)}$$

(We, who have the extra perspective of base ten, may check this base-three computation by translating the problem into our customary base: $102_3 = 11_{ten}$, and $21_3 = 7$; thus the problem was "really" $11 + 7 = 18$. The answer $200_3$ is right, for 18 is 2 nines.)

How does the base-three pupil subtract? First, to each basic addition fact there corresponds a subtraction fact, which he will memorize. For instance, $2 + 2 = 11$ (base three), implies that

$$11 - 2 = 2 \quad \text{(base three).}$$

Then he solves this type of problem:

$$121 - 12 = ? \quad \text{(base three)}$$

as follows

$$121 = 100 + 20 + 1 \quad \text{(base three),}$$
$$12 = \phantom{100 + {}} 10 + 2 \quad \text{(base three).}$$

Unable to "subtract 2 from 1," he rewrites $20 + 1$ as $10 + 11$ (base three), and now has the easy subtraction problem:

$$121 = 100 + 10 + 11 \quad \text{(base three),}$$
$$-12 = \phantom{100 + {}} -(10 + 2) \quad \text{(base three).}$$

Since $11 - 2 = 2$ (base three), he sees that

$$121 - 12 = 100 + 2 = 102 \quad \text{(base three).}$$

(We, in the base-ten world, may check his work: $121_3 = 16_{ten}$, $12_3 = 5_{ten}$, $102_3 = 11_{ten}$, and $16 - 5$ indeed equals 11.)

How does the base-three pupil multiply? First he will deal with numbers that have lots of 0's, say, in a problem such as

$$1000 \times 10000 \quad \text{(base three).}$$

He knows that three equals 10, hence

$$1000 = 10 \cdot 10 \cdot 10 \quad \text{(base three),}$$

and

$$10000 = 10 \cdot 10 \cdot 10 \cdot 10 \quad \text{(base three).}$$

Thus

$$1000 \cdot 10000 = 10 \cdot 10 \cdot 10 \cdot 10 \cdot 10 \cdot 10 \cdot 10 = 10000000 \quad \text{(base three).}$$

He learns that to multiply such numbers "you simply count up the 0's". (Any resemblance to the same rule in base ten is not a coincidence; both are special cases of the law of exponents, $a^m \cdot a^n = a^{m+n}$.)

He next may compute $20_3 \cdot 100_3$. He will begin by rewriting this as

$$2 \cdot 10 \cdot 100 \quad \text{(base three),}$$

which he knows, by counting up 0's, is

$$2 \cdot 1000 \quad \text{(base three)},$$

or simply

$$2000 \quad \text{(base three)}.$$

He may then be confronted with a somewhat more involved problem, such as $20_3 \cdot 200_3$, which he rewrites as

$$2 \cdot 10 \cdot 2 \cdot 100 \quad \text{(base three)},$$

or

$$2 \cdot 2 \cdot 10 \cdot 100 \quad \text{(base three)}.$$

He has now memorized $2 \cdot 2 = 11$ and knows that $10 \cdot 100 = 1000$. The problem has become: compute

$$11 \cdot 1000.$$

Writing 11 as $10 + 1$, he reduces this to the simpler problem

$$10 \cdot 1000 + 1 \cdot 1000 \quad \text{(base three)},$$

which is

$$10000 + 1000 \quad \text{(base three)},$$

or, finally,

$$11000 \quad \text{(base three)}.$$

(We may again use our base-ten knowledge to check whether

$$20_3 \cdot 200_3 = 11000_3.$$

We have $20_3 = 6$, $200_3 = 18$, and $11000_3 = 81 + 27 = 108$, and 6 times 18 is indeed 108.) Although we may sympathize with the child who must write such a long expression as 11,000, while we get away with 108, he may in turn respond, "But look at the size of your multiplication table. I'm sure I would get 6 times 9 mixed up with 7 times 8."

And it will not be much longer before the base-three pupil will compute

$$212 \cdot 21 \quad \text{(base three)}$$

with this scheme

$$
\begin{array}{r}
212 \\
\times\ 21 \\
\hline
212 \\
1201 \\
\hline
12222
\end{array}
\quad \text{(base three)}.
$$

This looks just as odd to us as

$$
\begin{array}{r}
23 \\
\times\ 7 \\
\hline
161
\end{array}
\quad \text{(base ten)}
$$

looks to him; yet it is the same problem, as the reader may check.

Perhaps in the third grade, the base-three pupil meets this problem:

$$0.01 \times 0.001 \quad \text{(base three)}.$$

If he is too hasty he may just count up the 0's to the right of the dot and say, "Easy, it's just 0.0001." He should be praised for making an analogy with the case $10 \cdot 100 = 1000$, but the analogy has produced the wrong answer.

What do $0.01_3$ and $0.001_3$ mean? We have

$$0.01 = \frac{1}{\text{three} \times \text{three}} = \frac{1}{10 \cdot 10} \quad \text{(base three)}$$

and

$$0.001 = \frac{1}{10 \cdot 10 \cdot 10} \quad \text{(base three)}.$$

Thus the product is

$$\frac{1}{10} \cdot \frac{1}{10} \cdot \frac{1}{10} \cdot \frac{1}{10} \cdot \frac{1}{10},$$

which is recorded by putting a "1" in the *fifth* place to the right of the dot. Hence there should be four 0's to the right of the dot,

$$0.01 \cdot 0.001 = 0.00001.$$

His guess is one "0" short.

In the world of base two, matters become even simpler. These are the addition and multiplication tables that the child would have to memorize:

| + | 0 | 1 |
|---|---|---|
| 0 | 0 | 1 |
| 1 | 1 | 10 |

| × | 0 | 1 |
|---|---|---|
| 0 | 0 | 0 |
| 1 | 0 | 1 |

(base two).

Here is the computation of $11001_2 \cdot 1011_2$. Notice that all it involves is addition and "shoving to the left" the right number of places:

```
        11001
      ×  1011
        11001    (base two)
        11001
       11001
      100010011
```

The reader may check that this corresponds to $25 \times 11 = 275$ in base ten.

Although there is no likelihood that the decimal system will be replaced by the binary system in daily life, it should be pointed out that the binary system is the one best suited to electronic computing machines. First, since only two digits, 0 and 1, are required, each natural number can be recorded by a set of transistors, some conducting (corresponding to 1), some not conducting (corresponding to 0). Second, once the machine is wired to do addition, it can be made to multiply merely by wiring it to shove numbers to the left.

Some 4000 years ago the Egyptians used a system for representing numbers between 0 and 1 that resembles our technique for bases 2, 3, and 10. They represented a rational between 0 and 1 as a sum of rationals of the form $1/N$, the so-called **unit fractions**, such that no unit fraction was repeated. For example, they wrote

$$\frac{2}{31} = \frac{1}{20} + \frac{1}{124} + \frac{1}{155}.$$

The reader may check that the representation is correct.

There are several differences between the Egyptian system and the decimal system. First, their representation for a rational number always stops, whereas the decimal representation need not (consider $\frac{1}{3} = 0.3333 \ldots$). Second, there are many ways to represent a number in the Egyptian method, whereas the decimal system provides at most two representations (as an example of two representations, consider $\frac{1}{2} = 0.5000 \ldots = 0.49999 \ldots$). For example, in the Egyptian system, there are many representations for $\frac{2}{5}$, including these three:

$$\frac{1}{3} + \frac{1}{15} = \frac{1}{4} + \frac{1}{7} + \frac{1}{140} = \frac{1}{5} + \frac{1}{6} + \frac{1}{30}.$$

The Egyptian system has practically nothing to recommend it. It is not easy to determine when one number is larger than another. (Which is larger, $\frac{1}{3} + \frac{1}{11} + \frac{1}{231}$ or $\frac{1}{4} + \frac{1}{7} + \frac{1}{140}$?) By what routine do we represent in the Egyptian style the product or sum of two numbers? (For example, how do we represent $\frac{2}{5} \times \frac{2}{5}$ or $\frac{2}{5} + \frac{2}{5}$?) Perhaps it was their cumbersome technique for representing numbers that prevented the Egyptians from ever developing mathematics beyond the solution of equations of the simplest type.

As a matter of fact, it is not obvious that every rational number between 0 and 1 *has* a representation as a sum of distinct unit fractions. This question interested Fibonacci, a mathematician who lived in Pisa when that city's famous tower was begun. In his *Liber Abacci*, published in 1202, written in medieval Latin and never translated, Fibonacci described a simple technique for constructing such a representation. Before proving that it always works, we will illustrate the technique by applying it to the rational $\frac{4}{13}$.

*Step 1.* Select the largest unit fraction less than or equal to $\frac{4}{13}$; that is, $\frac{1}{4}$. Subtract $\frac{1}{4}$ from $\frac{4}{13}$:

$$\frac{4}{13} - \frac{1}{4} = \frac{3}{52}.$$

*Step 2.* Select the largest unit fraction less than or equal to $\frac{3}{52}$; that is, $\frac{1}{18}$. Subtract $\frac{1}{18}$ from $\frac{3}{52}$:

$$\frac{3}{52} - \frac{1}{18} = \frac{2}{936}.$$

*Step 3.* Select the largest unit fraction less than or equal to $\frac{2}{936}$; that is, $\frac{1}{468}$. Subtract $\frac{1}{468}$ from $\frac{2}{936}$:

$$\frac{2}{936} - \frac{1}{468} = 0.$$

Having reached 0 we stop. Since we have reached 0 by successive subtractions of $\frac{1}{4}$, $\frac{1}{18}$, and $\frac{1}{468}$, we see that

$$\frac{4}{13} = \frac{1}{4} + \frac{1}{18} + \frac{1}{468}.$$

This is how Fibonacci describes his method:

> The rule for this category is that you divide the larger number by the smaller; and when the division itself is not even, look to see between what two numbers the division falls. Then take the larger part, and subtract it, and keep the remainder.

The reader will observe that the same technique of repeated subtractions is also used to find the decimal or binary representation of $\frac{4}{13}$. But, whereas the decimal or binary representation might go on and on and never stop, Fibonacci's representation always stops. The reader may apply Fibonacci's technique to $\frac{15}{19}$. He will see that it provides the representation $\frac{15}{19} = \frac{1}{2} + \frac{1}{4} + \frac{1}{26} + \frac{1}{988}$.

We shall prove that Fibonacci's technique always stops (though not necessarily in three steps) when applied to a positive rational less than 1.

Observe the successive rationals we met in applying Fibonacci's technique to $\frac{4}{13}$. These rationals were $\frac{4}{13}$, $\frac{3}{52}$, $\frac{2}{936}$, and 0. Though the denominators grow swiftly, the numerators 4, 3, 2, and 0 shrink. This suggests a way of proving that the process must always stop when applied to a rational. All we have to do is prove that the numerators shrink at each step. (This would imply, for example, that Fibonacci's technique, when applied to $\frac{19}{53}$, would have to stop in fewer than 19 steps, since the numerator is 19.) Thus we need only prove this

LEMMA. *If A and B are natural numbers, and A is less than B, and $1/N$ is the largest unit fraction less than or equal to $A/B$, then the numerator $NA - B$ of the rational $(NA - B)/BN$ (which is $A/B - 1/N$) is less than A.*

PROOF. If $A$ divides $B$, then $A/B$ can be expressed as a unit fraction, $1/N$. We have

$$\frac{A}{B} - \frac{1}{N} = 0,$$

and the numerator $NA - B$ mentioned in the lemma is simply 0. This establishes the lemma if $A$ divides $B$.

It remains to prove that $NA - B$ is less than $A$ if $A$ does not divide $B$. Let us first note how we find $N$ in practice. Consider, for example, $\frac{4}{13}$. To find the largest unit fraction less than $\frac{4}{13}$ we can rewrite $\frac{4}{13}$ as follows:

$$\frac{4}{13} = \frac{1}{\frac{13}{4}}.$$

But since $\frac{13}{4} = 3 + \frac{1}{4}$, the largest unit fraction less than $\frac{4}{13}$ has as denominator the first natural number larger than $3 + \frac{1}{4}$; in short, the next natural number after 3.

Similarly, to find the largest unit fraction less than or equal to $A/B$ (if $A$ does not divide $B$), we write

$$B = QA + R$$

(where $R$, the remainder when $B$ is divided by $A$, is a positive number, since $A$ does not divide $B$), and then observe that the largest unit fraction less than $A/B$ is

$$\frac{1}{Q+1}.$$

Carrying out the subtraction,

$$\frac{A}{B} - \frac{1}{Q+1},$$

we obtain

$$\frac{A(Q+1) - B}{B(Q+1)}.$$

We wish to prove that the new numerator,

$$A(Q+1) - B,$$

is less than $A$. Recalling that $B = QA + R$, we have

$$A(Q+1) - B = A(Q+1) - (QA + R)$$
$$= AQ + A - QA - R$$
$$= A - R.$$

Since $R$ is not 0, $A - R$ is indeed less than $A$, and the lemma is proved for the case in which $A$ does not divide $B$. This completes the proof.

Fibonacci's technique suggests this more difficult problem, one that was first posed in 1956 and which several mathematicians have worked on in vain: Can we use a technique like Fibonacci's to represent a rational with an odd denominator as a sum of distinct unit fractions with odd denominators?

For example, let us try to represent $\frac{2}{7}$ as a sum of distinct unit fractions with odd denominators. Were it not for the restriction that the unit fractions must have odd denominators, we would subtract $\frac{1}{4}$. Instead, we must subtract $\frac{1}{5}$, the largest unit fraction with odd denominator that is not larger than $\frac{2}{7}$. We obtain

$$\frac{2}{7} - \frac{1}{5} = \frac{3}{35}$$

as the next rational to work on. Notice that the new numerator, 3, is larger than the old numerator, 2. This means that the lemma that helped Fibonacci will not help us.

But let us continue undaunted. The largest unit fraction less than or equal to $\frac{3}{35}$ is $\frac{1}{12}$. But since 12 is even, we must use $\frac{1}{13}$. We obtain

$$\frac{3}{35} - \frac{1}{13} = \frac{4}{455}.$$

Again the numerator has grown, though not so quickly as the denominator.

Let us next consider $\frac{4}{455}$. The largest unit fraction less than or equal to $\frac{4}{455}$ is $\frac{1}{114}$. But since 114 is even, we must use $\frac{1}{115}$. We obtain

$$\frac{4}{455} - \frac{1}{115} = \frac{5}{52325}.$$

Again the numerator has grown. But this time, since the numerator divides the denominator, we can rewrite $\frac{5}{52325}$ as a unit fraction, $\frac{1}{10465}$. Thus our modified version of Fibonacci's technique stops and, incidentally, has found for us this representation of $\frac{2}{7}$ as a sum of distinct unit fractions with odd denominators:

$$\frac{2}{7} = \frac{1}{5} + \frac{1}{13} + \frac{1}{115} + \frac{1}{10465}.$$

The reader may check that this same method produces the following representations:

$$\frac{2}{9} = \frac{1}{5} + \frac{1}{45}, \quad \frac{2}{5} = \frac{1}{3} + \frac{1}{15}, \quad \frac{5}{13} = \frac{1}{3} + \frac{1}{21} + \frac{1}{273}.$$

The interesting point is that in each of these cases (and in many more that have been done on computing machines) the process stops. Whether the process always stops is not known, though it is known that such fractions are the sum of distinct unit fractions with odd denominators. (See R 6 or R 7.)

We have seen in this chapter various ways of representing numbers. In our usual system we represent numbers as a sum of powers of 10 and permit each power to occur at most nine times in the sum. In the binary system, we use powers of 2 and permit each power to appear at most once. We should realize that the decimal system is not the only possible method of representing numbers. And, in particular, when we speak of a number such as $\pi$ or $\sqrt{2}$, we do not need to think of its representation in the decimal system or in any other system.

Too many people feel that a number is not well described until they have its decimal representation. It is more precise to think of $\pi$ as the ratio between the circumference and diameter of a circle than as $\frac{22}{7}$ or 3.1 or 3.14 or 3.1416, all of which are just rough estimates of $\pi$.

And we should all be reminded, by the cumbersome Egyptian representation of rationals, that our usual representation of numbers as the sum of rationals whose denominators are powers of one fixed number is a device of great simplicity.

### Exercises

1. Fill in the blanks.
   (a) In base ___ the only digits are 0, 1, and 2.
   (b) In base five the digits are _____.

2. Express each of the integers from 1 through 10 in base two.

3. Check that
   (a) $17_{ten} = 122_3$.
   (b) $27_{ten} = 1000_3$.
   (c) $11_{ten} = 102_3$.
   (d) $16_{ten} = 10000_2$.
   (e) $31_{ten} = 11111_2$.
   (f) $100_{ten} = 1100100_2$.

4. Express $200_{ten}$ in (a) base two, (b) base three, (c) base four.

5. Express in base ten (a) $1011_2$, (b) $1011_3$, (c) $100_5$, (d) $200_5$, (e) $2000_5$.

6. Express (a) $2^5 + 2^2 + 1$ in base two, (b) $2 \cdot 3^5 + 3^2 + 2 \cdot 3 + 1$ in base three, (c) $9 \cdot \text{ten}^4 + 4 \cdot \text{ten} + 2$ in base ten.

7. Express $160_{ten}$ in (a) base two, (b) base three.

8. Express (a) $31_5$ in base two, (b) $41_8$ in base three. (*Suggestion:* First express them in base ten.)

9. Express in base ten these base-twelve numbers: (a) $t$, (b) $100$, (c) $2e$.

10. Write in the binary notation (a) $\cancel{IIII}\ \cancel{IIII}\ \cancel{IIII}\ II$, (b) 15, (c) 16, and (d) 21.

11. Write in the binary notation (a) 23, (b) 41, (c) 42, (d) 99.

12. Write in the decimal notation (a) $111_2$, (b) $110_2$, (c) $10101_2$.

13. Write in the decimal notation (a) $1011_2$, (b) $11111_2$, (c) $100001_2$.

14. Which is larger?
    (a) $222_3$ or $1001_3$
    (b) $101000_2$ or $1112_3$

15. Which is larger?
    (a) 9057 or 8999
    (b) $1000_2$ or $1011_2$
    (c) 74 or $1010011_2$
    (d) $1001_{ten}$ or $1001_2$

16. Express in the form $0.$___ (a) $\frac{5}{8}$ in base ten, (b) $\frac{5}{8}$ in base two, (c) $\frac{5}{8}$ in base four, (d) $\frac{5}{8}$ in base eight.

17. Translate into a base-ten decimal (a) $\frac{3}{4}$, (b) $0.11_2$, (c) $0.4_5$, (d) $0.21_3$.

18. How are the bases two, ten, and twelve used in a yardstick?

19. If you have four pennies, four nickels, and four quarters, what amounts up to $1.24 can you pay?

20. Express (a) $\frac{11}{16}$ in the form $0.\underline{\quad}_2$, (b) $\frac{2}{3} + \frac{1}{27}$ in the form $0.\underline{\quad}_3$, (c) $\frac{4}{25} + \frac{2}{125}$ in the form $0.\underline{\quad}_5$.

21. What is the next integer after (a) $99_{\text{ten}}$, (b) $44_5$, (c) $11_2$? Express each answer in the same base as that of the given number.

22. Express in the form $a/b$, where $a$ and $b$ are to be written in base ten, (a) $1.111_2$, (b) $1.111_3$, (c) $1.111_4$, (d) $1.111_{\text{ten}}$. (See p. 83 for the meaning of the underline.)

23. Express in the form $\underline{\quad}.\underline{\quad}_2$: (a) $\frac{173}{4}$, (b) $\frac{17}{16}$.

24. Check that $111_5 = 11111_2$.

25. When a natural number $A$ is written in base three, how can you decide whether it is divisible by (a) 3, (b) 9, (c) 2, (d) 4?

26. Which is larger, $2415_6$ or $2415_7$?

27. Fill in at least the first three digits in
    (a) $\frac{2}{5} = 0.\underline{\quad}_3$,      (b) $\frac{5}{7} = 0.\underline{\quad}_2$,      (c) $\frac{3}{32} = 0.\underline{\quad}_2$,      (d) $\frac{1}{2} = 0.\underline{\quad}_5$.

28. On a ruler divided into at least sixteenths show $0.1011_2$.

29. Make the base-four addition and multiplication tables.

30. Make the base-five addition and multiplication tables.

31. Compute (a) $11_2 + 1$, (b) $11_3 + 1$, (c) $44_5 + 1$.

32. Using only the arithmetic of the binary system, compute:
    (a) $1010_2 + 101_2$,      (b) $1110_2 + 1001_2$,      (c) $1011_2 + 11111_2$.

    Check your answers by translating the problem into the decimal system.

33. Compute (a) $\begin{array}{r} 201_3 \\ +120_3 \end{array}$      (b) $\begin{array}{r} 122_3 \\ +212_3 \end{array}$      (c) $\begin{array}{r} 403_5 \\ +124_5 \end{array}$

    (d) Check your answers in base ten.

34. Compute (a) $100_5 \cdot 1000_5$, (b) $0.001_5 \cdot 0.0001_5$.

35. The guess that the pupil made on p. 248 was off by one place. Will it always be off by that amount?

36. Compute (a) $200_5 \cdot 3000_5$, (b) $0.002_5 \cdot 0.0003_5$.

37. Compute in base three (a) $122_3 \cdot 201_3$, (b) $1221_3 \cdot 21_3$. (c) Check your answers in base ten.

38. Using only the binary system, compute
    (a) $111_2 \cdot 10_2$,      (b) $1011_2 \cdot 101_2$,      (c) $1010_2 \cdot 11_2$,      (d) $111_2 \cdot 111_2$.

39. Using only the binary system, compute
    (a) $101_2 \cdot 1011_2$,      (b) $111_2 \cdot 1000_2$,      (c) $1111111_2 \cdot 1111_2$.
    (d) Check your answers in base ten.

40. Show that $0.010101_2 = \frac{1}{3}$, using the technique described on p. 84. (Recall that $100_2 = 4$.)

41. Express in base twelve (a) $100_{\text{ten}}$, (b) $5 \times 7$, (c) $t + 5$, (d) $3e$, (e) $\frac{1}{3}$ (as a base-twelve "decimal"), (f) $\frac{1}{2}$, (g) $\frac{1}{4}$.

42. Since $(1.4)^2 = 1.96$ and $(1.5)^2 = 2.25$, we see that $\sqrt{2}$ is between 1.4 and 1.5. Prove in a similar fashion that $\sqrt{2}$ is between (a) 1.41 and 1.42, (b) 1.414 and 1.415.

43. See E 42.
    (a) Show that $\sqrt{2}$ is between $1.1_3$ and $1.2_3$.
    (b) Show that $\sqrt{2}$ is between $1.10_3$ and $1.11_3$.
    (c) Fill in the next digit, $\sqrt{2} = 1.10\_\cdots$, base three.

44. See E 42 and E 43. Fill in the three digits: $\sqrt{2} = 1.\_\_\_\cdots$ base two.
    As E 42–44 suggest, the symbol $\sqrt{2}$ is an adequate description of "that positive number whose square is 2," just as $\frac{1}{3}$ is an adequate description of "that number which, when multiplied by three, yields one." There is no more new information in $\sqrt{2} = 1.414\cdots$ than in $\frac{1}{3} = 0.010101_2$.

45. See E 42. Clearly $\sqrt{3}$ is between 1 and 2. Using the technique suggested by E 42, find the first three digits to the right of the "decimal point" in the representation of $\sqrt{3}$ in (a) base ten, (b) base two, (c) base three.

46. Compute in base five (a) $24_5 \cdot 32_5$, (b) $412_5 \cdot 304_5$. (c) Check your answers in base ten.

47. Express the products in base twelve: (a) $9 \cdot 7$, (b) $t \cdot t$ (where $t = $ ten).

48. Show that $0.010101_2 = \frac{1}{3}$, using geometric series, as described in Appendix E.

49. Compute (a) $101101_2 \times 10101_2$, (b) $1111_2 \times 1111_2$. (c) Check your answer in base ten.

50. In which bases are these computations?
    (a) $3 + 4 = 12$, (b) $12 + 12 = 30$, (c) $3 \times 4 = 10$.

51. In which bases will you see the familiar sight $2 \times 3 = 6$?

52. Is $121212_{(3)}$ odd or even?

53. Check that the following Egyptian representations are correct:

    (a) $\dfrac{2}{31} = \dfrac{1}{20} + \dfrac{1}{124} + \dfrac{1}{155}$,  (b) $\dfrac{2}{17} = \dfrac{1}{12} + \dfrac{1}{51} + \dfrac{1}{68}$.

54. Apply Fibonacci's technique to (a) $\frac{5}{11}$, (b) $\frac{8}{19}$.

55. Apply Fibonacci's technique to (a) $\frac{5}{13}$, (b) $\frac{7}{11}$, (c) $\frac{7}{16}$.

56. Apply Fibonacci's technique to (a) $\frac{4}{9}$, (b) $\frac{7}{50}$, (c) $\frac{5}{23}$.

57. Show that the modified Fibonacci technique provides the following representations of rationals with odd denominator as a sum of unit fractions with odd denominators:

    (a) $\dfrac{3}{7} = \dfrac{1}{3} + \dfrac{1}{11} + \dfrac{1}{231}$,  (b) $\dfrac{5}{11} = \dfrac{1}{3} + \dfrac{1}{9} + \dfrac{1}{99}$.

●

58. Using the formula for the sum of a geometric series (Theorem 2 of Appendix E), prove that
    (a) 999 is one less than 1000.

(b) 9999 is one less than 10,000.

(c) 99,999 is one less than 100,000.

59. Using the formula for the sum of a geometric series (Theorem 2 of Appendix E), prove that

(a) $111_2$ is one less than $1000_2$.

(b) $1111_2$ is one less than $10000_2$.

(c) $11111_2$ is one less than $100000_2$.

60. In the base ten we have the rule for "casting out 9's." Develop the analogous rule for "casting out 2's" for the base 3. In particular, devise a shortcut for determining whether a number written in the base 3 notation is even or odd.

61. (a) Using Theorem 3 of Appendix E, show that $0.689999999\ldots = 0.690000000\ldots$.

(b) Similarly, show that $0.01111111 \ldots = 0.1000000000 \ldots$ (in the base 2).

62. Using geometric series (see Appendix E), prove

(a) that $0.10000 \ldots$ (base 2 notation) is equal to $0.111111 \ldots$ (base 3 notation), where the base 3 representation of numbers less than 1 is analogous to the base 2 and base 10 representations for numbers less than 1;

(b) that $0.3333333 \ldots$ (base 10 notation) is equal to $0.10000000 \ldots$ (base 3 notation).

63. (a) Prove that $0.10000 \ldots$ (base 3 notation) is equal to $0.010101 \ldots$ (0 and 1 continuing to alternate) (base 2 notation).

(b) Prove that $0.10101010 \ldots$ (0 and 1 continuing to alternate) (base 3 notation) is equal to $0.011$ (base 2 notation).

64. (a) What is the remainder when $9053_{\text{ten}}$ is divided by 10?

(b) What is the remainder when $2001_3$ is divided by 3?

(c) What is the remainder when $10011_2$ is divided by 2?

65. Some rationals, such as $\frac{1}{2}, \frac{3}{8}, \frac{2}{5}$, have a decimal representation that "stops" (repeating block is 0), while others, such as $\frac{1}{3}$, have a decimal representation that never stops.

(a) Determine, by the usual "long division," the decimal representation of at least ten rationals.

(b) Which stop? Which do not stop?

66. Let $A/B$ be a rational with $(A, B) = 1$, and assume its decimal representation stops.

(a) Prove that $B$ divides some power of 10.

(b) Prove that in the expression of $B$ as the product of primes only the primes 2 and 5 can appear.

67. Let $A/B$ be a rational with $(A, B) = 1$, where no prime other than 2 or 5 divides $B$. Prove that the decimal representation of $A/B$ stops. (*Hint:* Prove that there exist natural numbers $C$ and $n$ such that $A/B = C/10^n$.)

68. Let $A/B$ be a rational with $(A, B) = 1$. Prove that if there is a prime other than 2 or 5 dividing $B$, then the decimal representation of $A/B$ does not stop.

69. Which fractions $A/B$, where $(A, B) = 1$, have a "decimal representation" in base 2 that stops? Which do not? Explain your answer.

70. (a) In which bases does the "decimal" representation of $\frac{1}{2}$ terminate?
    (b) Similarly for $\frac{1}{6}$? Prove your answer.

71. If $A$ and $B$ are both odd, what can be said about the number of unit fractions with odd denominator whose sum is $A/B$?
    (a) Experiment.
    (b) Make a conjecture.
    (c) Prove it.

72. Prove that whenever $A$ is odd and $B$ is even, then $A/B$ is not the sum of a finite number of unit fractions with odd denominators.

73. Each time the (unmodified) Fibonacci technique is applied to a rational, the "new" numerator is less than the "old" numerator. (The lemma asserts this.) Prove that in the first step of the modified Fibonacci technique the "new" numerator is less than twice the "old" numerator.

74. Prove that if $A$ is odd and $B$ is even, then the modified Fibonacci technique applied to $A/B$ never stops. (*Hint:* See E 72.)

75. (a) Express $0.123_4$ in the form $A/B$, where $A$ and $B$ are expressed in base ten.
    (b) Express $0.123_4$ as a base-ten decimal.

76. Let $A$ and $B$ be positive integers, neither of which is divisible by 3.
    (a) In base three, what can the units digit of $A$ be?
    (b) In base three, what can the units digit of $A^2$ be?
    (c) In base three, what can the units digit of $2B^2$ be?
    (d) Can $A^2 = 2B^2$?
    (e) Use (d) to prove that $\sqrt{2}$ is irrational.

77. In base ten, the divisibility rules for 3 and 9 are essentially the same: If the sum of the digits of the number $A$ is divisible by the number (3 or 9), then so is $A$ itself. If numbers are written in base nine, for what divisors would there be a similar test for divisibility (that is, a test involving the sum of digits)?

78. Express $\frac{3}{7}$ as a base-two "decimal." Carry out the division

$$\begin{array}{r} 0.\overline{\qquad\qquad} \\ 111 \overline{\smash{)}11.0000\ldots} \end{array}$$

    as in Chapter 6.

● ●

79. Is it possible for

$$\overbrace{111\cdots1_3}^{m \text{ 1's}} \quad \text{to equal} \quad \overbrace{111\cdots1_2}^{n \text{ 1's}},$$

    other than in the case $m = 1$ and $n = 1$?

80. We outline a way of multiplying any two natural numbers that uses only multiplication and division by two. We illustrate it by computing $35 \times 56$. First find the quotient when 2 is divided into 35, namely 17. Repeat the same process on 17, obtaining 8. Continue till you reach 1. (In this case we obtain 35, 17, 8, 4, 2, 1.)

Pair off with these numbers those that you obtain from 56 by repeatedly multiplying by 2, in this manner:

| 35 | 56 |
|----|------|
| 17 | 112 |
| 8 | 224 |
| 4 | 448 |
| 2 | 896 |
| 1 | 1792 |

Next cross out all entries in the right-hand column that correspond to *even* entries in the left-hand column. Add the remaining numbers, 56, 112, and 1792, in the right-hand column. Their sum, 1960, is the product $35 \times 56$.

(a) Compute $47 \times 72$ by this method.

(b) Compute $64 \times 8$ by this method.

(c) Prove that this method always works.

81. What are the last two digits of the base-three expression for $17^{100}$?

### References

1. B. L. van der Waerden, *Science Awakening*, Noordhoff Ltd., Groningen, Holland, 1954. (The first two chapters discuss in detail the art of computing and the representation of numbers in ancient times. In particular, pp. 15–30 are devoted to Egyptian arithmetic; pp. 37–40, to the base 60; pp. 51–61, to the Hindu numerals and the positional system.)

2. O. Ore, *Number Theory and Its History*, McGraw-Hill, New York, 1948. (The history of various notational systems is given on pp. 1–24—his "decadic" is the same as our "decimal." Pages 34–37 present a second method for translating from one base to another.)

3. T. Dantzig, *Number, the Language of Science*, Macmillan, New York, 1954. (Chapter 1 and pp. 253–260 discuss the representation of numbers.)

4. O. Ore, *Graphs and Their Uses*, Random House, New York, 1963. (The game of nim is studied with the aid of base 2, pp. 73–75.)

5. M. Dunton and R. Grimm, Fibonacci on Egyptian fractions, *Fibonacci Quarterly*, vol. 4, 1966, pp. 339–354. (This translation of a few pages of the *Liber Abacci* is part of a projected translation of the entire work into English by a number theorist and a classicist.)

6. R. Breusch, A special case of Egyptian fractions, *American Mathematical Monthly*, vol. 61, 1954, pp. 200–201.

7. *R. L. Graham, On finite sums of unit fractions, *Proceedings London Mathematical Society*, vol. 14, 1964, pp. 193–207.

*Chapter* **16**

# TYPES OF NUMBERS

The discovery that $\sqrt{2}$ is not rational led us to separate the numbers into two types, rational and irrational, just as we had separated the natural numbers into the odd and the even. In this chapter we will meet a more profound duality, one that, in a very precise sense, generalizes the duality of the rationals and the irrationals.

Moreover, we will be forced to introduce the complex numbers—numbers that do not usually correspond to points on the line. Numbers that do correspond to points on the line, and which have been adequate for the first fifteen chapters, we call **real numbers**.

We should point out that the material on complex numbers, pp. 267–274, can be read independently of the rest of the chapter.

That $\sqrt{2}$ is not the quotient of two natural numbers can be put another way: there is no rational $M/N$ that, when substituted for $X$ in the equation $1X^2 = 2$, gives us a true statement. Looking at $X$ as a place where numbers, not cars, can be parked, we can say that whenever a rational is parked at $X$ we obtain a false statement. But when we place $\sqrt{2}$ (or, for that matter, $-\sqrt{2}$) at $X$ we obtain a true statement.

We will summarize these remarks in the statement, the *equation $X^2 = 2$ has the roots $\sqrt{2}$ and $-\sqrt{2}$*, and neither of these roots is a rational. It is customary to write equations with only a 0 to the right of the equal sign. Thus we could also say that the equation $X^2 - 2 = 0$ has two roots and that neither of them is rational.

We can consider an even simpler equation, where $X$, but not $X^2$, appears; for example, the equation $3X - 4 = 0$. Whenever we place an integer at $X$ we always obtain a false statement, since 3 does not divide 4. But if we substitute $\frac{4}{3}$ for $X$ we get a true statement: $3(\frac{4}{3}) - 4 = 0$. Thus $\frac{4}{3}$ is a *root* of the equation $3X - 4 = 0$. The equation $3X - 4 = 0$ has just one root, and that root is rational.

As long as we deal with equations of the form $AX + B = 0$, where $A$ and $B$ are integers (and $A$ is not 0), we can find a root among the rationals. Indeed, such equations have only one root, namely, $-B/A$. But, as we saw with $X^2 - 2 = 0$, to find a root for an equation of the type $AX^2 + B = 0$ we might have to look outside of the rationals.

But what happens when we meet an equation in which both $X^2$ and $X$ appear? For example, can we find a root for the equation

(1) $$X^2 + 6X + 9 = 0?$$

This particular equation can be rewritten as

(2) $$(X + 3)^2 = 0,$$

since $(X + 3)(X + 3) = X^2 + 6X + 9$. (Appendix C discusses such operations.)

The only number whose square is 0 is the number 0. Thus, the only number that can replace $X$ in Equation (2) and yield a true statement is the number $-3$. Hence Equation (1) has exactly one root, $-3$. [As a check, we substitute $-3$ for each $X$ of Equation (1); this gives us the statement $(-3)^2 + 6(-3) + 9 = 0$, which is true.]

Let us try to find roots for the equation

(3) $$X^2 + 6X + 2 = 0,$$

which is the same type as (1). The reader is invited to substitute various numbers wherever $X$ appears in (3) and see whether he comes upon one that leads to a true statement.

But rather than conduct a random search for a root of (3), let us first add 9 to both sides in order to make it resemble (1), which turned out to be so easy. We then obtain

$$X^2 + 6X + 9 + 2 = 9,$$

or

$$(X + 3)^2 + 2 = 9.$$

Subtracting 2 from both sides gives us

$$(X + 3)^2 = 7.$$

If a number $R$ is to be a root of (3), then $(R + 3)^2$ must be 7; that is,

$$R + 3 = \sqrt{7},$$

or

$$R + 3 = -\sqrt{7}.$$

The reader may check that the two numbers thus suggested, $-3 + \sqrt{7}$ and $-3 - \sqrt{7}$, are roots of (3) by showing that each of them leads to a true statement when substituted for each $X$ in (3).

Next, let us consider equations of the type $AX^3 + BX^2 + CX + D = 0$, where $A$, $B$, $C$, and $D$ are given real numbers. Here $X^3$ enters the scene. For example, let us see whether the equation

(4) $$8X^3 - 12X^2 - 2X + 3 = 0$$

has a root.

We are trying to find a number $R$ such that $8R^3 - 12R^2 - 2R + 3$ is 0. Let us try a few guesses, say $-2, -1, 0, 1, 2$, and 3. When we replace $X$ in (4) by $-2$ we obtain

$$8(-2)^3 - 12(-2)^2 - 2(-2) + 3,$$

which equals

$$-64 - 48 + 4 + 3,$$

or $-105$. Thus $-2$ is *not* a root of (4). Replacing $X$ by the various guesses yields these results

| $X$ | $8X^3 - 12X^2 - 2X + 3$ |
|---|---|
| $-2$ | $-105$ |
| $-1$ | $-15$ |
| 0 | 3 |
| 1 | $-3$ |
| 2 | 15 |
| 3 | 105 |

Note that when $X$ is replaced by 3, or by larger numbers, $8X^3 - 12X^2 - 2X + 3$ takes on large positive values, for then $X^3$ is much larger than both $X^2$ and $X$. Similarly, when $X$ is replaced by negative numbers that are large in size, $8X^3 - 12X^2 - 2X + 3$ takes on negative values that are large in size.

A graph of the expression renders these observations visible to the eye:

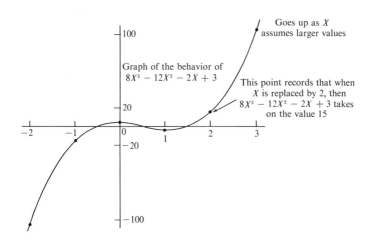

Goes up as $X$ assumes larger values

Graph of the behavior of $8X^3 - 12X^2 - 2X + 3$

This point records that when $X$ is replaced by 2, then $8X^3 - 12X^2 - 2X + 3$ takes on the value 15

As $X$ "wanders" from the negative numbers that lie far to the left toward the large positive numbers, there will be at least one point where the curve crosses the axis of real numbers; that is, where $8X^3 - 12X^2 - 2X + 3$ becomes 0. (A glance at the graph shows that there are in fact three roots. They turn out to be $-\frac{1}{2}$, $\frac{1}{2}$, and $\frac{3}{2}$.)

The same reasoning proves

THEOREM 1. *Any equation of the type $AX^3 + BX^2 + CX + D = 0$, where A, B, C and D are given real numbers and where A is not 0, has at least one real root.*

It can happen that some equations of the type appearing in Theorem 1 have only one real root. For example, the equation $X^3 = 0$ has just one real root; namely, 0.

Before we go further it is advisable to define a few terms. Expressions such as $3X - 4$, $X^2 + X + 1$, $X^3 - 2X^2 + X - 2$, and $X^5 + 5X - 5$ we shall call **polynomials.** A polynomial is a sum of terms of the form $SX^n$, where $S$ is a real number and $n$ is a natural number. If $S$ is 1, as in $1X^5$, we just write $X^5$. The $S$ is called the **coefficient** of $X^n$. The largest $n$ for which the corresponding coefficient is not 0 is called the **degree** of the polynomial. A polynomial of degree 0 is thus of the form $SX^0$, where $S$ is not 0. We will write such a polynomial simply as $S$; for example, instead of $5X^0$, we will write 5. The polynomial that has all of its coefficients equal to zero we call the **zero polynomial** and write it 0; we do not assign it a degree. Thus $5X + 4$ has degree 1 and $6X^{10} - 5X + 2$ has degree 10.

We may add, subtract, multiply, and divide polynomials just as we do integers. As an illustration of addition, we have

$$(6X^2 + 1) + (-5X^2 + X + 7) = X^2 + X + 8.$$

The computation of the product of polynomials involves the same bookkeeping as the multiplication of natural numbers. For instance, let us compute the product $(7X + 3)(8X^2 - 5X + 2)$:

$$
\begin{array}{r}
8X^2 - \phantom{1}5X + 2 \\
\times \phantom{8X^2 - 5}7X + 3 \\
\hline
24X^2 - 15X + 6 \quad \leftarrow \text{from the "3"} \\
56X^3 - 35X^2 + 14X \phantom{+6} \quad \leftarrow \text{from the "7}X\text{"} \\
\hline
56X^3 - 11X^2 - \phantom{1}X \phantom{1}+ 6 \quad \leftarrow \text{the product}
\end{array}
$$

We illustrate division by finding the quotient and remainder when $X^2 - 3X + 1$ is divided into $X^4 + 6X^3 - 1$, which is short for $X^4 + 6X^3 + 0X^2 + 0X - 1$. We arrange the "long division" this way:

$$
\begin{array}{r}
X^2 + \phantom{0}9X + \phantom{0}26 \quad \leftarrow \text{quotient} \\
X^2 - 3X + 1\overline{\smash{\big)}\,X^4 + 6X^3 + \phantom{0}0X^2 + \phantom{0}0X - \phantom{0}1} \\
\underline{X^4 - 3X^3 + \phantom{0}X^2} \\
9X^3 - \phantom{00}X^2 + \phantom{0}0X \\
\underline{9X^3 - 27X^2 + \phantom{0}9X} \\
26X^2 - \phantom{0}9X - \phantom{0}1 \\
\underline{26X^2 - 78X + 26} \\
69X - 27 \quad \leftarrow \text{remainder}
\end{array}
$$

Just as in ordinary arithmetic, this can be checked by a multiplication. The reader may verify that

$$X^4 + 6X^3 - 1 = (X^2 + 9X + 26)(X^2 - 3X + 1) + (69X - 27).$$

The vocabulary of the arithmetic of integers, including "divisor," "multiple," and "factorization," carries over to the arithmetic of polynomials; we will not pause to spell them out.

*An equation of the form $P = 0$, where $P$ is a polynomial of degree n, is called an* **equation of degree** *n.* We may say, for example, that Theorem 1 concerns equations of degree 3.

The same reasoning that proved Theorem 1 also proves

THEOREM 2. *Any equation of odd degree has at least one real root.*

Having met so often the contrast between odd and even, we would expect

THEOREM 3. *For any even natural number n larger than 0, there is an equation of degree n that has no real roots.*

PROOF. For any real number $a$, whether $a$ is negative or positive, $a^2$, $a^4$, $a^6$, and so on are never negative. Thus an equation of the form $X^n + 1 = 0$, with $n$ even, cannot have any real roots.

*Any real number that is a root of an equation of the type $P = 0$, where $P$ is a polynomial whose coefficients are integers (not all 0), we will call* **algebraic** (since they are the numbers we meet while doing algebra). The demand that the coefficients be integers is crucial; otherwise any number $a$ would be algebraic since it is a root of the equation $X + (-a) = 0$.

For example, $\frac{3}{4}$ is algebraic because it is a root of the equation $4X - 3 = 0$. Similarly, $\sqrt{2}$ is algebraic because it is a root of $X^2 - 2 = 0$, and $-3 + \sqrt{7}$ is algebraic because it is a root of $X^2 + 6X + 2 = 0$, as we saw earlier. Clearly the rationals are just algebraic numbers of a special type—namely, those algebraic numbers which are roots of equations of degree 1.

*Any real number that is not algebraic we will call* **transcendental.** The question comes to mind whether every real number is algebraic, and hence no number is transcendental. That is, given a number $a$, is there a polynomial $P$ with integral coefficients such that, when the $X$'s in $P$ are replaced by $a$, we obtain the value 0? This is not an easy question. Recall how difficult it was to prove that $\sqrt{2}$ is irrational, that is, not the root of any equation of the form $P = 0$, *where P is a polynomial of degree 1, with integral coefficients.* To show that a number is transcendental, we would have to show that it is not the root of *any* equation of the form $P = 0$, no matter how high we allow the degree of $P$ to be, where $P$ is a polynomial whose coefficients are integers (not all 0).

In Chapter 18 we will prove, using set theory, that there are transcendentals. For that proof we will need

THEOREM 4. *An equation of degree n has at most n real roots.*

This should come as no surprise. Any equation of degree 1, $AX + B = 0$, has one root; namely, $-B/A$. The equation $X^2 + 6X + 9 = 0$ has one root, and $X^2 + 6X + 2 = 0$ has two roots, as we have already seen. The equation $X^2 + 1 = 0$ has no real roots, since no real number has its square equal to $-1$.

For the proof of Theorem 4 we will need

LEMMA 1. *If r is a root of $P = 0$, then there is a polynomial $Q$ with real coefficients such that $P = Q \times (X - r)$.*

The proof will be easier to follow if we first try a specific case. As the reader may check, 2 is a root of the equation $3X^2 - 10X + 8 = 0$. We will build a polynomial $Q$ such that $3X^2 - 10X + 8 = Q \times (X - 2)$. Another way of putting it is this: we will build a polynomial $Q$ such that when we subtract $Q \times (X - 2)$ from $3X^2 - 10X + 8$ we will obtain 0.

We will build $Q$ bit by bit in order to get rid of the $X^2$, and then the $X$. First, let us get rid of the $3X^2$. To do this we subtract $3X(X - 2)$ from $3X^2 - 10X + 8$. That is, we form

$$(3X^2 - 10X + 8) - 3X(X - 2),$$

which reduces to $3X^2 - 10X + 8 - 3X^2 + 6X$, or simply $-4X + 8$.

Next we get rid of $-4X$ in the polynomial $-4X + 8$. To do this we subtract $-4(X - 2)$ from $-4X + 8$. That is, we form

$$-4X + 8 - (-4)(X - 2),$$

which reduces to $-4X + 8 + 4X - 8$, or simply 0.

Not only did we get rid of the $-4X$, but, simultaneously, we got rid of the 8. This, as we will see in the proof of the lemma, is no coincidence.

Collecting our subtractions we see that

$$3X^2 - 10X + 8 - 3X(X - 2) - (-4)(X - 2) = 0,$$

or

$$3X^2 - 10X + 8 - (3X - 4)(X - 2) = 0.$$

Thus

(5) $$3X^2 - 10X + 8 = (3X - 4)(X - 2).$$

We have found our $Q$, namely $3X - 4$, and the reader can check it by multiplying out the right side of (5). [Relation (5) yields a true statement no matter what we replace $X$ by. For example, if we replace $X$ by 3, we obtain $5 = 5$; if we replace $X$ by 2, we obtain $0 = 0$; and, if we replace $X$ by $\frac{4}{3}$, we again obtain $0 = 0$.]

The reader is advised to use the same technique to find a polynomial $Q$ such that

$$2X^3 - 3X^2 - X - 2 = Q(X - 2).$$

(This will involve three subtractions.)

Now we are ready for the

PROOF OF LEMMA 1. Let $r$ be a root of $P = 0$. By repeated subtractions as above we can find a polynomial $Q$ such that when $Q(X - r)$ is subtracted from the polynomial $P$, all terms involving $X$ will cancel. (Or, to put it another way, we may divide $X - r$ into $P$ by "long division" and obtain a remainder that is of degree 0, or the zero-polynomial.) We want to show that for this $Q$, the term in $P$ that has no $X$ will also cancel, as it did in the preceding example.

In any case, we know that there is a real number $s$ such that

(6) $$P - Q(X - r) = s.$$

We want to prove that $s = 0$. For the first time we will use the fact that $r$ is a root of $P = 0$. Let us replace each $X$ in $P$, $Q$, and $X - r$ of (6) by $r$. Since (6) yields a true statement whenever we replace $X$ by any real number, it will yield a true statement when $X$ is replaced by $r$. When $X$ is replaced by $r$, then $P$ takes on the value 0; $Q$ takes on some value that we shall call $q$; and $X - r$ takes on the value $r - r$, which is 0. Thus from (6) we deduce that

$$0 - (q \times 0) = s.$$

Thus $s = 0$, as we had hoped. This proves the lemma.

We are now ready to give a

PROOF OF THEOREM 4. For equations of degree 1, the theorem is clearly true. Let us consider equations of degree 2. If the equation $P = 0$ has no roots, then

Theorem 4 certainly is true for this equation. If $P = 0$ has a root, let $r$ be one of the roots. According to the lemma, there is a polynomial $Q$ such that $P = Q(X - r)$. Since $P$ has degree 2, $Q$ must have degree 1.

Now let $r'$ be another root, different from $r$, of the equation $P = 0$. That is, when $r'$ replaces each of the $X$'s in $P$, then $P = 0$. If we replace every $X$ in $Q$ by this same $r'$, then $Q$ will assume some value that we will denote by $q$. We will now show that $q = 0$.

When $X$ is replaced by $r'$ at each of its appearances in $P$, $Q$, and $X - r$ in the equation

$$P = Q(X - r),$$

we obtain

$$0 = q(r' - r).$$

Since $r'$ is different from $r$, we see that $r' - r$ is not 0. But $q(r' - r)$ is 0. Thus $q$ must be 0. That is, $r'$ is a root of the equation $Q = 0$. Thus every root of the equation $P = 0$, other than $r$, is also a root of the equation $Q = 0$. But the equation $Q = 0$ has just one root, since $Q$ is of degree 1. Thus the equation $P = 0$ has at most two roots, hence Theorem 4 is proved for equations of degree 2.

If $P = 0$ is an equation of degree 3, the reasoning is similar, and makes use of the fact that an equation of degree 2 has at most 2 roots. In detail, if $r$ is a root of the equation $P = 0$, then according to Lemma 1 there is a polynomial $Q$ (this time of degree 2) such that $P = Q \times (X - r)$.

Just as in the previous case, we can prove that any root $r'$ of $P = 0$, other than $r$, must also be a root of $Q = 0$. Since we have already proved that the equation $Q = 0$ has at most two roots, we then conclude that the equation $P = 0$ can have at most three roots. The same argument then applies, step by step, to equations of degree 4, then 5, then 6, and so on, which proves Theorem 4.

Theorem 4 tells us that an equation $P = 0$ has only a limited supply of roots. Theorem 3 tells us that there are equations without any real roots whatsoever; for example, the equation $X^2 + 1 = 0$ has no roots. This is inconvenient, since such equations are met daily in mathematics and in its applications to the physical world, and it would be useful to have roots for them on which we could operate as we do with the numbers of ordinary arithmetic.

We will now construct a mathematical system in which every equation will have at least one root. In this man-made structure we will define an operation $\oplus$, which will behave like ordinary addition, and an operation $\otimes$, which will behave like ordinary multiplication. These operations will satisfy the rules whose importance is emphasized in Chapter 12 and in Appendix C. We cite the more important ones as a reminder:

$$\left.\begin{array}{l} X \oplus Y = Y \oplus X \\ X \otimes Y = Y \otimes X \end{array}\right\} \text{commutativity};$$

$$\left.\begin{array}{l} X \oplus (Y \oplus Z) = (X \oplus Y) \oplus Z \\ X \otimes (Y \otimes Z) = (X \otimes Y) \otimes Z \end{array}\right\} \text{associativity};$$

and the rule that relates $\otimes$ to $\oplus$,

$$X \otimes (Y \oplus Z) = (X \otimes Y) \oplus (X \otimes Z) \qquad \text{(distributivity)}.$$

The first hints that man could build such a mathematical system appeared in the sixteenth century, but it was not till the early part of the nineteenth century that it was accepted as a legitimate part of mathematics.

We now define this structure that will enable us to find roots for all equations, including the equation $X^2 + 1 = 0$. We shall build it on the whole plane, whereas the real numbers, with their addition and multiplication, are built only on the line. (Recall that the real numbers include the transcendental numbers and the algebraic numbers, and that the algebraic numbers include the rational numbers, which in turn include the integers. The number system we will now construct will include all the reals.) We will not attempt to give the history behind the definitions of $\oplus$ and $\otimes$; we will simply show that these definitions provide a useful mathematical structure.

If $X$ and $Y$ are any two points in the plane, we shall define $X \oplus Y$ as the fourth vertex of the parallelogram, two of whose sides are $0X$ and $0Y$:

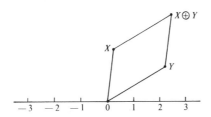

Since the parallelogram determined by $0$, $X$, and $Y$ is the same as the parallelogram determined by $0$, $Y$, and $X$, then $X \oplus Y = Y \oplus X$. It is not so clear, however, that $X \oplus (Y \oplus Z) = (X \oplus Y) \oplus Z$ for any three points $X, Y, Z$ in the plane; thus the reader is urged to draw the (four) necessary parallelograms and check that $\oplus$ is associative for a specific $X$, $Y$, and $Z$. He will be delighted to see the point $X \oplus (Y \oplus Z)$ coincide with the point $(X \oplus Y) \oplus Z$.

If $0$, $X$, and $Y$ happen to be on a line, then the parallelogram collapses, and its fourth vertex lies on the same line as $0$, $X$, and $Y$. For example, as the reader may check, $2 \oplus 1 = 3$, and $4 \oplus (-4) = 0$. The reader may check (with a few diagrams) that whenever $X$ and $Y$ happen to be on the line of real numbers, then $X \oplus Y$ coincides with $X + Y$. Therefore the operation $\oplus$ can be thought of as an extension to the whole plane of our everyday addition. Or we might prefer to think that our everyday addition is but a small part of the operation $\oplus$ defined throughout the plane.

Next we define the second half of the structure, the operation $\otimes$. Observe first that a point $X$ in the plane (different from 0) determines an angle in this manner, measured counterclockwise from the ray of positive real numbers.

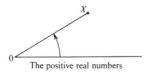

The positive real numbers

In particular, any point on the vertical line through 0, and above 0, determines an angle of 90°. Every negative real number determines the angle 180°. Every positive real number determines the angle 0°. Finally, every point directly below 0 determines the angle 270°.

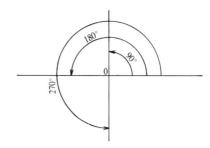

Of course, if we add any multiple of 360° to an angle, we describe the same direction; thus the negative real numbers, for example, also determine the angles 540°, 900°, 1260°, . . . .

If $X$ and $Y$ are points in the plane, and if both differ from 0, then we define $X \otimes Y$ in this manner. *The point $X$ lies on a circle whose center is 0. Let us call the radius of this circle r. The point $X$ determines an angle $A$, as shown in this diagram:*

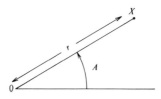

*Similarly, Y will be located at a certain distance, say, s, from 0, and will determine a certain angle, say, B. Here is how we will draw X ⊗ Y. We must tell how far and in what direction X ⊗ Y is to be from 0. The point X ⊗ Y will be on the circle with center 0 and radius r × s* (this makes use of ordinary multiplication) *and will have the angle A + B* (this makes use of ordinary addition). *If X or Y happens to be 0, then we define X ⊗ Y to be 0.*

*The reader can check that X ⊗ Y is simply the point obtained from Y by rotating Y about 0 by the angle of X, and then magnifying or shrinking the result by a factor equal to the radius of X.*

As an example, let us find X ⊗ Y for the X and Y in this diagram:

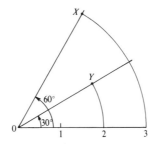

Since $r$ is 3 and $s$ is 2, $X \otimes Y$ is somewhere on the circle with center 0 and radius 6. Where on this circle is $X \otimes Y$? The angle of $X \otimes Y$ is $60° + 30°$; that is, $90°$. Thus $X \otimes Y$ lies on the line through 0 that is perpendicular to the line of real numbers:

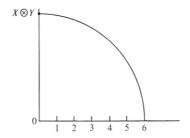

That $X \otimes Y = Y \otimes X$ for any $X$ and $Y$ in the plane follows from the commutativity of ordinary multiplication ($r \times s = s \times r$) and of ordinary addition ($A + B = B + A$). For similar reasons, $X \otimes (Y \otimes Z) = (X \otimes Y) \otimes Z$ for any three points in the plane.

Next we must show that

$$X \otimes (Y \oplus Z) = (X \otimes Y) \oplus (X \otimes Z)$$

for any $X$, $Y$, and $Z$ in the plane. We prove this first for the special case in which $X$ is on the circle of radius 1. For any point $X'$ in the plane (since $r$, the radius of $X$, is 1), $X \otimes X'$ and $X'$ are at the same distance from 0. In fact $X \otimes X'$ is obtained from $X'$ by rotating $X'$ about 0 through the angle determined by $X$. This fact we will now use in the three cases where $X'$ is $Y$, $Z$, or $Y \oplus Z$.

Consider the parallelogram defining $Y \oplus Z$:

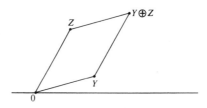

Now rotate this whole parallelogram by the angle of $X$, whatever it may be:

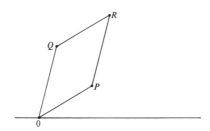

The rotation moves $Y$ to $P$, $Z$ to $Q$, and $Y \oplus Z$ to $R$. Since we are considering only the case in which $X$ lies on the circle of radius 1, we have

(7)           $P = X \otimes Y, \quad Q = X \otimes Z, \quad R = X \otimes (Y \oplus Z).$

But inspection of parallelogram $0PRQ$ shows that $R = P \oplus Q$. Combining this with (7), we obtain

(8)                          $X \otimes (Y \oplus Z) = (X \otimes Y) \oplus (X \otimes Z),$

thus proving distributivity for the special case in which $X$ is on the circle of radius 1.

In case $X$ is not on the circle of radius 1 we must magnify (or shrink) parallelogram $0PRQ$ by the distance from 0 to $X$. (Why?) Since the magnification of a parallelogram is still a parallelogram of the same shape (think of a parallelogram magnified by a slide projector), we see that (8) holds for any $X$, $Y$ and $Z$ in the plane. Thus distributivity is proved.

Just as we saw that addition on the line is but a small part of the operation $\oplus$, we will now show that multiplication on the line is but a small part of the operation $\otimes$. For example, let us compute $2 \otimes 3$ and see how it compares with $2 \times 3$. Clearly $2 \otimes 3$ is somewhere on the circle of radius 6 and center 0. Since the angle of 2 is 0° and the angle of 3 is 0°, the angle of 2 ⊗ 3 is (0 + 0)°, which is 0°. Thus 2 ⊗ 3 is 6, which is the same as 2 × 3.

As another check, let us see whether (−1) ⊗ (−1) is equal to (−1) × (−1); that is, 1. Since −1 is on the circle of radius 1 and center 0, then (−1) ⊗ (−1) is somewhere on the circle of radius 1 × 1 and center 0; that is, the circle of radius 1 and center 0. Where does (−1) ⊗ (−1) lie on this circle? The angle of −1 is 180°; thus the angle of (−1) ⊗ (−1) is (180 + 180)°; that is, 360°. Thus (−1) ⊗ (−1) is the point 1, on the line of real numbers. In a similar manner, the reader may check that 2 ⊗ (−3) is −6, and then go on to show why $X ⊗ Y$ is $X × Y$ whenever both $X$ and $Y$ happen to lie on the real line.

The reader who is suspicious of a multiplication and addition of points in the plane should be reminded that up to the sixteenth century, mathematicians thought that all arithmetic should be done with that part of the real line to the right of 0. If the root of an equation happened to be negative, they called it "fictitious" and threw it away as an absurdity. But today we are as accustomed to working on that part of the real line to the left of 0 as we are to reading thermometers that record temperatures below 0.

The main object of our construction on the plane is to provide roots for equations that have no real roots. In particular, we hope that the equation $X^2 + 1 = 0$ now has a root. Let us see whether there is a point $B$ in the plane such that $(B ⊗ B) ⊕ 1 = 0$; that is, a point $B$ such that $B ⊗ B = −1$.

On what size circle would $B$ have to lie, and at what angle? We know that −1 is on the circle of radius 1 and has the angle 180°, or in general, (180 + multiple of 360)°. Therefore if $B$ is on the circle of radius $r$, we must have $r × r = 1$. Thus $r = 1$. Next, what is the angle of $B$? Since the angle of $B ⊗ B$ is twice the angle of $B$ and is to be 180°, we see that $B$ may have the angle 90°. Thus we have found a root for the equation $X^2 = 0$ directly above 0 as in this picture:

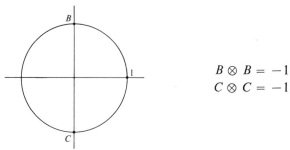

$$B ⊗ B = −1$$
$$C ⊗ C = −1$$

Moreover, as the reader may check, it is also true that $C ⊗ C = −1$. Thus the plane provides at least two roots of $X^2 ⊕ 1 = 0$. The reader may check that the plane provides no more roots of $X^2 ⊕ 1 = 0$.

It is the custom to give $B$ the name $i$. The symbol $i$ stands for "imaginary," and is a memorial to the uncertainties of the mathematicians who first leaped off the tight rope of the real line into the mysteries of the plane. This might remind us of the "X" in X-ray, which records Roentgen's perplexity when he first noticed X-rays in 1895; little could he know then that they are just part of a whole scale of radiation that includes heat and light as special cases.

*In order to feel more at home with the operations $\oplus$ and $\otimes$ let us first of all write $+$ instead of $\oplus$ and $\times$ instead of $\otimes$. After all, $\oplus$ and $+$ do agree on the real line, and so do $\otimes$ and $\times$.* For example, instead of the cumbersome $3 \otimes i$ we will write $3 \times i$, or just $3i$. As the reader may check, $3i$ appears on the plane in this position:

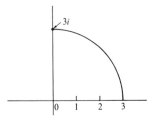

With this agreement, we summarize our observations concerning $i$ in this important

THEOREM 5. *The equation $X^2 + 1 = 0$ has two roots; they are called $i$ and $-i$.*

The equation $X^5 - 1 = 0$, which has only one root on the line of real numbers, namely 1, has five roots in the plane, regularly spaced on the circle of radius 1, as shown in this diagram:

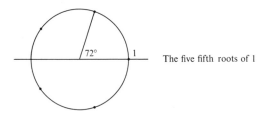

The five fifth roots of 1

The reader may check that they are roots by using the definition of multiplication of points in the plane.

For any real number $b$, the product $bi$ is a point on the vertical line through 0 (above 0 if $b$ is positive, below 0 if $b$ is negative). Using this observation we will prove that every point $P$ in the plane can be written in the form

$$a + bi,$$

where $a$ and $b$ are real numbers. To do this, we simply draw the parallelogram (in fact, rectangle) determined by $P$ in this manner:

By the definition of $\oplus$, which we have agreed to write $+$, $P = A + B$. But $A$ is a real number, which we can call $a$, and $B$ is of the form $bi$ (that is, $b \otimes i$) for some real $b$. Thus

$$P = a + bi.$$

The fact that every point in the plane can be written in the form $a + bi$ gives us a second way of describing these points. In our first method we gave the angle determined by the point and the distance from the point to 0. Now we can give $a$ and $b$, which relate the point to the line of reals and to the vertical line through 0. For example, the point $-3 + 4i$ is 3 units to the left of the vertical line through 0, and 4 units above the line of reals.

This raises a problem. How can we compute the product of two points given in this new manner? For example, how can we find the real numbers $a$ and $b$ such that

$$(3 + 4i) \times (1 + i) = a + bi?$$

One method is to go back to the geometric definition of multiplication of points in the plane. Using the Pythagorean Theorem, the reader will find that the point $3 + 4i$ lies on the circle of radius 5, and that the point $1 + i$ lies on the circle of radius $\sqrt{2}$, which is approximately 1.41. Thus $(3 + 4i) \times (1 + i)$ lies on the circle whose radius is about 7.05. To determine the angle of $(3 + 4i) \times (1 + i)$ the reader should draw the two points $3 + 4i$ and $1 + i$ and add their angles with the aid of a compass or protractor.

If the reader is careful he will find that $(3 + 4i) \times (1 + i)$ seems to be very close to $-1 + 7i$. That is, the $a$ and the $b$ we were seeking seem to be integers, despite the presence of $\sqrt{2}$ and the numerically inconvenient angle determined by the point $3 + 4i$. But at this moment, because of our approximation of $\sqrt{2}$ and because the accuracy of our drawings is always limited by the thickness of the pencil point, we are not sure that $a$ is $-1$ and $b$ is 7.

Let us compute $(3 + 4i) \times (1 + i)$ in a different way, without drawing any pictures. Using distributivity, we can reduce our problem to two simpler problems, since

$$(3 + 4i) \times (1 + i) = [(3 + 4i) \times 1] + [(3 + 4i) \times i].$$

As the reader may easily check by going back to our definition of $\otimes$ in the plane,

$$(3 + 4i) \times 1 = 3 + 4i.$$

Using commutativity of multiplication in the plane, we have

$$(3 + 4i) \times i = i \times (3 + 4i).$$

Then, by distributivity,

$$i \times (3 + 4i) = i \times 3 + i \times 4i.$$

By going back to the definition of multiplication in the plane, the reader may check that $i \times 3 = 3i$ and that $i \times 4i = -4$. Thus

$$(3 + 4i)(1 + i) = 3 + 4i + 3i - 4,$$

which the reader may simplify to $-1 + 7i$. Our suspicion that $a$ and $b$ are integers is correct.

As the preceding computation shows, the arithmetic of the operations $\oplus$ and $\otimes$ in the plane closely resembles that of our usual addition and multiplication of real numbers. In fact, since the operations $\oplus$ and $\otimes$ satisfy the basic rules of arithmetic, computation with these operations differs from computation with ordinary addition and multiplication simply in the appearance of the symbol $i$ and the extra rule, $i^2 = -1$.

*When we consider the plane together with the operations $\oplus$ and $\otimes$, we call the points of the plane* **complex numbers.** The word "complex" is more awesome than instructive. If history, like a sweater, could be unraveled and knitted again, we might call the points of the plane merely "numbers," and then treat those points that happen to lie on the horizontal line through 0 as a special type of number. Presumably, the word "complex" refers to the fact that a complex number is the sum of a real number and a number on the vertical line through 0 (the line of so-called imaginary numbers).

The real numbers are sufficient to provide every equation of odd degree with at least one root (Theorem 2) and to provide some equations of even degree with roots. The algebraic significance of the complex numbers lies in the fact that every equation with real coefficients has at least one root among the complex numbers. (Of course, all the roots might be real, such as the root of $3X + 1 = 0$, but recall that every real number is also a complex number.) Indeed, far more is true, as we see in the

FUNDAMENTAL THEOREM OF ALGEBRA. *Any equation of degree greater than zero with complex coefficients has at least one root among the complex numbers.*

The first proof of this theorem was given in 1799 by Gauss. We will not give a proof, but only mention that the result depends both on algebra and topology. (See R 7 for a sketch of the proof.)

However, we will display the power of this theorem and show the usefulness of the complex numbers by proving

THEOREM 6. *Any polynomial with real coefficients can be expressed as the product of such polynomials whose degrees are either 1 or 2.*

The theorem may be startling, for it implies that any polynomial with real coefficients and of degree at least 3 can be "factored." Theorem 2 and Lemma 1 already told us that a polynomial of odd degree could be factored. As instances of Theorem 6 we cite

$$X^4 + 1 = (X^2 + \sqrt{2}\,X + 1)(X^2 - \sqrt{2}\,X + 1)$$

and

$$X^4 + X^2 + 1 = (X^2 + X + 1)(X^2 - X + 1),$$

both of which may be checked by multiplication.

Before we begin the proof of Theorem 6, we introduce the notion of the conjugate of a complex number. The **conjugate** of the complex number $A = a + bi$ is the complex number $a - bi$. It is denoted $\bar{A}$ or $\overline{a + bi}$. Geometrically, $\bar{A}$ is simply the reflection of $A$ across the axis of real numbers.

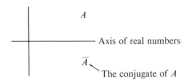

A glance at the diagram shows that

$$A + \bar{A} \qquad \text{and} \qquad A\bar{A}$$

are real numbers. Furthermore, for any two complex numbers $A$ and $B$ we have

(9) $$\overline{A + B} = \bar{A} + \bar{B}.$$

To see this, just reflect the parallelogram used to define $A + B$ across the axis of real numbers. Similarly,

(10) $$\overline{A \times B} = \bar{A} \times \bar{B}.$$

Observe that for a real number, $a$, we have

(11) $$\bar{a} = a.$$

The next lemma generalizes the fact that both $i$ and $\bar{i}$ (which is $-i$) are roots of the equation $X^2 + 1 = 0$.

LEMMA 2. *If the complex number R is a root of the equation P = 0, where P is a polynomial with real coefficients, then $\overline{R}$ is also a root of that equation.*

PROOF. We illustrate the argument for a polynomial picked at random, say $P = 5X^4 - 6X^3 + 8X + \sqrt{2}$. Since $R$ is a root of $P = 0$, we have

$$5R^4 - 6R^3 + 8R + \sqrt{2} = 0.$$

Now let us "conjugate" both sides of that equation, obtaining

$$\overline{5R^4 - 6R^3 + 8R + \sqrt{2}} = \overline{0}.$$

By (*11*), $0 = \overline{0}$; by (*9*), we conclude that

$$\overline{5R^4} - \overline{6R^3} + \overline{8R} + \overline{\sqrt{2}} = 0.$$

By (*10*), $\overline{5R^4} = \overline{5}\,\overline{R^4}$; by (*11*), $\overline{5} = 5$. Thus

$$\overline{5R^4} = 5\overline{R}^4.$$

Applying the same reasoning to the other terms, we obtain

$$5\overline{R}^4 - 6\overline{R}^3 + 8\overline{R} + \sqrt{2} = 0.$$

Thus $\overline{R}$ is also a root of $P = 0$. Since the same argument can be applied to any polynomial, the lemma is proved.

LEMMA 3. *Let A be any complex number. Then, when $(X - A)(X - \overline{A})$ is multiplied out, the resulting product,*

$$X^2 - (A + \overline{A})X + A\overline{A},$$

*has only real coefficients.*

PROOF. This is just a restatement of the observation that $A + \overline{A}$ and $A\overline{A}$ are real.

LEMMA 4. *If P is a polynomial with real coefficients and R is a root of the equation P = 0, and if R is not real, there is a polynomial Q with real coefficients such that*

$$P = Q(X - R)(X - \overline{R}).$$

PROOF. By Lemma 2, $\overline{R}$ is also a root of the equation $P = 0$. By the same reasoning as used in the proof of Theorem 4, we know that there is a polynomial $Q$ (with complex coefficients) such that

$$P = Q(X - R)(X - \overline{R}).$$

By Lemma 2, $Q$ is the quotient of two polynomials with real coefficients, namely $P$ and $(X - R)(X - \overline{R})$. Hence the coefficients of $Q$ are real, and the lemma is proved.

Now we come to the

PROOF OF THEOREM 6. Let $P$ be any polynomial whose coefficients are real. If the degree of $P$ is 1 or 2, there is nothing to prove. If the degree of $P$ is 3, we know by Theorem 1 that it has a real root $r$, and then by Lemma 1, that there is a polynomial $Q$ with real coefficients such that $P = Q(X - r)$. Thus Theorem 6 is true for degree 3.

Rather than treat each degree separately, we will proceed as we did in the proof of the five-color theorem, where we made the coloring of a given map depend on the coloring of a map with fewer countries.

Let us imagine that we have proved Theorem 6 for all polynomials $P$ whose degree is, say, at most 99. We will then show that Theorem 6 holds for polynomials of degree 100.

Let $P$ be a polynomial of degree 100. By the Fundamental Theorem of Algebra, the equation $P = 0$ has at least one complex root $R$. There are two possibilities:

(i) $R$ is real,

(ii) $R$ is not real, hence $\overline{R}$ is different from $R$.

Consider case (i) first. By Lemma 1, there is a polynomial $Q$ with real coefficients such that

$$(12) \qquad P = Q(X - R).$$

But $Q$ has degree 99 and, by our assumption, can be expressed as the product of polynomials with real coefficients and of degree at most 2. Combining this with $(12)$ we see that Theorem 6 holds for $P$.

Now to case (ii). By Lemma 4, $P$ is the product of a polynomial $Q$ and the polynomial $(X - R)(X - \overline{R})$,

$$(13) \qquad P = Q(X - R)(X - \overline{R}).$$

But $Q$ must have degree 98, and hence is the product of polynomials of degree at most 2. Combining this with $(13)$ shows that Theorem 6 holds for $P$.

Thus if Theorem 6 holds through degree 99, it follows that it holds for degree 100. The same reasoning then shows that it holds for degree 101. In the same manner, beginning at degree 3, we conclude that it holds for degree 4; hence for degree 5; hence for degree 6; and so on. Theorem 6 is proved.

We have already pointed out the resemblance of this proof to that of the five-color theorem. But there is a deeper similarity between this proof and that of the theorem in Chapter 5 which asserts that every prime is special. Both introduce and depend on structures that are in no way referred to in the statement of the theorem itself. Recall that the notions "prime" and "special" refer only to multiplication, not to addition. Yet the proof that every prime is special leans heavily on addition. And now we have proved a theorem about polynomials with real coefficients by entering the realm of complex numbers. This may be inevitable, for no one has ever found a proof that stays within the real numbers.

One more comment on the proof is in order. Theorem 6 concerns the product of polynomials; it does not refer to the notion "root of an equation." Yet this notion, expressed in Lemmas 1 and 4, is the heart of the proof. No one has managed to avoid it.

The complex numbers were not built in a day. Man began with the counting numbers, 1, 2, 3, 4, 5, . . . . But to find roots for equations of the type $AX = B$, he had to introduce all the positive rationals (not just those with denominator equal to 1). Zero was introduced to simplify the representation of numbers (for example, consider how important 0 is in distinguishing 37 from 307 and 30070). To find a root for equations of the type $X + A = 0$, he had to invent the negative numbers. To find a root for $X^2 - 2 = 0$, he had to enter the realm of irrationals. But to guarantee that any equation with rational coefficients has a root, he had to build the complex numbers. In searching for roots of equations, man need never go beyond the complex numbers.

Though the complex numbers first served only the pure mathematician, in less than a century after their acceptance in algebra, they were applied to the physical world. In 1893, Steinmetz utilized the complex numbers to simplify the theory of alternating currents. Unlike direct current, such as a battery provides, an alternating current fluctuates in magnitude and direction—customarily sixty times a second.

In direct current, a constant voltage $E$, working against a resistance $R$, produces a constant current $I$. These three quantities are related by the equation

$$E = IR,$$

which involves the multiplication of real numbers. (This equation, which says that "voltage drop is proportional to current" is the basis of Chapter 8.)

A coil rotating within a constant magnetic field produces an alternating current. As Steinmetz wrote in 1893,

> The current rises from zero to a maximum; then decreases again to nothing, reverses and rises to a maximum in the opposite direction; decreases to zero, again reverses and rises to a maximum in the first direction—and so on.
>
> Thus in all calculations with alternating current, instead of a simple mechanical value of direct current theory, the investigator had to use a complicated function of time to represent the alternating current. The theory of alternating current apparatus thereby became so complicated that the investigator never got very far. . . .
>
> The idea suggested itself at length of representing the alternating current by a single complex number. . . . This proved the solution of the alternating current calculation.

It gave to the alternating current a single numerical value, just as to the direct current, instead of the complicated function of time of the previous theory; and thereby it made alternating current calculations as simple as direct current calculations.

The introduction of the complex number has eliminated the function of time from the alternating current theory, and has made the alternating current theory the simple algebra of the complex number, just as the direct current theory leads to the simple algebra of the real number.

We will sketch how Steinmetz, as has been said, "generated electricity out of the square root of $-1$."

To describe the motor, which produces the varying voltage, he uses the single complex number **E**. The distance from **E** to 0 shall be the *maximum voltage* the machine is producing. The angle of **E** is determined by the *initial position* of the rotating coil.

Corresponding to the alternating current is the complex number **I**. The distance from **I** to 0 is the *maximum current* flowing through the circuit. The angle of **I** records the *lag* in the current—it does not necessarily reach its maximum when the voltage does.

Within the typical circuit are, in addition to the motor: a resistance, a capacitor (for storing electrical charge), and a fixed coil (which produces a magnetic field when current passes through it). With the resistance is associated the real number $R$; with the capacitor, the real number $X_C$; with the fixed coil, the real number $X_L$. Steinmetz introduced the single complex number

$$\mathbf{R} = R + (X_L - X_C)i,$$

the *complex impedance*, and summarized the basic behavior of alternating currents in the single equation

$$\mathbf{E} = \mathbf{I} \otimes \mathbf{R},$$

where $\otimes$ refers to the multiplication of complex numbers. About two decades passed before electrical engineers completely switched over to the complex numbers.

The reader might expect us to construct operations $\oplus$ and $\otimes$ in space, as we did in the plane, which will extend the $\oplus$ and $\otimes$ of the complex numbers in the same sense that the $\oplus$ and $\otimes$ of complex numbers extend the $+$ and $\times$ of the real line.

It is easy to define an addition, $\oplus$, for points in space by using the same parallelogram technique that we used in defining $\oplus$ for the complex numbers; this new $\oplus$ is still commutative and associative. But it has been known for almost a century (Frobenius, 1878, and C. S. Pierce, 1881) that it is impossible to construct a $\otimes$ in space such that $\oplus$ and $\otimes$ together satisfy the usual rules of arithmetic, as given in Appendix C.

In 1843, Hamilton had managed to construct a $\oplus$ and a $\otimes$ in four-dimensional space (spaces of all dimensions are defined in Appendix F) that satisfied every rule of ordinary arithmetic except that the multiplication $\otimes$ was not commutative. In 1845, Cayley built a $\oplus$ and a $\otimes$ in eight-dimensional space that satisfied every rule of arithmetic except that $\otimes$ was neither commutative nor associative. The higher the dimension of the space, the more that had to be sacrificed. Note that when we go from the real numbers, built on a space of dimension 1, to the complex numbers, built on a space of dimension 2, we sacrifice the notion of "positive and negative" (that is, right and left of 0).

The reader may have noted that the numbers 1, 2, 4, and 8 of the preceding paragraph are powers of 2, and may ask, "What about other dimensions?" Frobenius and Pierce proved that only for dimensions 1, 2, and 4 can a $\oplus$ and a $\otimes$ be defined that will satisfy the rules of arithmetic, even if we drop the demand that $\otimes$ be commutative. In 1958, R. Bott, M. Kervaire, and J. Milnor, using the machinery of algebraic topology, proved that only for dimensions 1, 2, 4, and 8 can a $\oplus$ and a $\otimes$ be defined that satisfy the rules of arithmetic, even if we drop the demand that $\otimes$ be associative or commutative.

Yet, even in the line or in the plane, there are difficult problems. Among the more challenging problems of the nineteenth century was to decide whether $\pi$ (the ratio between the circumference and diameter of a circle) is an algebraic or a transcendental number. Hermite, who had already proved certain other numbers transcendental, stated in 1873, "I shall risk nothing on an attempt to prove the transcendence of the number $\pi$. If others undertake this enterprise, no one will be happier than I at their success, but believe me, my dear friend, it will not fail to cost them some efforts." Nine years later Lindemann proved that $\pi$ is transcendental.

No one knows, however, whether the number

$$\frac{1}{1} + \frac{1}{8} + \frac{1}{27} + \frac{1}{64} + \cdots,$$

formed by adding the reciprocals of all the cubes, and which is about 1.2, is rational or irrational.

Now that we have reached the complex numbers, let us show how the various types of numbers we have met in this book fit together. In this diagram the symbol $\supset$ is short for "contain."

(An irrational number can be either algebraic or transcendental.) The reader might well think of this book as a description of the various types of numbers and of some of their many applications.

In the next chapter we will apply polynomials and the complex numbers to geometric problems. In Chapter 18 we will prove that there are transcendental numbers.

### Exercises

1. (a) What is meant by a "root" of the equation $6X^2 - 7X - 10 = 0$?
   (b) Is 1 a root of that equation?
   (c) Is 2 a root?
   (d) Is $-1$ a root?
   (e) Show that $-\frac{5}{6}$ is a root.

2. How many of the numbers 0, 1, 2, 3, 4, 5 are roots of the equation $X^3 - 5X + 2 = 0$?

3. Find all real roots of (a) $X^2 + 4X + 4 = 0$, (b) $X^2 + 4X + 1 = 0$.

4. Check that 0, 1, and 2 are roots of $X^3 - 3X^2 + 2X = 0$. Can there be any more roots?

5. Which of the numbers 0, 1, $-1$, 2, $-2$, $\frac{1}{2}$ are roots of $2X^2 - 3X - 2 = 0$?

6. (a) What is the value of $5X^3 - 2X^2 + 6X - 7$ when $X$ is replaced by 0?
   (b) When $X$ is replaced by 1?
   (c) When $X$ is replaced by $\frac{1}{2}$?
   (d) Between what two numbers have we trapped a root of the equation
   $$5X^3 - 2X^2 + 6X - 7 = 0?$$

7. (a) Continuing with the equation of E 6, what is the numerical value of $5X^3 - 2X^2 + 6X - 7$ when $X$ is replaced by 0.8?
   (b) Same as (a) when $X$ is replaced by 0.9.
   (c) Between what two numbers have we trapped a root?
   (d) Which of these two numbers do you think is nearer the root?

8. (a) Are any of the numbers 1, 2, or 3 roots of the equation $2X^3 - X^2 + X - 5 = 0$?
   (b) Is there a root between 1 and 2?

9. (a) How does the table of values of $8X^3 - 12X^2 - 2X + 3$ on page 261 show us that the equation $8X^3 - 12X^2 - 2X + 3 = 0$ has at least three roots?
   (b) Check that $\frac{1}{2}$, $\frac{3}{2}$, and $-\frac{1}{2}$ are roots.

10. Show that 3 is a root of the equation $X^4 - 5X^3 + 10X + 24 = 0$.

11. Show that $-3 + \sqrt{7}$ is a root of the equation $X^2 + 6X + 2 = 0$.

12. Proceeding as we did with the equation $X^2 + 6X + 9 = 0$, find all the roots of
    (a) $X^2 - 6X + 9 = 0$ and (b) $X^2 + 3X + \frac{9}{4} = 0$.

13. Find all real roots of (a) $X^2 + 5 = 0$, (b) $X^2 + 14X + 49 = 0$.

14. Find all real roots of (a) $X^2 + 4X + 4 = 0$, (b) $X^2 + 2X + 1 = 0$.

15. Which of these equations has at least one real root? (a) $X^6 + 1 = 0$, (b) $X^6 + X^2 + 1 = 0$, (c) $X^7 + X + 3 = 0$, (d) Explain.

16. (a) Proceeding as we did with the equation $X^2 + 6X + 2 = 0$, find all the roots of $X^2 - 6X - 7 = 0$ and $X^2 - 2X - 4 = 0$.
    (b) Check that the roots you found in (a) lead to a true statement when substituted for $X$ in the equation.

17. (a) What is meant by an "algebraic" number?
    (b) Is 5 algebraic?
    (c) Is $\sqrt{7}$ algebraic?
    (d) Is $\frac{4}{5}$ algebraic?
    (e) Explain.

18. (a) Is $\sqrt[3]{2}$ algebraic?
    (b) Is $-\frac{3}{4}$ algebraic?
    (c) Is $-3 - \sqrt{7}$ algebraic?

19. What is the degree of the polynomials (a) $17X^3 - X + 1$, (b) $X^2 + X$, (c) $5X - 3$, (d) 19?

20. Give an example of a polynomial of degree 52.

21. Add the polynomials $6X^3 - X + 1$ and $5X^2 + X + 4$.

22. Add the polynomials $5X^4 + X - 2$ and $-5X^4 + \sqrt{2}\,X^3$.

23. Multiply the polynomials (a) $X - 5$ and $X + 5$, (b) $X^2 + 1$ and $X + 1$, (c) $X - 3$ and $X + 4$.

24. Multiply the polynomials (a) $X - \sqrt{2}$ and $X + \sqrt{2}$, (b) $X + (3 + \sqrt{7})$ and $X + (3 - \sqrt{7})$.

25. Multiply the polynomials $X^2 + \sqrt{2}\,X + 1$ and $X^2 - \sqrt{2}\,X + 1$.

26. Let $P$ be a polynomial of degree 3 and $Q$ be a polynomial of degree 5. What can be said about the degree of (a) $P + Q$, (b) $P \times Q$?

27. Find the quotient and remainder when you divide $X + 1$ into $X^3 + 7X^2 + 5X - 1$.

28. Find the quotient and remainder when you divide $X^2 + X + 1$ into $X^5 + 3X^4 + 5X^3 + 11X^2 + X + 2$.

29. Find the quotient and remainder when you divide $X^3 + 1$ into $X^6 + 4X^2 + 1$. (First write the latter as $X^6 + 0X^5 + 0X^4 + 0X^3 + 4X^2 + 0X + 1$.)

30. When you divide a polynomial of degree 100 by a polynomial of degree 21, what will you be able to say in advance about the degrees of the quotient and the remainder?

31. What resemblance is there between polynomials and the decimal system of notation?

32. (a) Is every natural number a rational number?
    (b) Is every rational number an algebraic number?
    (c) Is every irrational number a transcendental number?

33. The "real impedance" of an alternating current circuit is defined as
$$\sqrt{R^2 + (X_L - X_C)^2}.$$
How is it related to the "complex impedance"?

34. Does the equation $5X^{19} - \sqrt{3}\,X + 0.42 = 0$ have any real roots?

35. (a) Check that 0 is a root of the equation $5X^7 - 2X^3 + X^2 - 6X = 0$.
    (b) Noting that $X - 0$ is simply $X$, find a polynomial $Q$ such that
    $$5X^7 - 2X^3 + X^2 - 6X = Q \times (X - 0),$$
    thereby illustrating Lemma 1.

36. If $r$ is 0, then Lemma 1 is very easy to prove. Prove the lemma for this special case.

37. (a) Check that 2 is a root of $3X^2 - 5X - 2 = 0$.
    (b) Find a polynomial $Q$ such that $3X^2 - 5X - 2 = Q \times (X - 2)$.

38. (a) Check that $-1$ is a root of $2X^2 + 5X + 3 = 0$.
    (b) Find a polynomial $Q$ such that $2X^2 + 5X + 3 = Q \times (X + 1)$, thereby illustrating Lemma 1.

39. Fill in the blanks: If $P$ is a polynomial of degree 201 and has real coefficients, then the equation $P = 0$ has anywhere from ____ to ____ roots.

40. Fill in the blanks: If $P$ is a polynomial of degree 202 and has real coefficients, then the equation $P = 0$ has anywhere from ____ to ____ real roots.

41. Fill in the blank: If 7 is a root of the equation $P = 0$, then there is a polynomial $Q$ such that _____.

42. Is there a polynomial $Q$ such that $X^{11} - 10X^7 + X = Q \times (X - 2)$?

43. Is there a polynomial $Q$ such that (a) $2X^5 - 6X^3 + X - 18 = Q \times (X - 2)$?
    (b) $2X^5 - 6X^3 + X - 18 = Q \times (X - 1)$?

44. (a) Is $\pi$ equal to $\frac{22}{7}$?
    (b) Is $\pi$ equal to 3.1416?

45. Is it possible to add $\pi$ several times, and then add to this sum several $\pi^2$, then several $\pi^3$ and have a sum that is an integer?

46. (a) Draw three points $X$, $Y$, and $Z$ in the plane.
    (b) Find the point $X \oplus (Y \oplus Z)$ by drawing the two necessary parallelograms.
    (c) Find the point $(X \oplus Y) \oplus Z$ by drawing the two necessary parallelograms.
    (d) Compare the results of (b) and (c) (they should be the same).

47. Choose three points $X$, $Y$, and $Z$ in the plane, such that $X$ lies within the circle of radius 1 (for convenience). Using the geometric definitions of $\oplus$ and $\otimes$, compute $X \otimes (Y \oplus Z)$ and $(X \otimes Y) \oplus (X \otimes Z)$. What algebraic rule tells us that the two results should coincide?

48. Using the geometric definition of $\otimes$, show that
    (a) $(-2) \otimes 3 = -6$,    (b) $(-2) \otimes (-4) = 8$,    (c) $2 \otimes 3 = 6$.

49. There are no real roots of the equation $X^2 + 1 = 0$. Draw the two complex roots.

50. (a) How many real roots are there of the equation $X^3 - 1 = 0$?
    (b) Using the geometric definition of $\otimes$, show that the three complex numbers in this diagram are roots of $X^3 - 1 = 0$.

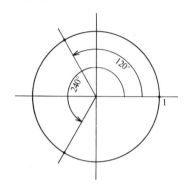

51. (a) Find the real roots of the equation $X^4 - 16 = 0$. (There are two.)
    (b) Draw the complex roots of the equation $X^4 - 16 = 0$. (There are four.)

52. (a) Using the geometric definition of multiplication of complex numbers, find the point $2i \times (3 + 4i)$.
    (b) Using only the rules of arithmetic, such as distributivity, compute $2i \times (3 + 4i)$, and draw the result. (Whenever you see $i^2$, replace it by $-1$.)
    (c) Compare the results of (a) and (b).

53. (a) Using the geometric definition of multiplication of complex numbers, compute and draw $(2 + i) \times (1 + 2i)$.
    (b) Compute $(2 + i) \times (1 + 2i)$ arithmetically, and draw the result.
    (c) Compare the two points drawn in (a) and (b).

54. (a) Using the geometric definition of multiplication of complex numbers, compute and draw $(1 + i) \times (-1 + i)$.
    (b) Compute $(1 + i) \times (-1 + i)$ arithmetically, and draw the result.
    (c) Compare the two points drawn in (a) and (b).

55. (a) Using the geometric definition of multiplication of complex numbers, compute and draw $(3 + 4i) \times (4 + 3i)$.
    (b) Compute $(3 + 4i) \times (4 + 3i)$ arithmetically, and draw the result.
    (c) Compare the points drawn in (a) and (b).

56. Prove that $(2 \otimes i) \otimes (3 \otimes i) = -6$
    (a) by drawing the points $2 \otimes i$ and $3 \otimes i$ and using only the geometric definition of $\otimes$;
    (b) by using the commutativity and associativity of $\otimes$ and the relation $i^2 = -1$ (draw no pictures).

57. Draw the two complex roots of the equation $X^2 + 4 = 0$.

58. Draw the two complex roots of the equation $X^2 - i = 0$. (Think of it as $X^2 = i$.)

59. Draw a complex number $X$ such that $\overline{X} = X$.

60. (a) Check that $3 + i$ is a root of the equation $X^2 - 6X + 10 = 0$.

(b) Which lemma guarantees that $3 - i$ is also a root?

(c) Check that $3 - i$ is a root.

61. (a) Multiply $X - (2 + 5i)$ and $X - (2 - 5i)$.

(b) Which lemma guarantees that the coefficients of the product are real?

62. Compute (a) $\overline{3 + 5i}$, (b) $(3 + 5i)(\overline{3 + 5i})$, (c) $(3 + 5i) + (\overline{3 + 5i})$.

63. (a) Select and draw a point $P$ arbitrarily in the plane.

(b) Draw the point $Q$ such that $Q \oplus P = 0$.

●

64. Prove the converse of Lemma 1. That is, prove that if $r$ is a real number and if $P$ and $Q$ are polynomials satisfying $P = Q \times (X - r)$, then $r$ is a root of the equation $P = 0$.

65. (a) How would you modify Lemma 1 if we allow complex coefficients?

(b) Prove that Theorem 6 remains true if we allow complex coefficients.

66. How would you define a divisor of a polynomial?

67. Prove that $(a + bi) \otimes (c + di) = (ac - bd) \oplus (ad + bc)i$. (*Hint:* Begin by using distributivity.)

● ●

68. Prove that $1 + \sqrt{2}$ is algebraic by finding an equation with integral coefficients of which it is a root.

69. Find an equation (of degree 4) with integral coefficients of which $\sqrt{2} + \sqrt{3}$ is a root. This will show that $\sqrt{2} + \sqrt{3}$ is algebraic. It can be proved that the sum and the product of any two algebraic numbers are algebraic.

70. (a) Prove that if $r$ is a root of the equation $AX^3 + BX^2 + CX + D = 0$, then $1/r$ is a root of the equation $DX^3 + CX^2 + BX + A = 0$.

(b) Using (a), prove that if $r$ is algebraic, so is $1/r$.

71. (a) Prove that if $r$ is a root of the equation $AX^3 + BX^2 + CX + D = 0$, then $r/2$ is a root of the equation $8AX^3 + 4BX^2 + 2CX + D = 0$.

(b) Using the idea suggested by (a), prove that if $r$ is algebraic, so is $r/2$.

72. Prove that every point on the vertical line through 0 can be written in the form $b \otimes i$ for some real number $b$.

73. Any point $X$ (other than 0) determines a point $S$ on the circle of radius 1 and a positive real number $r$, as shown by this diagram.

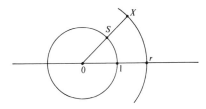

Prove that $S \otimes r = X$.

74. Exercise 67 suggests a direct and short way of defining the complex numbers, $\oplus$ and $\otimes$—a way that avoids geometry but exploits the real numbers, with their $+$ and $\times$: Simply define a complex number to be a pair of real numbers $(a, b)$, and define $(a, b) \oplus (c, d)$ to be $(a + c, b + d)$, and $(a, b) \otimes (c, d)$ to be $(ac - bd, ad + bc)$.
   (a) Which is the $(a, b)$ that we should name $-1$?
   (b) Which is the $(a, b)$ that we should name $i$?

75. Consider only polynomials with real coefficients. Assume that every polynomial of even degree has a divisor of degree 1 or 2. Show that Theorem 6 would follow.

76. (a) Show that the polynomial $X^3 - 2$ is not the product of polynomials of smaller degree that have only rational coefficients.
   (b) Express $X^3 - 2$ as the product of polynomials of smaller degree that have real coefficients.
   (c) Express $X^3 - 2$ as the product of polynomials of degree 1 whose coefficients are complex.

77. Does the equation $2X^3 - iX^2 + (17 + i)X - 2 = 0$ have any
   (a) real roots?      (b) complex roots?

78. Prove that any polynomial with complex coefficients can be expressed as the product of polynomials of the first degree (with complex coefficients).

79. Draw the five complex roots of the equation $X^5 - 1 = 0$. Denote by $F$ the one that has angle 72°.
   (a) Show that the other roots are $F^2$, $F^3$, $F^4$, and $F^5 = 1$.
   (b) Show that $F + F^4$ is real by drawing the appropriate parallelogram. Estimate $F + F^4$ to two decimal places by measuring it. (It is convenient to make the radius of the circle 10 centimeters and use the metric system.)
   (c) Treat $F^2 + F^3$ similarly.
   (d) Using (b) and (c), estimate $1 + F + F^2 + F^3 + F^4$.

80. Consider the number defined as follows. We draw two perpendicular lines as in (14) and the curve traced out by all points $P$ such that the rectangle with diagonal $OP$ has area 1 and horizontal side at least 1. The curve and one such rectangle are shown in (14).

(14)

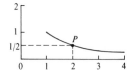

Then we draw the endless staircase above the curve (14), determined in this manner:

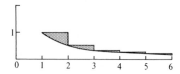

Let $A$ be the total area of the endless shaded region.

(a) With the aid of the endless supply of rectangles shown in this diagram

(15)

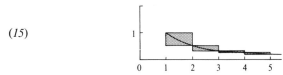

show that $A$ is less than 1.

(b) Show that $A$ is larger than $\frac{1}{2}$.

The number $A$, which is between 0.577 and 0.578, is known as Euler's constant. It has been in use for over two centuries, yet no one knows whether it is rational or irrational, or whether it is algebraic or transcendental.

81. Consider polynomials with real coefficients. Define such a polynomial to be "prime" if it is not the product of such polynomials of smaller degrees. (Thus $X^2 + 1$ and $6X + 2$ are primes.)

(a) What degree can a prime polynomial have?

(b) Show that the Euclidean algorithm, page 65, holds for polynomials, the main change being that in the equation $B = QA + R$, either the "remainder" is the 0-polynomial or the degree of $R$ is less than the degree of $A$.

(c) Show that every polynomial is the product of prime polynomials.

(d) Define "special" polynomial.

(e) Show that the analog of Lemma 4, page 65, is also valid for polynomials.

(f) Prove that every prime polynomial is special.

(g) State and prove the analog of the Fundamental Theorem of Arithmetic.

82. Carry out the analog of E 81, except part (a), for polynomials whose coefficients are all rational. (For instance, $X^3 - 2$ is now a prime polynomial.)

83. A Gaussian integer is a complex number of the form $a + bi$, where $a$ and $b$ are "ordinary" integers—that is, integers in the usual sense of the word.

(a) Show that an ordinary integer $m$ can be expressed as the sum of the squares of two ordinary integers if it can be written as the product of a Gaussian integer with its conjugate,

$$m = (a + bi)\overline{(a + bi)}.$$

(b) Use (a) to prove that if the integers $m$ and $n$ can both be expressed as the sum of two squares, then so can their product $mn$. (For instance, $13 = 2^2 + 3^2$, $25 = 3^2 + 4^2$, and $13 \cdot 25 = 325 = 6^2 + 17^2$.)

84. This is an era of specialization. At a certain dairy each delivery man has a certain frequency at which he will deliver milk. There is one milkman who will deliver every 2nd day; call him the 2-milkman. There is one who will deliver every 3rd day; call him the 3-milkman. There are a 4-milkman, a 5-milkman, a 6-milkman, and so on. Indeed there is an endless supply of milkmen, each specializing in one of the possible frequencies.

An immortal baby is born who requires exactly one quart of milk delivered every single day of his life. The dairy can meet his demands by scheduling the 2-milkman for days 0, 2, 4, 6, 8, . . . , forever; the 4-milkman for days 1, 5, 9, 13, 17, . . . , and so on; the 8-milkman for days 3, 11, 19, 27, 35, . . . . Generally, the $2^n$-milkman delivers, beginning when the baby is $2^{n-1} - 1$ days old.

But that scheme involves an *infinite* number of milkmen. Can the company keep the baby happy with only a *finite* number of milkmen?

(a) Show that a schedule that has the $M_1$-milkman begin on day $D_1$, the $M_2$-milkman on day $D_2$, ..., the $M_k$-milkman on day $D_k$, would permit us to express the "polynomial of infinite degree"

$$1 + X + X^2 + X^3 + \cdots$$

in the form

$$X^{D_1} + X^{D_1+M_1} + X^{D_2+2M_1} + \cdots$$
$$+ X^{D_2} + X^{D_2+M_2} + X^{D_2+2M_2} + \cdots$$
$$\cdots\cdots\cdots\cdots\cdots\cdots\cdots\cdots\cdots\cdots\cdots\cdots$$
$$+ X^{D_k} + X^{D_k+M_k} + X^{D_k+2M_k} + \cdots$$

For convenience, let us assume that $M_1, M_2, \ldots, M_k$ increase in size from left to right.

(b) Using the formula for the sum of a geometric series, show that if the baby were being supplied by a finite number of milkmen, then

$$\frac{1}{1-X} = \frac{X^{D_1}}{1 - X^{M_1}} + \frac{X^{D_2}}{1 - X^{M_2}} + \cdots + \frac{X^{D_k}}{1 - X^{M_k}}$$

(c) Show that the equation in (b) cannot hold, by considering the values of both sides of the equation when $X$ is replaced by complex numbers on the circle of radius 1 that have an angle near $(360/M_k)°$.

(d) Consider the problem of providing the baby exactly two quarts a day by a finite number of these milkmen, each of whom will only leave one quart at a delivery.

85. A polyomino is a generalization of a domino. It consists of several squares of a checkerboard wired rigidly together. The shaded four squares, for instance,

make up a polyomino, consisting of 6 adjacent squares from which the second and fifth have been deleted. Copies of this polyomino can tile a 7 by 12 rectangle. The diagram below shows 21 such polyominoes, numbered from 1 through 21; a square numbered $N$ is covered by the polyomino numbered $N$.

| 15 | 8  | 16 | 8  | 8  | 13 | 8  | 13 | 13 | 19 | 13 | 21 |
|----|----|----|----|----|----|----|----|----|----|----|----|
| 5  | 14 | 5  | 5  | 17 | 5  | 12 | 18 | 12 | 12 | 20 | 12 |
| 15 | 4  | 16 | 4  | 4  | 11 | 4  | 11 | 11 | 19 | 11 | 21 |
| 15 | 14 | 16 | 7  | 17 | 7  | 7  | 18 | 7  | 19 | 20 | 21 |
| 3  | 14 | 3  | 3  | 17 | 3  | 10 | 18 | 10 | 10 | 20 | 10 |
| 15 | 2  | 16 | 2  | 2  | 9  | 2  | 9  | 9  | 19 | 9  | 21 |
| 1  | 14 | 1  | 1  | 17 | 1  | 6  | 18 | 6  | 6  | 20 | 6  |

We now show how polynomials in two letters $X$ and $Y$ can be applied to problems concerning tiling by polyominoes. Specifically, we outline a proof of this

THEOREM. *It is impossible to tile any rectangle with polyominoes formed by deleting the 3rd and 5th of 7 adjacent squares*

$\left(\right.$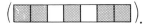$\left.\right)$.

First of all, by a "polynomial in $X$ and $Y$" we mean a sum of terms of the form $SX^n Y^m$, where $S$ is real and $m$ and $n$ are natural numbers. For instance, $XY + 5X^3Y^2 - 6XY^4 + X$ and $1 + 8XY^5$ are polynomials in $X$ and $Y$. These polynomials can be added and multiplied in the expected—purely mechanical—way.

(a) Assume that an $a$ by $b$ rectangle can be tiled with copies of the polyomino

.

Show that this implies that there are polynomials $P$ and $Q$ in the letters $X$ and $Y$ such that

(16) $\quad (1 + X + X^2 + \cdots + X^{a-1})(1 + Y + Y^2 + \cdots + Y^{b-1})$
$$= P(1 + X + X^3 + X^5 + X^6) + Q(1 + Y + Y^3 + Y^5 + Y^6).$$

(*Hint:* Use $X^m Y^n$ to record the square of the $a$ by $b$ rectangle $m$ squares to the right and $n$ squares above the lower left corner square. (The side of length $a$ is horizontal.)

(b) Using the formula for the sum of a geometric series, show that

$$1 + X + \cdots + X^{a-1} = \frac{X^a - 1}{X - 1}$$

and

$$1 + Y + \cdots + Y^{b-1} = \frac{Y^b - 1}{Y - 1}.$$

(Consider $X$ and $Y$ to be real numbers other than 1.)

(c) To show that the tiling is impossible, show that (16) cannot hold. To do this, first show that the left side of (16) is *not* 0 when $X$ and $Y$ are replaced by any number right of $-1$ but left of 0 on the number line. Then show that the expression $1 + X + X^3 + X^5 + X^6$ has a root between $-1$ and 0. Finish the proof.

86. What is the similarity between E 84 and E 85?

87. See E 85. We outline a problem first solved by N. G. de Bruijn. Let $a$, $b$, and $n$ be positive integers. What must be true about $n$, $a$, and $b$ in order that an $a$ by $b$ rectangle can be cut into 1 by $n$ rectangles?

(a) Show that if $n$ divides $a$ or $n$ divides $b$, then the $a$ by $b$ rectangle can be cut into 1 by $n$ rectangles.

(b) Does the converse of (a) hold?

(c) Generalize (a) and (b) to filling an $a$ by $b$ by $c$ box with 1 by 1 by $n$ bricks.

88. By the size or "absolute value" of a real number $r$, we shall mean $r$ if $r$ is positive, and $-r$ if $r$ is not positive. Thus the size of 3 is 3, the size of $-3$ is also 3, and the size of 0 is 0. Clearly, the size of a real number is its distance from 0. Let $|r|$ denote the size of $r$.
    (a) Let $s$ be any real number. Prove that there is an integer $M$ such that $|s - (M/13)|$ is less than $\frac{1}{26}$. (*Hint:* Draw the rationals with denominator 13.)
    (b) Let $s$ be any real number and $N$ any natural number bigger than 0. Prove that there is an integer $M$ such that $|s - (M/N)|$ is less than $1/2N$.

89. In E 88 we saw that any real number can be approximated closely by a rational if we allow the denominator of the rational to be large. In 1955, K. F. Roth obtained a profound result on the approximation of algebraic numbers by rationals, of which the following is a consequence: If $s$ is an irrational algebraic number, then there are only a finite number of rationals $M/N$ with the property that $|s - (M/N)|$ is less than $2/N^3$. This suggests that if a number $s$ can be approximated "too" closely by rationals, then $s$ must be transcendental. This will be exploited in E 90. Show that Roth's theorem does not extend to rational $s$.

90. Using E 89, prove that

$$ s = \left(\frac{1}{10}\right)^3 + \left(\frac{1}{10}\right)^9 + \left(\frac{1}{10}\right)^{27} + \cdots + \left(\frac{1}{10}\right)^{3^n} + \cdots $$

is transcendental. The following steps outline the proof.
    (a) Prove that $s$ is irrational.
    (b) Prove that $|s - [(\frac{1}{10})^3 + \cdots + (\frac{1}{10})^{3^n}]|$ is less than $2/(10^{3^n})^3$.
    (*Hint:* Observe that $|s - [(\frac{1}{10})^3 + \cdots + (\frac{1}{10})^{3^n}]|$ is equal to $(\frac{1}{10})^{3^{n+1}} + (\frac{1}{10})^{3^{n+2}} + \cdots$, which is less than the geometric series with first term $(\frac{1}{10})^{3^{n+1}}$ and ratio $\frac{1}{2}$.)
    (c) Now use Roth's theorem to prove that $s$ must be transcendental.

91. Consider $N^2$ dots arranged regularly as an $N$ by $N$ square array ($N$ rows of $N$ equally spaced dots). What is the largest number of these dots you can choose such that no three of them lie on a straight line? Call this number $D(N)$.
    (a) Why is $D(N)$ not larger than $2N$?
    (b) Show that $D(2) = 4$, $D(3) = 6$, and $D(4) = 8$.
    (c) Examine $D(5)$ and $D(6)$.
    (d) The formula for $D(N)$ is not yet known. However, if $N$ is prime, it is known that $D(N)$ is at least as large as $N$. To show this, label the $N^2$ dots $(a, b)$ where $a$ and $b$ go from 0 to $N - 1$. Dot $(a, b)$ is $a$ dots to the right and $b$ dots up from the bottom left dot $(0, 0)$. Let $R(i) =$ remainder when the integer $i$ is divided by $N$. Then the $N$ dots

    $$(0, R(0)), (1, R(1^2)), (2, R(2^2)), \ldots, (N, R(N^2))$$

    have no three on a line. [A proof of this uses analytic geometry and the fact that a polynomial of degree two whose coefficients are in the field of integers (mod $N$) has at most two roots.] Apply this technique for $N = 7$ and $N = 11$.

92. (This generalizes E 59 of Chapter 6.) Can you find three dots $A$, $B$, and $C$ among the endless array described in E 59 of Chapter 6 such that $\angle ABC$ is 60°?

Exercises 27–30 of Chapter 17 develop trigonometry from the complex number, and may be studied now.

## References

1. B. L. van der Waerden, *Science Awakening*, Noordhoff Ltd., Groningen, Holland, 1954. (The Babylonian treatment of equations of degree 2 is described on pp. 69–70.)

2. L. Weisner, *Introduction to the Theory of Equations*, Macmillan, 1949. (The complex numbers are developed on pp. 1–19.)

3. E. T. Bell, *The Development of Mathematics*, McGraw-Hill, New York, 1945. (Pages 155–164, 215–218 discuss the complex numbers. How the attitude toward the negative and complex numbers has changed through the centuries is shown on pp. 172–180. For a discussion of the work of Pierce and Frobenius and for further references see pp. 249–251. The Fundamental Theorem of Algebra is discussed on p. 178.)

4. E. T. Bell, *Men of Mathematics*, Simon and Schuster, New York, 1937. (See p. 464, on which the remark of Hermite appears, for a discussion of the geometric significance of the transcendence of $\pi$. Remarks on Gauss's proof of the Fundamental Theorem of Algebra are made on pp. 332–334.)

5. *I. Niven, *Irrational Numbers*, Carus Mathematical Monograph 11, Mathematical Association of America, distributed by Wiley, New York. (On p. 20 the irrationality of $\pi$ and $\pi^2$ is proved with the aid of the elementary integral calculus. Chapter 7 presents Liouville's proof, of 1844, for the existence of transcendentals.)

6. D. J. Struik, *A Concise History of Mathematics*, Dover, New York, 1948. (A concise history of the work of Tartaglia and Ferrari on roots of equations of degree at most 4 is given on pp. 109–116.)

7. R. Courant and H. Robbins, *What Is Mathematics?*, Oxford University Press, New York, 1941. [A sketch of a proof of the Fundamental Theorem of Algebra (for the reader familiar with trigonometry) is presented on pp. 269–271. Liouville's proof of the existence of transcendentals is to be found in pp. 104–107.]

8. O. Veblen and J. W. Young, *Projective Geometry*, vol. 1, Ginn, Boston, 1938. (Addition for points on the line is defined geometrically on pp. 141–142; see, in particular, the footnote on p. 142. On pp. 144–145 the same is done for multiplication; see, in particular, the footnote on p. 145. The resulting addition and multiplication form a field, as defined in Appendix C.)

9. A. Seidenberg, *Lectures in Projective Geometry*, Van Nostrand, New York, 1962. (Geometry is built out of algebra on pp. 34–35.)

10. J. B. Roberts, *The Real Number System in an Algebraic Setting*, W. H. Freeman and Company, San Francisco, 1961. (Beginning with the natural numbers, this book constructs the integers, then the rationals, and, finally, all the real numbers.)

11. *L. E. Dickson, *Linear Algebra*, Hafner, New York. (On p. 14 the 8-dimensional algebra of Cayley is defined.)

12. *K. F. Roth, Rational approximations to algebraic numbers, *Mathematika*, vol. 2, 1955, pp. 1–20. (For a history of this problem see the review by E. R. Kolchin in *Mathematical Reviews*, vol. 17, 1956, p. 242.)

13. *B. L. van der Waerden, *Modern Algebra*, vol. 1, Ungar, New York, 1949. (A proof of the Fundamental Theorem of Algebra is given on pp. 225–228.)

14. *C. C. MacDuffee, *An Introduction to Abstract Algebra*, Wiley, New York, 1956. (On pp. 117–121 the Fundamental Theorem of Arithmetic is extended to the complex numbers of the form $a + bi$, where $a$ and $b$ are integers. As exercise 2 on p. 121 shows, Theorem 51.4 of p. 120 can be used to prove that every prime natural number of the form $4k + 1$ can be expressed uniquely as the sum of the squares of two natural numbers.)

15. T. Dantzig, *Number, the Language of Science*, Macmillan, New York, 1954. (Chapters 8 and 9 discuss the real numbers; Chapter 10, the complex numbers; Chapter C, roots and their relation to geometry.)

16. J. B. Kelly, Polynomials and polyominoes, *American Mathematical Monthly*, vol. 73, 1966, pp. 464–471.

17. S. W. Golomb, *Polyominoes*, Scribner's, New York, 1965. (Questions of tiling by polyominoes lead the reader into geometry, combinatorics, and symmetry.)

*Chapter* **17**

# CONSTRUCTION
# BY STRAIGHTEDGE
# AND COMPASS

In Chapter 7 we used the distinction between the rational and irrational numbers to answer the geometric question, "Which rectangles can be tiled with congruent squares?" In Chapter 8 the algebra of electrical networks showed that a 1 by $\sqrt{2}$ rectangle cannot be tiled by squares even if we allow them to be of different sizes. Now we will apply polynomials and the complex numbers to another geometric problem, one that goes back to the time of Euclid, known as "construction by straightedge and compass." All that we need from Chapter 16 are the notions of polynomial and root, and the arithmetic of complex numbers.

Now let us turn to the problem. We have at our disposal an *unmarked* stick, called a "straightedge," with which to draw lines, and an *uncalibrated* compass that can be set in any position:

Straightedge

Compass

(We may imagine that we have an endless set of compasses so that, if need be, we can draw circles as large as we please.) By using the straightedge repeatedly we can draw line segments as long as we please. The general problem is this: *What can we draw (construct) with a straightedge and compass?*

For instance, we can cut a given line segment in half, as this picture shows:

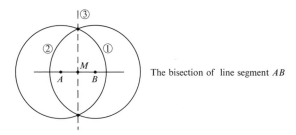

The bisection of line segment $AB$

This picture and the circled numbers describe the following steps in the bisection of line segment $AB$.

*Step 1.* Place the point of the compass at $A$. Set the compass such that the distance between the point and the pencil point is greater than half the length of $AB$. With this setting draw a circle whose center is at $A$.

*Step 2.* With the *same* setting of the compass draw a circle whose center is at $B$.

*Step 3.* Draw a line through the two points at which the circles meet. (They will meet because their radius was purposely chosen large enough so that they would.) The point $M$ at which this line meets the original line segment $AB$ is the midpoint of $AB$.

That $M$ *is indeed the midpoint of $AB$* can be proved by elementary geometry, but we will not concern ourselves with this aspect of the construction. The three-step construction proves

THEOREM 1. *It is possible to bisect any given line segment by straightedge and compass.*

The construction of the midpoint illustrates some of the rules in a permissible construction. For instance, we may set the compass in such a way that its two points coincide with two known points. But we may also set the compass at an arbitrary undetermined opening—as long as the final result does not depend on our knowledge of the exact size of this opening.

The diagram that describes the bisection of a line segment incidentally shows how to construct an angle of 90° by straightedge and compass. But we can do better, as is shown by

THEOREM 2. *Given a line $L$ and a point $P$, it is possible to construct a line through $P$ perpendicular to $L$ by straightedge and compass.*

PROOF. Line *L* and point *P* are given.

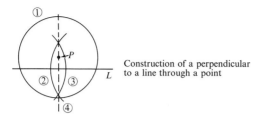

Construction of a perpendicular to a line through a point

*Step 1.* Draw a circle whose center is at *P*, and whose radius is arbitrary (but large enough so that the circle intersects line *L* in two distinct points).

*Step 2.* Draw a circle whose radius is larger than that drawn in Step 1 and whose center is the right-hand point at which the first circle meets *L*.

*Step 3.* Draw a circle whose radius is the same as that used in Step 2 and whose center is the left-hand point at which the circle drawn in Step 1 meets *L*.

*Step 4.* Draw the line through the two points at which the circles drawn in Steps 2 and 3 meet. This is the desired line; the proof is complete.

We have seen how to bisect any given line segment. Can we bisect any given angle? The answer is "yes," as is shown by

THEOREM 3. *It is possible to bisect any given angle by straightedge and compass.*

PROOF. Angle *AOB* is given

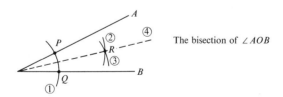

The bisection of ∠ *AOB*

*Step 1.* Draw a circle of arbitrary radius and center at *O*.

*Step 2.* Draw circles of equal size and centered at *P* and *Q*. (The radius in these steps need not be the same as that used in Step 1.)

*Step 3.* Let a point of intersection of the two circles of Step 2 be denoted by *R*; then draw the line through *O* and *R*.

This ∠ *AOR* is half as large as ∠ *AOB*.

By repeated bisection we may cut any given line segment or any given angle into quarters, eighths, sixteenths, and so on. Can we trisect any line segment (cut into three congruent segments) with straightedge and compass? Can we trisect any given angle with straightedge and compass?

Before we consider these questions let us prove some lemmas that will be useful later.

LEMMA 1. *It is possible to construct by use of compass only a copy of any given line segment (prescribing in advance the line on which the segment is to lie, and one of its ends.)*

PROOF. The given line segment is *AB*. The given line, on which we are to construct a copy of *AB*, is *L*. The prescribed end of the copy is *C*. As the diagram shows, one setting of the compass suffices (and the straightedge is not needed).

In the single step we draw a circle whose radius is the length of the line segment *AB*. Then *CD* is a copy of *AB*.

LEMMA 2. *Given a line segment of length 1, it is possible to construct a line segment of length $(\sqrt{5} - 1)/4$ by straightedge and compass.*

PROOF. Let *AB* be the given line segment of length 1.

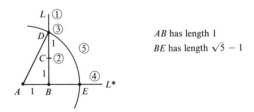

*AB* has length 1
*BE* has length $\sqrt{5} - 1$

The diagram shows how to construct a line segment of length $\sqrt{5} - 1$. With the aid of two bisections (Theorem 1), we would then obtain a line segment of length $(\sqrt{5} - 1)/4$.

*Step 1.* Construct the line *L* perpendicular to *AB* that passes through *B* (Theorem 2).

*Step 2.* Construct *C* on the line *L* such that *BC* is a copy of *AB* (Lemma 1).

*Step 3.* Construct *D* on the line *L* such that *CD* is a copy of *BC*.

*Step 4.* Draw the line L* on which the segment *AB* lies.

*Step 5.* Draw the circle whose radius is equal to the length of *AD* and whose center is *A*. This circle meets the line L* at a point that we label *E*. The segment *BE* has the desired length, $\sqrt{5} - 1$. (To see this, use the Pythagorean theorem to show that *AD*, hence *AE*, has length $\sqrt{5}$.)

LEMMA 3. *It is possible to construct a copy of any given angle (prescribing in advance a line on which one of the two sides of the angle is to lie, and its vertex.)*

PROOF. The given angle is ∠ *AOB*. The given line, on which one side of the copy is to lie, is *L*. The prescribed vertex is *P*.

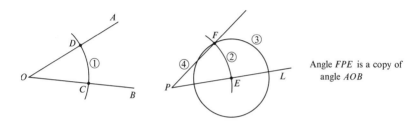

Angle *FPE* is a copy of angle *AOB*

*Step 1.* Draw any circle whose center is *O*. (It meets the two sides of the given angle at *C* and *D*.)

*Step 2.* With the *same* setting of the compass as in Step 1, draw a circle with center at *P*. The point at which it meets *L* we call *E*.

*Step 3.* Set the compass at the distance from *C* to *D*, and draw a circle with that radius and with its center at *E*. This circle meets the circle constructed in Step 2 at two points, one of which we denote by *F*.

*Step 4.* Draw the line through *P* and *F*.

The ∠ *FPE* is a copy of ∠ *AOB*.

LEMMA 4. *It is possible to construct through a given point the line that is parallel to a given line.*

PROOF. Let the given point be *P* and the given line be *L*.

Construction of L* through *P*, parallel to *L*

*Step 1.* Select an arbitrary point $Q$ on line $L$.

*Step 2.* Draw the line through $P$ and $Q$.

*Step 3.* Using Lemma 3, copy the angle between line $L$ and the line $PQ$ (prescribing its vertex $P$ and one side, namely the line $PQ$). The second side of this angle, the line $L^*$, is parallel to $L$.

With Lemma 4 in our possession we are ready to answer the first of our two trisection questions, with

THEOREM 4. *It is possible to trisect any given line segment by straightedge and compass.*

PROOF. The given line segment is $AB$.

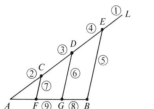

Trisection of line segment $AB$

*Step 1.* Draw an arbitrary line $L$ through $A$.

*Step 2.* Choose an arbitrary point $C$ on $L$.

*Step 3.* Draw $D$ on $L$ such that $CD$ is a copy of $AC$ (Lemma 1).

*Step 4.* Draw $E$ on $L$ such that $DE$ is a copy of $CD$ (Lemma 1).

*Step 5.* Draw the line through $BE$.

*Step 6.* Draw lines through $D$ and $C$ parallel to $EB$ (Lemma 4).

The points $F$ and $G$, where the lines constructed in Step 6 meet $AB$, divide $AB$ into three congruent segments, $AF$, $FG$, and $GB$. We have shown how to trisect an arbitrary line segment.

The same technique used in the proof of Theorem 4 shows how to cut a given line segment into any number of congruent line segments. Combining Lemma 1 and Theorem 4, we see that starting with a line segment of length 1, we can construct a line segment whose length is any prescribed positive rational number. As Lemma 2 shows, we can also construct some of irrational length. Theorem 6 will show that we cannot construct a line segment of *any* preassigned length.

Let us return to the second question, "Can we trisect any given angle with straightedge and compass?"

For instance, can we trisect an angle of 90°? In other words, can we construct an angle of 30° with straightedge and compass? This is not difficult, as is shown by the proof of

THEOREM 5. *It is possible to trisect an 90° angle by straightedge and compass.*

PROOF. We will simply construct a 30° angle. (With the aid of Lemma 3, we could then copy it to trisect a given 90° angle.) First, construct an equilateral triangle as follows:

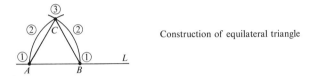

Construction of equilateral triangle

*Step 1.* Draw an arbitrary line *L*, and choose two points, *A* and *B*, on it.
*Step 2.* Set the compass at the radius *AB*, and draw two circles, one with center *A* and one with center *B*.

Denote by *C* one of the two points where the circles meet. Because the three sides of triangle *ABC* are equal, the three angles of the triangle are also equal. Since the sum of the angles in a triangle is 180°, we see that each of the angles is 60°. Bisecting any of them (Theorem 3) yields a 30° angle. This ends the proof.

A 45° angle can also be trisected by straightedge and compass; simply use Theorem 3 to bisect the 30° angle constructed in Theorem 5 and then copy the resulting 15° angle on a side of the given angle. However, it was shown in the nineteenth century that there is no general technique for trisecting any given angle. This is a consequence of the much stronger assertion stated in

THEOREM 6. *It is impossible to trisect a 60° angle by straightedge and compass.*

Though we will not be able to give a complete proof of Theorem 6, we will reduce it to an algebraic statement whose truth is much easier to accept.

Before we develop the machinery for approaching Theorem 6, we first show another way of looking at it. If we could trisect a 60° angle, we could construct a 20° angle. By copying one 20° angle next to another (Lemma 3) we could construct a 40° angle. With the aid of this angle we could construct a regular

9-gon by drawing nine of these 40° angles with a common vertex, and then drawing a circle whose center is at the vertex:

Thus Theorem 6 is equivalent to: *It is impossible to inscribe a regular 9-gon in a circle by straightedge and compass.*

Let us look at the problem for a moment in this alternative way. In this form Theorem 6 raises a new question: For which numbers *n*, where *n* = 3, 4, 5, . . . , is the regular *n*-gon constructible by straightedge and compass? From here on, we shall use "constructible" to mean constructible by straightedge and compass. The case *n* = 6 (the hexagon) is constructible, since, as we saw, an angle of 60° is constructible. The equilateral triangle illustrates the case *n* = 3. The construction of a 90° angle disposes of the case *n* = 4 (the square). Bisection of the 90° angle takes care of *n* = 8 (the octagon). Thus, for the integers *n* from 3 to 10, we see that for *n* = 3, 4, 6, 8 the regular *n*-gon is constructible. Theorem 6 treats *n* = 9. The cases *n* = 5 (pentagon) and *n* = 10 (decagon) are equivalent, for if one is constructible, so is the other; if one is not, neither is the other. The case *n* = 7, like *n* = 9, is also not constructible; the proof, similar to that which we will give for Theorem 6, will be omitted. Though we will only sketch a proof of Theorem 6, we now provide a complete proof of

THEOREM 7. *It is possible to construct an angle of 72° [ = (360/5)°], hence a regular pentagon, by straightedge and compass.*

PROOF. Let *F* be that complex number on the unit circle whose angle is 72°. Then, by the geometric definition of complex multiplication, 1, *F*, $F^2$, $F^3$, $F^4$ are the vertices of a regular pentagon.

(*1*)

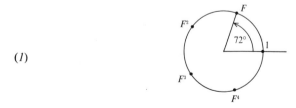

We will construct *F* indirectly, for first we will construct the point *Q* directly below it on the horizontal line through the center, and then use

Theorem 2 to find $F$ itself. The distance from $Q$ to the center we shall call $a$.

(2)

By the formula for the sum of a geometric progression,

$$1 + F + F^2 + F^3 + F^4 = \frac{F^5 - 1}{F - 1}.$$

Since $F^5 = 1$, we conclude that

(3) $$1 + F + F^2 + F^3 + F^4 = 0.$$

Inspection of diagram (2) and recollection of the geometric definition of addition of complex numbers shows that

(4) $$F + F^4 = 2a.$$

Squaring both sides of (4) yields

(5) $$F^2 + 2F^5 + F^8 = 4a^2.$$

But

$$F^5 = 1 \quad \text{and} \quad F^8 = F^5 \cdot F^3 = F^3.$$

Thus (5) reduces to

$$F^2 + 2 + F^3 = 4a^2,$$

which tells us that

(6) $$F^2 + F^3 = 4a^2 - 2.$$

Combining (3), (4), and (6) we obtain

$$0 = 1 + F + F^2 + F^3 + F^4 = 1 + (F + F^4) + (F^2 + F^3)$$
$$= 1 + 2a + 4a^2 - 2$$
$$= 4a^2 + 2a - 1.$$

Thus we see that the number $a$ is a root of the equation

(7) $$4X^2 + 2X - 1 = 0.$$

But what are the roots of this equation, which is so similar to equation (3) of Chapter 16? To deal with it, we first divide by 4, obtaining

$$X^2 + \frac{1}{2}X - \frac{1}{4} = 0.$$

Then we add $(\frac{1}{4})^2 = 1/16$ to both sides:

(8) $$X^2 + \frac{1}{2}X + \left(\frac{1}{4}\right)^2 - \frac{1}{4} = \frac{1}{16}.$$

Now (8) can be written as

$$\left(X + \frac{1}{4}\right)^2 - \frac{1}{4} = \frac{1}{16}$$

or

$$\left(X + \frac{1}{4}\right)^2 = \frac{1}{4} + \frac{1}{16} = \frac{5}{16}.$$

Thus for any root $R$ of (7) we know that

$$R + \frac{1}{4} = \sqrt{\frac{5}{16}} \quad \text{or} \quad R + \frac{1}{4} = -\sqrt{\frac{5}{16}};$$

that is, the two roots of (7) are

(9) $$-\frac{1}{4} + \frac{\sqrt{5}}{4} \quad \text{and} \quad -\frac{1}{4} - \frac{\sqrt{5}}{4}.$$

Since $a$ is a root of (7), $a$ must equal one of the two numbers in (9). Moreover, since $a$ is positive and $-1/4 - \sqrt{5}/4$ is negative, we conclude that

$$a = -\frac{1}{4} + \frac{\sqrt{5}}{4},$$

or simply

$$a = \frac{\sqrt{5} - 1}{4}.$$

By Lemma 2, $a$ is constructible by straightedge and compass. Hence we can construct the point $F$, and Theorem 7 is proved.

As we mentioned, we will not be able to complete the proof of Theorem 6, which concerns the trisection of a 60° angle. But we will be able to get well into the argument.

THEOREM 6. *It is impossible to trisect a 60° angle by straightedge and compass.*

OUTLINE OF PROOF. In view of previous remarks, we will merely show that a regular 9-gon cannot be constructed by straightedge and compass.

Let $G$ be the complex number on the unit circle whose angle is 40°. The nine points $1, G, G^2, G^3, \ldots, G^8$ are the vertices of a regular 9-gon. If we could construct the point $G$, Theorem 2 would enable us to construct the point $P$ directly below it on the horizontal line through the center. The distance from $P$ to the center we will call $c$.

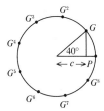

We will find a polynomial of which $c$ is a root. This will allow us to bring the techniques of algebra to bear on the problem. We begin with the observation that

(10) $$G + G^8 = 2c,$$

an equation that simply reflects our definition of the addition of complex numbers. Squaring both sides of the equation yields

(11) $$G^2 + 2G^9 + G^{16} = 4c^2.$$

But $G^9 = 1$ and $G^{16} = G^7 \cdot G^9 = G^7 \cdot 1 = G^7$. Thus we may rewrite (11) as

$$G^2 + 2 + G^7 = 4c^2.$$

Hence

(12) $$G^2 + G^7 = 4c^2 - 2.$$

Multiplying (10) and (12) together results in

$$(G + G^8)(G^2 + G^7) = 2c\,(4c^2 - 2),$$

hence

(13) $$G^3 + G^8 + G^{10} + G^{15} = 8c^3 - 4c.$$

But $G^{10} = G^9 \cdot G = G$, and $G^{15} = G^9 \cdot G^6 = G^6$.

Thus (13) tells us that

(14) $$G^3 + G^8 + G + G^6 = 8c^3 - 4c.$$

Together, (10) and (14) yield

$$G^3 + 2c + G^6 = 8c^3 - 4c$$

or

(15) $$G^3 + G^6 = 8c^3 - 6c.$$

We see that $G^3$ has angle $120°$ and that $G^6$ is below it. Inspection of the parallelogram formed by the two equilateral triangles in this picture

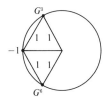

shows that

$$G^3 + G^6 = -1.$$

From (15) we obtain the equation

$$-1 = 8c^3 - 6c,$$

or simply

$$8c^3 - 6c + 1 = 0.$$

Hence $c$ is a root of the equation

(16)                    $$8X^3 - 6X + 1 = 0.$$

We conclude that $c$ is algebraic, but that information is not enough to insure that it can be constructed. The numbers that we can construct are a special type of algebraic number, like $(\sqrt{5} - 1)/4$, that can be expressed in terms of *square roots* and the four arithmetic operations (addition, multiplication, subtraction, and division). *It can be proved by algebraic means* that no root of (16) is obtainable by repeated applications of those five processes. For this reason the 20° angle cannot be constructed by straightedge and compass. This concludes the sketch of the proof.

It was Gauss who, at the age of 17, determined which regular $n$-gons are constructible by straightedge and compass. With the aid of the complex numbers he showed that for an $n$-gon to be constructible these conditions must hold: When $n$ is factored into primes *no prime other than 2 appears more than once*, and if a prime $P$ other than 2 does appear then $P$ *must be* 1 *more than a power of two*. (Thus a 5-, 17-, or 85-gon is constructible, but a 9-, 13-, or 25-gon is not.)

If the reader reviews this chapter, he will find a variety of mathematical fields represented. Geometry provided the Pythagorean Theorem and would justify the various constructions, such as bisection or trisection of a line segment. Algebra provides the formula $1 + X + X^2 + X^3 + X^4 = (X^5 - 1)/(X - 1)$ and a technique for learning that the roots of the equation $4X^2 + 2X - 1 = 0$ are $-1/4 + \sqrt{5}/4$ and $-1/4 - \sqrt{5}/4$. The complex numbers, part of algebra, provide a means of translating certain geometric questions into algebraic ones. Finally, the completion of the proof of Theorem 6—which shows why no root of the equation $8X^3 - 6X + 1 = 0$ is expressible by repeated square roots—belongs to the theory of vector spaces, which is part of algebra.

Truly the orbits of the mathematical universe are marvelously and inextricably intertwined.

**Exercises**

1. Given line segments of lengths $a$ and $b$, construct by straightedge and compass a line segment of length $a + b$.

2. Given line segments of length 1, $a$, and $b$, show that the construction below provides a line segment $CD$ of length $ab$.

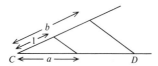

3. Given line segments of length 1, and $a$, and $b$, show that the construction below provides a segment $CD$ of length $a/b$.

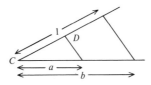

4. (a) Cut a line segment into seven segments of equal length, with straightedge and compass.
   (b) Cut the same line segment into ten segments of equal length in a similar fashion.
   (c) Using (a) and (b) show which is larger; 5/7 or 7/10.

5. We cannot construct a line segment whose length is the same as the circumference of a given circle (since $\pi$ is transcendental). Show how to construct line segments the sum of whose lengths is as close as we please to the circumference.

6. (a) Show why any angle $ACB$ inscribed in a circle in such a way that $A$ and $B$ are the ends of a diameter is 90°.
   (b) Show that the length of $CD$ in the diagram below is $\sqrt{a}$. (*Hint:* Why are $\triangle ADC$ and $\triangle CDB$ similar?)

   (c) Given line segments of lengths 1 and $a$, show how to construct a line segment of length $\sqrt{a}$.
   (d) Illustrate this by constructing $\sqrt{3}$ and $\sqrt{5}$.

7. Show how to cut any given line segment into five congruent line segments by straightedge and compass.

8. Say that you drew a circle but forgot where its center is. How could you find the center by straightedge and compass? (*Hint:* Construct two lines that pass through the center.)

9. Construct a regular pentagon by straightedge and compass.

10. We outline a way of showing that $1 + F + F^2 + F^3 + F^4 = 0$ (where $F$ is the complex number appearing in the proof of Theorem 7) that does not use the formula for the sum of a geometric series.

(a) Let $S = 1 + F + F^2 + F^3 + F^4$. Show that $FS = S$.

(b) From (a) deduce that $S = 0$.

11. (a) Draw an angle of $40°$, and measure the length $c$ referred to in the proof of Theorem 6.

(b) For your value of $c$, is $8c^3 - 6c - 1$ near $0$?

12. Which of these angles can be constructed by straightedge and compass? Which cannot? (a) $20°$, (b) $10°$, (c) $12°$, (d) $3°$, (e) $36°$, (f) $22\frac{1}{2}°$, (g) $15°$, (h) $1°$, (i) $1\frac{1}{2}°$.

13. The algebraic result on which the proof of Theorem 6 rests is this

THEOREM. *Let* $A, B, C, D$ *be integers. If the equation* $AX^3 + BX^2 + CX + D = 0$ *has no rational root, then none of its roots can be constructed with straightedge and compass.*

(a) Show that no root of equation $X^3 - 2 = 0$ is rational.

(b) Deduce that it is impossible to construct with straightedge and compass the edge of a cube twice as large in volume as that of a given cube.

14. See E 13. Prove that there is no rational root of the equation $8X^3 - 6X - 1 = 0$. (*Hint:* Let the alleged root be $M/N$, where $M$ and $N$ are integers without a common divisor other than 1 or $-1$. Show that $8M^3 - 6MN^2 - N^3 = 0$ and that the only integers that satisfy this equation are $M = 0$ and $N = 0$.)

15. See E 13. We did not examine the 7-gon. We outline a treatment of this case. Let $H$ be the complex number on the unit circle that has angle $(360/7)°$. Let $S$ be the point directly below $H$ on the horizontal line through the center. The distance from $S$ to the center we call $d$. Now we proceed as in the case of the 9-gon.

(a) Show that $G + G^6 = 2d$.

(b) Show that $G^2 + G^5 = 4d^2 - 2$.

(c) Show that $G^3 + 2d + G^4 = 8d^3 - 4d$.

(d) Show that $8d^3 + 4d^2 - 4d - 1 = 0$.

(e) Complete the argument.

16. Carry out the following construction of a regular 6-gon, and explain why it works.

   *Step 1.* Draw a circle.

   *Step 2.* Mark at random a point $P_1$ on the circle.

   *Step 3.* Keeping the same setting of the compass, place the point of the compass at $P_1$ and draw a circle. Let $P_2$ be one of the intersections of this circle with the original circle.

   *Step 4.* Repeat Step 3, letting $P_2$ play the role of $P_1$. Call the resulting point $P_3$.

   *Step 5.* Construct $P_4$, $P_5$, and $P_6$ in the same manner.

   Points $P_1, P_2, P_3, P_4, P_5, P_6$ are vertices of a regular 6-gon. (Note that the straightedge is not used in this construction.)

17. The regular 10-gon is easily obtained from the regular 5-gon by bisection of the $72°$ angle. We outline a direct alternative argument for the constructibility of the 10-gon (and hence of the 5-gon).

   In this figure we have given a triangle $OAB$, in which $OA = 1 = OB$, $OC = x$, and $\angle AOB$ is $36°$. The dotted line $AC$ bisects $\angle OAB$.

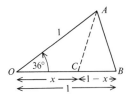

(a) Show that $\angle CAB$ is $36°$ and $BC = 1 - x$.
(b) Show that $\triangle AOB$ is similar to $\triangle CAB$.
(c) Deduce that $x/1 = (1 - x)/x$.
(d) Using (c) show that $x = (\sqrt{5} - 1)/2$.
(e) Use (d) to describe how to construct a $36°$ angle and hence a regular 10-gon.

●

18. Is it possible with straightedge and compass to
    (a) cut the circumference of a circle into five congruent arcs?
    (b) cut a line segment into five congruent line segments?
    (c) cut the circumference of a circle into 11 congruent arcs?
    (d) cut a line segment into 11 congruent line segments?

19. (a) Show that any odd integer $N$ is the difference of two squares.
    (b) Use (a) to provide another way of constructing $\sqrt{N}$.
    (c) Illustrate by constructing $\sqrt{5}$.

20. What is wrong with the following alleged construction for trisecting $\angle AOB$? Draw a circle whose center is $O$. Call the two points at which the circle meets the sides of the angle $C$ and $D$. Construct point $X$ on the line segment $CD$ in such a way that $CX$ is one third of $CD$. Then $\angle COX$ is one third of $\angle AOB$. Are all these statements true? If not, where is the error?

21. On a protractor angles are shown at $1°$ intervals, in particular the nonconstructible angle of $40°$. Why does this not contradict Theorem 6?

22. Archimedes trisected an angle as follows:

    *Step 1.* Draw a circle whose center is at the vertex $O$ of the angle. Call its radius $r$. Let the two sides of the angle meet the circle at points $A$ and $B$.

    *Step 2.* Mark off on the straightedge two points $P$ and $Q$ such that the distance from $P$ to $Q$ is $r$.

    *Step 3.* Place the straightedge in such a way that $P$ lies on the line through $O$ and $A$, $Q$ lies on the circle, and the straightedge passes through $B$.

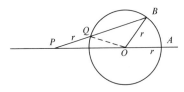

(a) Show that the angle formed by the straightedge and the line through $O$ and $A$ is one-third the size of $\angle AOB$.

(b) Why doesn't (a) contradict Theorem 6?

● ●

23. We are given a circle, whose center is $O$ and whose radius is $r$, and a point $P$ outside the circle. The point $Q$ on the segment $OP$ such that

$$OP \cdot OQ = r^2$$

is called the *inverse* of $P$ with respect to the circle.

(a) Show that $Q$ is inside the circle.
(b) Justify the following construction of $Q$, given $O$, $P$, and the circle:

*Step 1.* Draw the circle passing through $O$ whose center is $P$.
*Step 2.* Label the points at which that circle meets the given circle $R$ and $S$.
*Step 3.* Draw the circle through $O$ whose center is $R$.
*Step 4.* Draw the circle through $O$ whose center is $S$.

The circles drawn in Steps 3 and 4 meet at $O$ and at a second point. This second point is the inverse, $Q$, of $P$ with respect to the given circle.

24. We show how to bisect any given line segment $AB$ with compass only:

*Step 1.* Draw a circle through $A$, with center $B$.
*Step 2.* With the same setting as in Step 1, mark off three consecutive arcs on the circle, beginning at $A$. One end of the third arc is on the line through $A$ and $B$. Call it $P$.
*Step 3.* Draw the circle through $B$, with center $A$.
*Step 4.* Construct $Q$, the inverse of $P$ with respect to the circle constructed in Step 3.

Thus $Q$ is the midpoint of $AB$. Note that the construction of $Q$ (E 23) uses only the compass.

25. See E 24.
(a) Show how to shine a flashlight on three beads equally spaced on a wire, in such a way that their shadows on a flat surface are *not* equally spaced.
(b) Prove that it is impossible to bisect a line segment by straightedge alone.

26. Punch a small hole through the center of a jar lid or a circular piece of tagboard; switch on a lamp or flashlight near a table.
(a) Show, experimentally, that it is possible to tilt the lid in such a way that just one point of it touches the table and its shadow is a circle.

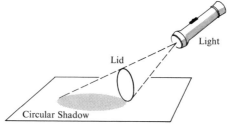

(It can be proved geometrically that this is possible.)

(b) Observe whether the spot of light from the hole is at the center of the shadow.
(c) In this diagram, $M$ and $M^*$ are midpoints of the segments $AB$ and $BC$:

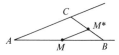

Prove that $MM^*$ is parallel to $AC$. Use this to prove that the spot mentioned in (b) is not at the center of the "shadow" circle.
(d) Using (a) and (c), show that it is impossible to construct the center of a given circle by straightedge alone.

In Exercises 27–30 we use the complex numbers to introduce and develop the branch of mathematics called trigonometry.

27. (a) Draw a circle of radius 10 centimeters. For convenience we will use 10 centimeters as a measure of length. Thus the circle has radius 1, and we call it "the unit circle."
(b) Draw a horizontal and a vertical line through its center.
(c) For any positive number $A$, consider the ray from the center that makes a (counterclockwise) angle of $A°$ with the right-hand portion of the horizontal line through the center. It meets the unit circle at a point $P$. Considered as a complex number, $P$ can be expressed in the form $a + bi$.

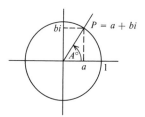

(d) We call $a$ the "cosine of $A$," and $b$ the "sine of $A$," and write

$$\cos A = a, \qquad \sin A = b.$$

For instance, it is not hard to see that

$$\cos 0 = 1, \qquad \sin 0 = 0; \qquad \cos 90 = 0, \qquad \sin 90 = 1.$$

Using your centimeter rule, find cos 20 and sin 20 to two decimal places.
(e) Using your centimeter rule, fill in this table, in which two entries have been made.

| $A$ | 0 | 10 | 20 | 30 | 40 | 60 | 80 | 100 | 180 |
|-----|---|----|----|----|----|----|----|-----|-----|
| $\cos A$ | | | | | | | | | $-1$ |
| $\sin A$ | | | | | | 0.87 | | | |

Note that the cosine and the sine of a number are numbers, just as the cube of a number is a number. The phrase "cosine of angle $AOB$" is short for "cosine of the number that tells the size of the angle $AOB$."

28. Show that for any number $A$, $(\cos A)^2 + (\sin A)^2 = 1$.

29. (a) Why is $(\cos A + i \sin A) \otimes (\cos B + i \sin B) = \cos(A + B) + i \sin(A + B)$, where $\otimes$ denotes the multiplication of complex numbers?
   (b) From (a) deduce that

$$\cos(A + B) = (\cos A)(\cos B) - (\sin A)(\sin B)$$
   and
$$\sin(A + B) = (\sin A)(\cos B) + (\cos A)(\sin B).$$

   These two equations are the essence of trigonometry.
   (c) Check the equations in (b) for $A = 20$ and $B = 40$.

30. From E 29(b) deduce that
   (a) $\cos 2A = (\cos A)^2 - (\sin A)^2$ $\quad [= 2(\cos A)^2 - 1$, by E 28],
   (b) $\sin 2A = 2(\cos A)(\sin A)$.

31. (a) Show, using the picture of the unit circle, that $\cos 45 = \sin 45$.
   (b) Combining (a) with E 29(b), deduce that $\cos 45 = \sqrt{2}/2$.

32. The three interior angles of an equilateral triangle are $60°$.
   (a) From this, and a picture of the unit circle, deduce that $\cos 60 = \frac{1}{2}$.
   (b) Deduce that $\sin 60 = \sqrt{3}/2$.

33. Prove that $\cos 3A = 4(\cos A)^3 - 3 \cos A$. (*Hint:* Write $3A$ as $2A + A$ and apply E 29(b) and E 30.)

34. (a) From E 32(a) and E 33, deduce that $-\frac{1}{2} = 4(\cos 40)^3 - 3 \cos 40$.
   (b) From (a) show that $\cos 40$ is a root of the equation $8X^3 - 6X + 1 = 0$.
   (c) Compare the argument leading up to (b) with the sketch of the proof of Theorem 6.

35. Show that $\sin 30 = \frac{1}{2}$.

36. (a) Use a picture to show that $\cos(A + 180) = -\cos A$.
   (b) Use E 29(b) to show that $\cos(A + 180) = -\cos A$.

37. See E 33. Check that $\cos 3A = 4(\cos A)^3 - 3 \cos A$ for (a) $A = 0$, (b) $A = 45$, (c) $A = 90$.

38. (a) Why is $[\cos A + (\sin A)i]^3 = \cos 3A + (\sin 3A)i$?
   (b) Use (a) to express $\cos 3A$ in terms of $\cos A$ and $\sin A$.

39. (a) Express $\cos 4A$ in terms of $\cos A$.
   (b) Express $\cos 8A$ in terms of $\cos A$.
   (c) Show why $\cos 140A$ can be expressed in terms of $\cos A$, but don't find the expression.
   (d) Show that $\cos 1$ is algebraic.

40. (a) Knowing that $\cos 45 = \sqrt{2}/2$, express $\cos 22\frac{1}{2}$ in terms of square roots.
   (b) Use (a) to estimate $\cos 22\frac{1}{2}$ to two decimal places.
   (c) Use a diagram and protractor to estimate $\cos 22\frac{1}{2}$.

41. How large can the sine of an angle be? How small?

42. In computing the energy of a radio signal it is necessary to evaluate sums such as

   (*17*) $\qquad (\cos 0)^2 + (\cos 1)^2 + (\cos 2)^2 + \cdots + (\cos 89)^2$

(a) Show that (*17*) has the same value as

(*18*)        $(\sin 0)^2 + (\sin 1)^2 + (\sin 2)^2 + \cdots + (\sin 89)^2$

(b) Show that the sum of (*17*) and (*18*) is 90.

(c) Deduce that (*17*) has the value, 45.

43. (a) Using a diagram, show that the distance from $\cos(A + B) + \sin(A + B)i$ to $1 + 0i$ equals the distance from $\cos A + (\sin A)i$ to $\cos B - (\sin B)i$.

   (b) Deduce from (a) that the equations in E 29(b) hold.

44. We have drawn our angles in the unit circle counterclockwise. It is convenient to consider clockwise too, and use negative numbers to describe them. Thus an angle of $-40°$ is drawn as follows:

The sine and cosine are defined as before. Thus

$$\cos(-40) = 0.77 \quad \text{and} \quad \sin(-40) = -0.64.$$

(a) With the aid of a diagram show that

$$\cos(-A) = \cos A \quad \text{and} \quad \sin(-A) = -\sin A.$$

(b) Using the geometric definition of complex multiplication, show that

$$[\cos A + (\sin A)i] \otimes [\cos(-B) + (\sin(-B)i] = \cos(A - B) + [\sin(A - B)]i.$$

(c) Deduce from (a) and (b) that

$$\cos(A - B) = \cos A \cos B + \sin A \sin B$$

and

$$\sin(A - B) = \sin A \cos B - \cos A \sin B.$$

(d) Check the equations in (c) for $A = 50$ and $B = 20$.

45. We have defined $\cos A$ and $\sin A$ by means of a circle, not only for our convenience, but also because that is the way they appear in many applications, such as calculus. For some applications in engineering and physics it is useful to relate cosine and sine to triangles (hence the origin of the word trigonometry: *tri* = three, *gonon* = angle, *meter* = measure).

   In a right triangle $ACB$, where $\angle ACB = 90°$, show that the cosine of $\angle BAC$ is $AC/AB$ and that the sine of $\angle BAC$ is $CB/AB$. (*Hint:* Note that $\triangle ACB$ is similar to the right triangle whose sides are $\cos \angle BAC$, $\sin \angle BAC$, and 1.)

46. See E 45. The substance of E 45 is usually summarized in the too terse equation

$$\text{cos of angle} = \frac{\text{adjacent}}{\text{hypotenuse}}, \quad \text{sin of angle} = \frac{\text{opposite}}{\text{hypotenuse}}$$

for a right triangle.

Use the table of E 27 and the above two equations to find

(a) the "adjacent" side if the hypotenuse is 20 and the angle is 50°,

(b) the "opposite" side if the angle is 30° and the hypotenuse is 100.

47. (a) Show that the area of the rectangle in this diagram is 4 sin *A* cos *A*.

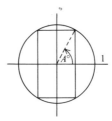

(b) Using (a), show that of all rectangles inscribed in a circle, the square has the largest area. [*Hint:* See E 30(b).]

48. Driving toward a high tower, a person notices that its top is 20° above the horizon. After getting a mile closer, he observes that the angle is now 30°. How high is the tower?

(a) Estimate the answer by drawing a scale model, where one inch equals one mile.

(b) Solve with the aid of the table in E 27.

49. A person wishes to measure the diameter of a circular pond without swimming across it, pulling a tape measure with his teeth. Instead he walks 100 feet from the pond, notices that the pond then occupies 60° of his vision. From this he determines the diameter. How?

50. Draw a circle whose radius is 10 centimeters and use it to estimate

(a) cos 21 and sin 21,

(b) cos 35 and sin 35,

(c) cos 56 and sin 56.

(d) Use (a), (b), (c) as a check for equations in E 29(b) in the case $A = 21$, $B = 35$.

51. A boy throws a ball with a speed of $v$ feet per second at an angle of $A°$ to the level ground. It can be shown that the horizontal distance the ball travels (neglecting air resistance) is

(*19*)                                   $(v^2/16) \sin A \cos A.$

(a) Show that (*19*) equals

(*20*)                                   $\left(\dfrac{v^2}{32}\right) \sin 2A.$

(b) Inspecting (*20*), decide at what angle the boy should throw the ball to obtain a maximum range.

(c) Using (*19*), show that the distance the ball travels for $A = 20°$ is the same as for $A = 70°$.

## References

All but R 3 are paperback.

1. W. W. Sawyer, *A Concrete Approach to Abstract Algebra*, W. H. Freeman and Company, San Francisco, 1959. (A very readable introduction to vector spaces which culminates in a complete proof that a 20° angle cannot be constructed with straightedge and compass.)

2. J. L. Kelley, *Algebra: A Modern Introduction*, Van Nostrand, Princeton, 1965.

3. R. Courant and H. Robbins, *What Is Mathematics?*, Oxford Univ. Press, New York, 1941. (Chapter 3, pp. 117–164, goes into the question of constructibility, including constructibility by compass alone.)

4. A. N. Kostovskii, *Geometrical Constructions Using Compasses Only*, Blaisdell, New York, 1962.

5. A. S. Smogorzhevskii, *The Ruler in Geometrical Constructions*, Blaisdell, New York, 1962.

6. B. I. Argunov and M. B. Balk, *Geometric Constructions in the Plane*, Heath, Boston, 1967.

7. S. I. Zetel, *Geometry of the Straightedge and Geometry of the Compasses*, Heath, Boston, 1967.

## Chapter 18

# INFINITE SETS

Infinity has fascinated man at least since the time of the Greeks. Who has not looked into the sky and wondered whether the stars are without end? Galileo even interrupts his *Dialogue on Two New Sciences* to record, in the year 1638, a conversation among three Renaissance gentlemen on some perplexing questions concerning infinity. We shall listen in on this conversation, examine it in detail, and, in the process of deciding which remarks are correct and which fallacious, present the fundamental discoveries of Georg Cantor, who, near the end of the nineteenth century, founded the mathematical theory of the infinite.

Now let us hear what the three gentlemen, Salviati, Simplicio, and Sagredo, have to say on this intriguing subject.

> *Simplicio:* Here a difficulty presents itself which appears to me insoluble. Since it is clear that we may have one line segment longer than another, each containing an infinite number of points, we are forced to admit that, within one and the same class, we may have something greater than infinity, because the infinity of points in the long line segment is greater than the infinity of points in the short line segment. This assigning to an infinite quantity a value greater than infinity is quite beyond my comprehension.
>
> *Salviati:* This is one of the difficulties which arise when we attempt, with our finite minds, to discuss the infinite, assigning to it those properties which we give to the finite and limited; but this I think is wrong, for we cannot speak of infinite quantities as being the one greater or less than or equal to another. To prove this I have in mind an argument, which, for the sake of clearness, I shall put in the form of questions to Simplicio who raised this difficulty.
>
> I take it for granted that you know which of the numbers are squares and which are not.*
>
> *Simplicio:* I am quite aware that a squared number is one which results from the multiplication of another number by itself: thus 4, 9, etc., are squared numbers which come from multiplying 2, 3, etc. by themselves.

---

* In this context, the word "number" is short for natural number, 1, 2, 3, 4, 5 $\cdots$. (Zero is not included.)

*Salviati:* Very well; and you also know that just as the products are called squares the factors are called roots; while on the other hand those numbers which do not consist of two equal factors are not squares. Therefore if I assert that all numbers, including both squares and non-squares, are more than the squares alone, I shall speak the truth, shall I not?

*Simplicio:* Most certainly.

*Salviati:* If I should ask further how many squares there are, one might reply truly that there are as many as the corresponding number of roots, since every square has its own root and every root its own square, while no square has more than one root and no root more than one square.

*Simplicio:* Precisely so.

*Salviati:* But if I inquire how many roots there are, it cannot be denied that there are as many as there are numbers because every number is a root of some square. This being granted we must say that there are as many squares as there are numbers because they are just as numerous as their roots, and all the numbers are roots. Yet at the outset we said there are many more numbers than squares, since the larger portion of them are not squares. Not only so, but the proportionate number of squares diminishes as we pass to larger numbers. Thus up to 100 we have 10 squares, that is, the squares constitute 1/10 part of all the numbers; up to 10,000 we find only 1/100th part to be squares; and up to a million only 1/1000th part; on the other hand in an infinite number, if one could conceive of such a thing, he would be forced to admit that there are as many squares as there are numbers all taken together.

*Sagredo:* What then must one conclude under these circumstances?

*Salviati:* So far as I see we can only infer that the totality of all numbers is infinite, that the number of squares is infinite, and that the number of their roots is infinite; neither is the number of squares less than the totality of all numbers, nor the latter greater than the former; and finally the attributes "equal," "greater," and "less" are not applicable to infinite, but only to finite, quantities. When, therefore, Simplicio introduces several lines of different lengths and asks me how it is possible that the longer ones do not contain more points than the shorter, I answer him that one line does not contain more or less or just as many points as another, but that each line contains an infinite number. Or if I had replied to him that the points in one line segment were equal in number to the squares; in another, greater than the totality of numbers; and in the little one, as many as the number of cubes, might I not, indeed, have satisfied him by thus placing more points in one line than in another and yet maintaining an infinite number in each? So much for the first difficulty.

*Sagredo:* Pray stop a moment and let me add to what has already been said an idea which just occurs to me. If the preceding be true, it seems to me impossible to say that one infinite number is greater than another. . . .

Before we analyze this conversation to see what is true and what false, let us make sure we agree on some fundamental terms to serve as tools in building a theory of infinity. *First of all, we shall use the word* **set** *to refer to any collection of objects or numbers.* We will say "a set of girls" rather than "a bevy of girls"; "a set of sheep" rather than "a flock of sheep"; "a set of fish" rather than "a school of fish." For example, in the foregoing conversation, Salviati mentioned the set of squares 1, 4, 9, 16, . . . and also the set of numbers 1, 2, 3, 4, . . . . (He and Cantor do not include 0 among the natural numbers, and we will go along with them throughout this chapter. The reader will see that Cantor's investigation of infinity could have been carried out with only slight changes if he had included 0 among the natural numbers.)

*The* **elements** *or* **members** *of a set are the individuals of which the set is composed.* For example, 9 is an element of the set of squares. The set that has no elements whatsoever is called the *empty set.* For instance, the set of real numbers whose square is $-1$ is the empty set. Two sets are *equal* if they have the same elements.

Let us now give a precise meaning to such phrases as "equal in number," "just as numerous," and "there are exactly as many." We will call two sets $S$ and $S'$ **conumerous** *if there is some way of pairing off the elements in $S$ with the elements in $S'$. Such a pairing of all the elements of $S$ with all the elements of $S'$ we will call a* **one-to-one correspondence** *between $S$ and $S'$.* (For instance, in a monogamous country the set $H$ of husbands is conumerous with the set $W$ of wives. The one-to-one correspondence that comes to mind pairs each husband with his wife.) Salviati was saying that the set of squares 1, 4, 9, . . . and the set of natural nmmbers are conumerous. This was his way of arranging one-to one correspondence between the two sets; each square is paired with its square root:

$$1, \quad 4, \quad 9, \quad 16, \quad \ldots, \quad N^2, \quad \ldots$$
$$\updownarrow \quad \updownarrow \quad \updownarrow \quad \updownarrow \qquad \updownarrow$$
$$1, \quad 2, \quad 3, \quad 4, \quad \ldots, \quad N, \quad \ldots$$

We will use the symbol $\leftrightarrow$ to indicate "is paired with." The word "conumerous" will be used usually instead of "equal in number," etc.

*If $S'$ and $S$ are two sets such that each element of $S$ is also an element of $S'$, we will say that $S$ is a* **subset** *of $S'$.* (For instance, every set is a subset of itself.) Thus the set of squares is a subset of the set of all natural numbers. What surprises the three gentlemen and ourselves is that a subset of a set $S$ can be conumerous with $S$ and yet fail to be all of $S$.

We will use braces, { }, as an abbreviation for "the set whose elements are." Thus {4, 7} is the set with two elements, namely, 4 and 7; and {1, 2, 3, . . .} is the set of natural numbers (0 being excepted in this chapter).

*By a* **proper subset** *of a set $S$ we mean any subset of $S$ other than $S$ itself.* For example, {1, 2, 3} has seven proper subsets: {1}, {2}, {3}, {1, 2}, {1, 3}, {2, 3}, and the empty set. As the reader may check, {1, 2, 3, 4} has fifteen proper subsets.

The set of natural numbers {1, 2, 3, . . .} (which, from now on, we will call $N$) is conumerous with many of its proper subsets, not just with the set of squares. As Salviati hinted, $N$ is conumerous with the set of cubes. (The reader may devise a convenient one-to-one correspondence.) Similarly, $N$ is conumerous with the set of all primes. It is a simple matter to arrange a one-to-one correspondence between the set $N$ of natural numbers and the set of all primes: pair 1 with the smallest prime, pair 2 with the next larger prime, pair 3 with the next, and so on. Thus, since 2 is the first prime, 3 the second, 5 the third, and so on, we have

$$
\begin{array}{ccccccccc}
1, & 2, & 3, & 4, & 5, & 6, & \ldots, & N, & \ldots \\
\updownarrow & \updownarrow & \updownarrow & \updownarrow & \updownarrow & \updownarrow & & \updownarrow & \\
2, & 3, & 5, & 7, & 11, & 13, & \ldots, & P_N, & \ldots
\end{array}
$$

where $P_N$ denotes the $N$th prime.

Similarly, the set $A$ of all points on the line segment from 0 to 2 is conumerous with the set $B$ of all points on the shorter line segment from 0 to 1. To justify this statement we must pair the elements of $A$ with the elements of $B$. A simple way of doing this is to match the typical element $a$ in $A$ to $a/2$, which is an element of $B$. (Observe that every element $b$ in $B$ is paired, by this rule, to some element in $A$, namely $2b$.)

Thus Simplicio's claim, "The infinity of points in the long line segment is greater than the infinity of points in the short line segment" is not justified, since the two sets are conumerous, even though one set is a proper subset of the other.

This phenomenon may seem strange to us, since it does not occur in daily life. After all, as the reader may check, the set $\{1, 2, 3, 4\}$ is not conumerous with any of its fifteen proper subsets. This contrast provides a fundamental distinction between the so-called finite sets and the infinite sets. *We will call a set* **finite** *if it is conumerous with a set of natural numbers* $\{1, 2, 3, 4, \ldots, n\}$ *for some natural number n, or is the empty set.* Thus, for example, the set of planets is finite, since it is conumerous with $\{1, 2, 3, 4, 5, 6, 7, 8, 9\}$. *Any set that is not finite we will call* **infinite.** For example, the set of primes is infinite (Theorem 2, Chapter 4). Whether the set of pairs of twin primes is finite or infinite is not known. (See Chapter 4.) The reader may easily convince himself that *no finite set is conumerous with a proper subset of itself.* But for infinite sets, as Salviati points out, this no longer holds. In fact, in 1887, Dedekind, rather than fight this seeming paradox, based his definition of infinite sets on it.

Although the number of elements in a finite set is greater than the number of elements in any proper subset, Simplicio should have been more careful than to jump to the conclusion that "The infinity of the points in the longer line segment is greater than the infinity of points in the shorter line segment." When we talk of infinite sets we must not use a word like "greater" until we have given it a precise meaning. Whether Salviati was right in saying, "We cannot speak of infinite quantities as being the one greater or less than or equal to another," or "The attributes 'equal,' 'greater,' and 'less' are not applicable to infinite, but only to finite, [sets]" was not settled until the year 1873.

Salviati and Sagredo seem to be of the opinion that any two infinite sets are conumerous. The evidence is quite persuasive: $N$, the set of squares, the set of cubes, and the set of primes are all conumerous. Indeed, we can prove

THEOREM 1. *Any infinite subset of N is conumerous with N.*

PROOF. Let $N'$ be an infinite subset of $N$. Then $N'$ has a smallest element that we call $n_1$ and pair with 1. We let $n_2$ be the smallest element other than $n_1$ in $N'$, and pair $n_2$ with 2. Next we let $n_3$ be the smallest element other than $n_1$ and $n_2$ in $N'$, and pair $n_3$ with 3. In this way we may continue, gradually pairing off all the elements of $N$ with elements of $N'$:

$$
\begin{array}{cccc}
1, & 2, & 3, & 4, \\
\updownarrow & \updownarrow & \updownarrow & \updownarrow \\
n_1, & n_2, & n_3, & n_4,
\end{array} \quad \cdots
$$

Since $N'$ is infinite, all the elements of $N$ can be paired by this rule. Moreover, each element of $N'$ will be paired eventually with some element of $N$. In fact, if $n'$ is an element of $N'$, then there are at most $n' - 1$ elements in $N'$ less than $n'$. Thus $n'$ will be paired by our rule; indeed, $n'$ will be paired to some element of $N$ no larger than $n'$.

Further evidence in favor of Salviati and Sagredo's opinion is our observation that the set of real numbers between 0 and 1 is conumerous with the set of real numbers between 0 and 2. Moreover, the borders of any two concentric circles are conumerous, as this diagram shows:

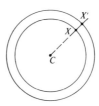

Here we pair $X$ with $X'$ if they are on the same ray from the center, $C$, of the two circles. (A similar argument applies to circles that are not concentric.)

We might also provide a geometric proof that any two line segments are conumerous by this rule:

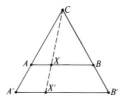

Place the two line segments $AB$ and $A'B'$ parallel to each other; denote by $C$ the point where the line through $A$ and $A'$ intersects the line through $B$ and $B'$, and marry $X$ to $X'$ if $C$, $X$, and $X'$ are on a line.

Salviati goes on to say, "The points in one line segment were equal in number to the squares," but doesn't bother to justify his statement. Perhaps he had in the back of his mind something like this:

*Salviati:* It is clear to me that the squares 1, 4, 9, . . . are conumerous with the set of all reals between 0 and 1. I picture two bags: one containing all the squares; the other, all the reals from 0 to 1. I put my left hand into the bag of squares and my right hand into the bag of reals, and I pair the two numbers I happen to select. These two numbers I remove from the bags. Now I repeat the process as often as necessary until both bags are empty.

*We:* But, Salviati, might you not empty one bag while the other is still far from empty? Your rule is so vague that it is impossible to predict what square will be matched with what real. Can you guarantee that you will not use up the squares before the reals, or the reals before the squares? What if it turns out that you pluck out the square 1 and the real 1? Next, you might pluck out the square 4 and the real $\frac{1}{2}$. After that, you might pluck out the square 9 and the real $\frac{1}{3}$, and so on. In this way, you might accidentally be pairing each square to a real of a very special form. Or what if you should have the misfortune of pairing each square with itself? We fear that many real numbers might end up not used at all.

*Salviati:* It is quite unlikely that I would be so unfortunate as to pluck out such an orderly set of reals.

*We:* But you must grant that it might happen. Besides, even if you live to be a hundred, you will never empty either bag. I would much prefer to see a more precise procedure for pairing off the squares to the reals from 0 to 1.

*Salviati:* At the moment I cannot think of one.

Thus, when the chips were down, Salviati did not convince us that the set of squares and the set of reals from 0 to 1 are conumerous. We still face the question: Are any of two infinite sets conumerous? If the answer is "yes," then we ought to be able to find, in particular, a rule that pairs off the natural numbers with the reals from 0 to 1. If the answer is "no," then we can introduce into infinite sets the notions of "greater" and "less," which we have for finite sets.

Before trying to devise a one-to-one correspondence between the natural numbers and the reals from 0 to 1 we might consider this simpler question: Can the set of natural numbers 1, 2, 3, . . . be married to the set of positive rationals? (Corollary 2 will treat the set of all rationals.)

It would seem highly unlikely that there could be such a one-to-one correspondence, since the set of positive rationals is formed by varying both the numerator $p$ and the denominator $q$. For example, the set of positive rationals with numerator 1 is conumerous with the set of natural numbers, since we can arrange this one-to-one correspondence:

$$1, \quad 2, \quad 3, \quad 4, \quad \ldots, \quad n, \quad \ldots$$
$$\updownarrow \quad \updownarrow \quad \updownarrow \quad \updownarrow \qquad\qquad \updownarrow$$
$$\frac{1}{1}, \quad \frac{1}{2}, \quad \frac{1}{3}, \quad \frac{1}{4}, \quad \ldots, \quad \frac{1}{n}, \quad \ldots$$

Yet, in 1873, Cantor discovered that the set of natural numbers and the set of positive rationals are conumerous. This is how he devised a one-to-one correspondence between the two sets. The reader will notice how he avoids pairing all the natural numbers with positive rationals of a fixed numerator.

First he placed the positive rationals on the plane, as indicated here:

$$
\begin{array}{cccccccc}
 & & & & \cdot & \cdot & \cdot & \cdot \\
 & & & & \cdot & \cdot & \cdot & \cdot \\
 & & & & \cdot & \cdot & \cdot & \cdot \\
(1) & & & & \frac{5}{1} & \frac{5}{2} & \frac{5}{3} & \frac{5}{4} & \cdot & \cdot & \cdot \\
 & & & & \frac{4}{1} & \frac{4}{2} & \frac{4}{3} & \frac{4}{4} & \cdot & \cdot & \cdot \\
 & & & & \frac{3}{1} & \frac{3}{2} & \frac{3}{3} & \frac{3}{4} & \cdot & \cdot & \cdot \\
 & & & & \frac{2}{1} & \frac{2}{2} & \frac{2}{3} & \frac{2}{4} & \cdot & \cdot & \cdot \\
 & & & & \frac{1}{1} & \frac{1}{2} & \frac{1}{3} & \frac{1}{4} & \cdot & \cdot & \cdot
\end{array}
$$

and then paired them with the natural numbers as indicated by this rule:

$$
\begin{array}{cccccc}
 & & \cdot \\
 & & \cdot \\
(2) & & 15 & \cdot & \cdot & \cdot \\
 & & 10 & 14 & \cdot & \cdot \\
 & & 6 & 9 & 13 & \cdot \\
 & & 3 & 5 & 8 & 12 \\
 & & 1 & 2 & 4 & 7 & 11 & \cdot
\end{array}
$$

Of course, expressions such as $\frac{6}{3}$, $\frac{4}{2}$, and $\frac{2}{1}$, which have equal value, all appear in ($1$). If we erase any rational not in its lowest terms (such as $\frac{4}{2}$ and $\frac{6}{3}$) from ($1$), then we can still arrange the one-to-one correspondence by placing no natural number at a dot from which the rational has been erased:

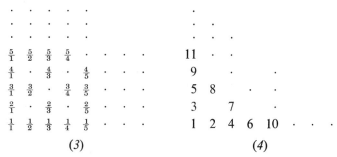

$$\begin{array}{cccccccc}
\frac{5}{1} & \frac{5}{2} & \frac{5}{3} & \frac{5}{4} & \cdot & \cdot & \cdot & \cdot \\
\frac{4}{1} & \cdot & \frac{4}{3} & \cdot & \frac{4}{5} & \cdot & \cdot & \cdot \\
\frac{3}{1} & \frac{3}{2} & \cdot & \frac{3}{4} & \frac{3}{5} & \cdot & \cdot & \cdot \\
\frac{2}{1} & \cdot & \frac{2}{3} & \cdot & \frac{2}{5} & \cdot & \cdot & \cdot \\
\frac{1}{1} & \frac{1}{2} & \frac{1}{3} & \frac{1}{4} & \frac{1}{5} & \cdot & \cdot & \cdot
\end{array}
\qquad
\begin{array}{ccccccc}
11 & \cdot & \cdot \\
9 & \cdot & & \cdot \\
5 & 8 & & \cdot & \cdot \\
3 & & 7 & & \cdot \\
1 & 2 & 4 & 6 & 10 & \cdot & \cdot
\end{array}$$

$$(3) \qquad\qquad\qquad (4)$$

Comparing (*3*) and (*4*), we see that the one-to-one correspondence begins as follows:

$$\begin{array}{ccccccccccc}
1 & 2 & 3 & 4 & 5 & 6 & 7 & 8 & 9 & 10 & 11 \quad \cdots \\
\updownarrow & \updownarrow & \updownarrow & \updownarrow & \updownarrow & \updownarrow & \updownarrow & \updownarrow & \updownarrow & \updownarrow & \updownarrow \\
\frac{1}{1} & \frac{1}{2} & \frac{2}{1} & \frac{1}{3} & \frac{3}{1} & \frac{1}{4} & \frac{2}{3} & \frac{3}{2} & \frac{4}{1} & \frac{1}{5} & \frac{5}{1} \quad \cdots
\end{array}$$

Note that the fractions are arranged systematically according to the sum of numerator and denominator; for instance, those with sum equal to 5 are $\frac{1}{4}$, $\frac{2}{3}$, $\frac{3}{2}$, $\frac{4}{1}$.

Calling a set **denumerable** *if it is conumerous with N, we can state*

THEOREM 2. *The set of positive rationals is denumerable.*

Of course the numbering in (*2*) is not the only one that would have done the trick. Rather than moving upward on the lines inclined at 45°, we could just as well have followed this path

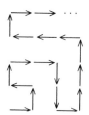

which places the positive integers as in this figure:

$$\begin{array}{cccc}
17 & 18 & \cdots \\
16 & 15 & 14 & 13 \\
5 & 6 & 7 & 12 \\
4 & 3 & 8 & 11 \\
1 & 2 & 9 & 10
\end{array}$$

We can use the idea of the proof of Theorem 2 to prove a more general result. *Let us begin by defining the* **union** *of sets* $S_1$, $S_2$, . . . *as the set of all elements that appear in at least one of* $S_1$, $S_2$, . . . . For example, the union of $\{1, 2\}$ and $\{1, 3\}$ is $\{1, 2, 3\}$. As another example, if $S_1$ is the set of positive rationals with numerator 1, if $S_2$ is the set of positive rationals with numerator 2, and, in general, if $S_n$ is the set of positive rationals with numerator $n$, then the union of $S_1$, $S_2$, $S_3$, . . . , $S_n$, . . . is simply the set of all positive rationals.

THEOREM 3. *If we have a denumerable collection of denumerable sets* $S_1$, $S_2$, $S_3$, . . . *(for convenience, no two of which share an element), then the union of all the* $S_1$, $S_2$, $S_3$, . . . , $S_n$, . . . *is also denumerable.*

PROOF. Let $S$ be the union of $S_1$, $S_2$, $S_3$, . . . ; that is, the set of elements belonging to at least one of $S_1$, $S_2$, $S_3$, . . . (in fact, exactly one, since no two share an element).

Since $S_1$ is denumerable we may place each of its elements on one of the dots of an endless line of equally spaced dots:

$(S_1)$

Since $S_2$ is denumerable, we may place its elements on the line of dots to the right of $S_1$:

$(S_1)$  $(S_2)$

Similarly, $S_3$ can be sprinkled to the right of $S_2$, and so on. Thus the set $S$, the union of $S_1, S_2, S_3, \ldots$, is paired, element by element, with the following endless arrangement of dots:

(5)

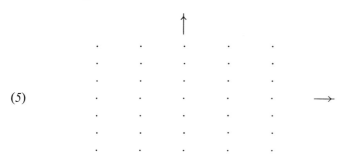

By the technique shown in (2), $S$ is denumerable. Theorem 3 is proved.

In a similar mood, we have

THEOREM 4. *The union, $S$, of a denumerable collection of finite sets $S_1$, $S_2$, $S_3$, ... is either finite or denumerable.*

PROOF. This time we are allowing the individual sets $S_1, S_2, S_3, \cdots$ to share elements. (Indeed, in the proof of Theorem 7 we will use Theorem 4 in a case where some of the $S_i$'s coincide.)

To prove that $S$ is either finite or denumerable we must present a one-to-one correspondence between $S$ and the set $\{1, 2, 3, \ldots, n\}$ for some $n$, or else between $S$ and the set $N$ of natural numbers. To arrange such a correspondence, we begin with $S_1$. Since $S_1$ is finite, it has, say, $n_1$ elements, and thus may be paired with the integers from 1 to $n_1$.

If $S_2$ is a subset of $S_1$ we go on to $S_3$. If $S_2$ is not a subset of $S_1$, then there will be among the elements of $S_2$ a finite set of $n_2$ elements not already in $S_1$. We then pair these elements with the next $n_2$ integers after $n_1$, namely $n_1 + 1$, $n_1 + 2, \ldots, n_1 + n_2$.

So far we have

$$\underbrace{1, 2, \ldots, n_1,}_{\substack{\text{paired to} \\ S_1}} \quad \underbrace{n_1 + 1, \ldots, n_1 + n_2}_{\substack{\text{paired to the part} \\ \text{of } S_2 \text{ not in } S_1}}$$

If $S_3$ is a subset of the union of $S_1$ and $S_2$ we go on to $S_4$. Otherwise, if $S_3$ has $n_3$ elements not already accounted for, we can pair them with the next $n_3$ integers after $n_1 + n_2$, namely $n_1 + n_2 + 1, n_1 + n_2 + 2, \cdots, n_1 + n_2 + n_3$.

So far we have

$$\underbrace{1, 2, \ldots, n_1}_{\substack{\text{paired to} \\ S_1}} \qquad \underbrace{n_1 + 1, \ldots, n_1 + n_2}_{\substack{\text{paired to the part} \\ \text{of } S_2 \text{ not in } S_1}} \qquad \underbrace{n_1 + n_2 + 1, \ldots, n_1 + n_2 + n_3}_{\substack{\text{paired to the part of } S_3 \text{ not} \\ \text{in the union of } S_1 \text{ and } S_2}}$$

If we continue in this manner, one of two things happens. We might meet an $m$ such that all the sets $S_{m+1}$, $S_{m+2}$, ... are subsets of the union of $S_1$, $S_2$, ..., $S_m$; in this case, $S$ is finite. On the other hand, we might meet no such $m$; in this case, $S$ is denumerable, for every element of $S$ is eventually paired with some element of $N$.

The following two corollaries illustrate the use of Theorems 3 and 4.

COROLLARY 1. *The set of all integers is denumerable; and so is the set of integers exclusive of zero.*

PROOF. Let $S_1$ be $\{1, -1, 0\}$; let $S_2$ be $\{2, -2\}$; let $S_3$ be $\{3, -3\}$; generally, let $S_n$ be $\{n, -n\}$ for any natural number $n$ larger than 1. Theorem 4 then asserts that the union of $S_1$, $S_2$, $S_3$, ... is denumerable (since this union is obviously not finite). But this union is precisely the set of all integers. This proves the first part of the corollary. Changing $S_1$ to $\{1, -1\}$ is all that is necessary to obtain a proof of the second part.

COROLLARY 2. *The set of rationals is denumerable.*

PROOF. *We take each rational to be represented in the form $A/B$, where $(A, B) = 1$ and $B$ is positive. Let $S_1$ consist of 0 and all rational numbers whose numerator is 1 or $-1$:*

$$S_1 = \{0, \tfrac{1}{1}, -\tfrac{1}{1}, \tfrac{1}{2}, -\tfrac{1}{2}, \tfrac{1}{3}, -\tfrac{1}{3}, \ldots\}$$

Let $S_2$ be the set of all rationals with numerator 2 or $-2$. Let $S_3$ be the set of all rationals with numerator 3 or $-3$. Let $S_n$ be, in general, the set of all rationals with numerator $n$ or $-n$ for any natural number $n$ larger than 1. By Corollary 1 and Theorem 4 each $S_n$ is denumerable. Thus, by Theorem 3, the union of $S_1$, $S_2$, $S_3$, ... is denumerable. But this union is precisely the set of rationals. This proves the corollary.

Theorems 2, 3, and 4 and their corollaries seem to suggest that any infinite set is denumerable. But we still have not been able to make any headway in this chapter in proving that the reals between 0 and 1 are denumerable. Perhaps Salviati has misled us. As regards this problem, Cantor wrote in a letter to Dedekind, dated November 29, 1873:

May I ask you a question, which has a certain theoretical interest for me but which I cannot answer; maybe you can answer it and would be so kind as to write to me about it. It goes as follows: take the set of all natural numbers $n$ and denote it $N$. Further, consider, say, the set of all positive real numbers $x$ and denote it $R$. Then the question is simply this: can $N$ be paired with $R$ in such a way that to every individual of one set corresponds one and only one individual of the other? At first glance, one says to oneself, "No, this is impossible, for $N$ consists of discrete parts and $R$ is a continuum." But nothing is proved by this objection. And much as I too feel that $N$ and $R$ do not permit such a pairing, I still cannot find the reason. And it is this reason that bothers me; maybe it is very simple.

Would not one at first glance be led to the conjecture that $N$ cannot be paired off with the set of all positive rational numbers $p/q$? And yet it is not hard to show . . . that such an ordering can be found.

The rationals are distributed on the line quite generously (see E 42 of Chapter 6)—between any two points, no matter how close they are to each other, there is an infinite set of rationals—and yet the set of rationals, as we saw, is denumerable. Intuition is an inadequate guide through the world of infinite sets. Cantor was asking Dedekind to prove rigorously that intuition is right when it says "the real numbers are not denumerable" and that Salviati, reaching into the two bags, is wrong.

Dedekind did not solve Cantor's problem. On December 2, 1873, Cantor wrote again to Dedekind:

I was extremely pleased to receive today your answer to my latest letter. I proposed my question to you for the following reason. I had asked it several years ago and had always remained in doubt whether the difficulty which it presents is a subjective one or whether it is inherent in the substance. As you write to me that you too are unable to answer it I may assume the latter. Besides, I would like to add that I have never seriously thought about it because it has no particular practical interest for me and I fully agree with you if you say that for this reason it doesn't deserve too much labor. Only it would be a beautiful result. . . .

Soon after, Cantor discovered a proof that the set of real numbers is not denumerable. On December 7, 1873, he wrote to Dedekind:

Recently I had time to follow up a little more fully the conjecture which I mentioned to you; only today I believe I have finished the matter. Should I have been deceived, I would not find a more lenient judge than you. I thus take the liberty of submitting to your judgment what I have written, in all the incompleteness of a first draft.

Thus, sometime between December 2 and December 7, 1873, Cantor had laid the cornerstone of the theory of infinity. In 1890 he discovered a second proof, much simpler than his proof of 1873. It is this second proof that we now give for

THEOREM 5 (Cantor's Theorem). *The set of real numbers is not denumerable.*

PROOF. We will argue very much as we did in proving that no finite list of primes can be complete. Instead of the Prime-manufacturing Machine of Chapter 4, we will require a "Real-manufacturing Machine," that will manufacture from any denumerable list of reals a real number not on that list.

Let us take any denumerable list of real numbers. For convenience, let us restrict our attention to real numbers between 0 and 1. In order to be able to treat all such lists, let us assume that each real number in a list is written in decimal form. We will denote the $j$th digit to the right of the decimal point in the $n$th number by $d_{n,j}$. Thus our list will look like this:

(6)
$$1 \longleftrightarrow 0.d_{1,1} \quad d_{1,2} \quad d_{1,3} \quad d_{1,4} \quad \ldots$$
$$2 \longleftrightarrow 0.d_{2,1} \quad d_{2,2} \quad d_{2,3} \quad d_{2,4} \quad \ldots$$
$$3 \longleftrightarrow 0.d_{3,1} \quad d_{3,2} \quad d_{3,3} \quad d_{3,4} \quad \ldots$$
$$\vdots$$
$$n \longleftrightarrow 0.d_{n,1} \quad d_{n,2} \quad d_{n,3} \quad d_{n,4} \quad \ldots$$
$$\vdots$$

The rule by which the particular numbers in the list are computed will not concern us. All that matters to us is that it is a denumerable list of real numbers; every natural number, no matter how large, is paired by the list to some real number. One example of such a list might be computed in a very orderly manner by using square roots:

(7)
$$1 \longleftrightarrow 0.9999 \ldots (= 1/\sqrt{1})$$
$$2 \longleftrightarrow 0.7071 \ldots (= 1/\sqrt{2})$$
$$3 \longleftrightarrow 0.5773 \ldots (= 1/\sqrt{3})$$
$$4 \longleftrightarrow 0.5000 \ldots (= 1/\sqrt{4})$$
$$\vdots$$
$$n \longleftrightarrow 0.d_{n,1} \quad d_{n,2} \quad d_{n,3} \quad d_{n,4} \ldots (= 1/\sqrt{n})$$
$$\vdots$$

In this list we have, for example, $d_{1,1} = 9$, $d_{1,2} = 9$, $d_{2,1} = 7$, $d_{2,2} = 0$.

Another quite orderly list consists of those rationals whose form is $1/n$:

(8)

$$1 \longleftrightarrow 0.9999\ldots \qquad\qquad \left(=\frac{1}{1}\right)$$

$$2 \longleftrightarrow 0.5000\ldots \qquad\qquad \left(=\frac{1}{2}\right)$$

$$3 \longleftrightarrow 0.3333\ldots \qquad\qquad \left(=\frac{1}{3}\right)$$

.
.
.

$$n \longleftrightarrow 0.d_{n,1}\ \ d_{n,2}\ \ d_{n,3}\ \ \ldots \qquad\qquad \left(=\frac{1}{n}\right)$$

.
.
.

The reader can clearly see that neither list contains all the positive reals; for example, $2/3$ is not in list (7), and $1/\sqrt{3}$ is not in list (8). What we will prove is that any list of the type (6), no matter how complicated the rule by which it was made, is not a complete roster of all real numbers between 0 and 1. For we will manufacture from any such list a real number between 0 and 1 that could not possibly be on the list.

In the construction of this unlikely number, the only digits that will concern us are those that lie on the diagonal to the right of the decimal point in (6). These are:

(9)

$$
\begin{array}{l}
d_{1,1}\\
\quad d_{2,2}\\
\qquad d_{3,3}\\
\qquad\quad d_{4,4}\\
\qquad\qquad\ddots
\end{array}
$$

The decimal representation of the number we will build will look like this:

$$0.b_1 b_2 b_3 \ldots b_n \ldots,$$

where $b_1,\ b_2,\ b_3,\ \ldots$ are each, of course, one of the ten integers from 0 to 9.

We will now give a rule for constructing the decimal representation of this number, which we will call $B$. The $n$th digit, $b_n$, will be 7 if $d_{n,n}$ in our original list is not 7, and will be 1 if $d_{n,n}$ happens to be 7. Thus, to begin, if we find that $d_{1,1}$ is 7, we take $b_1$ to be 1; and if $d_{1,1}$ is not 2, we take $b_1$ to be 7. Continuing in this way, we obtain $b_2,\ b_3,\ \ldots$ and thus define $B$.

For example, if we apply this procedure to (7), we find that $b_1 = 7$, $b_2 = 7$, $b_3 = 1$; and if we apply it to (8), we find that $b_1 = 7$, $b_2 = 7$, $b_3 = 7$.

Now we must show why $B = 0.b_1 b_2 b_3 \ldots$ cannot appear in list (6). First, it is not paired with 1, since $b_1$ is not $d_{1,1}$. Second, it is not paired with 2, since $b_2$ is not $d_{2,2}$. Third, it is not paired with 3, since $b_3$ is not $d_{3,3}$. As the reader may gather, $B$ appears nowhere in list (6), since it differs from the $n$th real number in the list in at least the $n$th digit to the right of the decimal point.

Since no denumerable list of positive real numbers is complete, it follows that the positive reals are not denumerable. This proves Cantor's Theorem.

By combining Theorems 2 and 5, the reader may give a new proof for

THEOREM 6. *There exist irrational real numbers.*

This is our third way of showing that not all numbers are rational. The first way (in Chapter 6), which depends on the Fundamental Theorem of Arithmetic, showed that $\sqrt{2}$, $\sqrt{3}$, $\sqrt{5}$, and $\sqrt{6}$, for example, are irrational. The second way (also in Chapter 6) showed that any decimal, such as $0.101001000100001\ldots$, that is not repeating, represents an irrational number.

But from Cantor's Theorem we can deduce much more than Theorem 6. In his letter of December 2, five days before he had proved Theorem 5, Cantor said concerning the question of whether the reals and the natural numbers are conumerous, "Provided that the answer is 'no,' we would this way have obtained a new proof of Liouville's theorem that there exist transcendental numbers." (Liouville in 1844 had already shown that there exist transcendentals, by exhibiting specific examples.) But Theorem 5 shows that Cantor's question, "Is the set of positive reals denumerable," must be answered "no." We will now use Theorem 5 to prove that there are transcendental numbers. To do this we will prove

THEOREM 7. *The set of algebraic numbers is denumerable.*

PROOF. Every polynomial $P$ whose coefficients are integers contributes only a finite number of algebraic numbers; namely, the roots of the equation $P = 0$. (See Theorem 4 of Chapter 16.)

If we could prove that the set of polynomials with integral coefficients is denumerable, we would then, by Theorem 4, be assured that the set of algebraic numbers is denumerable.

To do this we will in quite an indirect way make use of the denumerability of the positive rational numbers. We proceed as follows.

Each polynomial is completely described by its coefficients. Thus the polynomial $6 - 5X + X^3$, which is short for $6 - 5X + 0X^2 + 1X^3$, is described by the sequence

$$6, -5, 0, 1.$$

The polynomial $-5X^2 + 12X^4$ has the sequence

$$0, 0, -5, 0, 12.$$

The right-hand term in such a sequence is not 0, though the left-hand term may be 0. Moreover, any such sequence of integers describes some polynomial. For instance, the sequence $-2, 3, 1, 0, 2$ corresponds to the polynomial

$$-2 + 3X + 1X^2 + 0X^3 + 2X^4,$$

which is simply $-2 + 3X + X^2 + 2X^4$.

To prove that the set of polynomials with integral coefficients is denumerable, it suffices to prove that the set of such sequences of integers is denumerable. This we now do.

Every such sequence we will pair with a single positive rational number in the following manner. We make use of the set of primes in their usual order

$$2, 3, 5, 7, 11, 13, \ldots.$$

By way of illustration we associate with the sequence

$$6, -5, 0, 1$$

the rational number

$$2^6 \cdot 3^{-5} \cdot 5^0 \cdot 7^1,$$

which is

$$\frac{2^6 \cdot 7}{3^5}.$$

As a second illustration before stating the general rule, we associate with the sequence

$$-2, 3, 1, 0, 2$$

the rational number

$$2^{-2} \cdot 3^3 \cdot 5^1 \cdot 7^0 \cdot 11^2,$$

which is simply

$$\frac{3^3 \cdot 5 \cdot 11^2}{2^2}.$$

In general, then, we pair the sequence of integers

$$a_1, a_2, \ldots, a_n$$

(where $a_n$ is not 0) with the rational number

$$2^{a_1} 3^{a_2} \cdots P_n{}^{a_n},$$

where $P_n$ denotes the $n$th prime ($P_1 = 2$, $P_2 = 3$, $\ldots$).

Moreover, this procedure does not pair two different sequences with the same rational number. For instance, we can easily find the *unique* sequence paired with the rational number 10/117. We write

$$\frac{10}{117} = \frac{2 \cdot 5}{3^2 \cdot 13} = 2^1 3^{-2} 5^1 13^{-1} = 2^1 3^{-2} 5^1 7^0 11^0 13^{-1},$$

and then recover the sequence of exponents

$$1, \ -2, \ 1, \ 0, \ 0, \ -1$$

(which records the polynomial

$$1 - 2X + 1X^2 + 0X^3 + 0X^4 - 1X^5,$$

or simply

$$1 - 2X + X^2 - X^5).$$

Notice that we made use of the Fundamental Theorem of Arithmetic. Finally, the rational number 1 we pair off with a sequence consisting of one 0 (it corresponds to the zero-polynomial.)

Since, as Theorem 2 tells us, the set of positive rational numbers is denumerable, so is the set of sequences of the type considered; hence so is the set of all polynomials with integral coefficients. Thus the set of algebraic numbers is denumerable, and Theorem 7 is proved.

Combining Theorems 5 and 7 we obtain this strengthening of Theorem 6:

THEOREM 8. *There exist transcendental real numbers.*

Cantor's Theorem, which asserts that the set of real numbers is not conumerous with the set of natural numbers, shows that Salviati was not correct when he said, "The attributes of 'equal,' 'greater,' and 'less' are not applicable to infinite quantities." On the contrary, we can, following Cantor, give a very precise meaning to the statement "Set *A* is less numerous than set *B*" even when *A* and *B* are infinite.

*We will say that set A is* **less numerous** *than set B if A is conumerous with some subset of B, but not with B itself.* As an example, we know from Theorem 5 that the set of natural numbers is less numerous than the set of real numbers. *We will say that B is* **more numerous** *than A if A is less numerous than B.*

This notion of "less numerous," agreeing with our experience with finite sets, introduces a scale even among the infinite sets and raises questions that could not be considered before. For example, is there a set more numerous than the set of real numbers? Cantor showed that the answer is "yes." He even went on to prove that if *A* is any set, then the set *B* whose elements are the subsets of *A* is more numerous than *A*. (If *A* is finite, this is easy to see.) In particular, the set of all subsets of the real numbers is more numerous than the set of real numbers.

But Cantor raised a second question: Is every infinite subset of the reals either conumerous with the reals or else conumerous with the set of natural numbers?

The Continuum Hypothesis states Cantor's belief that the answer is "yes," but for almost a century it remained unproved. ("Continuum" is another name for the set of real numbers.) In 1963, P. Cohen settled this question by proving that the Continuum Hypothesis is not a consequence of the commonly accepted fundamental properties of sets. We will discuss this result—one of the most profound theorems of the twentieth century—in more detail in Chapter 19.

Following his discovery that not all infinite sets are conumerous, Cantor created an arithmetic that extends into the realm of infinite sets the familiar arithmetic of finite sets. First of all, he introduced symbols to denote the various sizes of infinite sets, just as the natural numbers 1, 2, 3, . . . denote the various sizes of finite sets. We will glance briefly at just two of these "infinite numbers."

Any denumerable set, Cantor would say, "has $\aleph_0$ elements." [The symbol $\aleph$ (alef) is the first letter of the Hebrew alphabet; the subscript, 0, distinguishes $\aleph_0$ (read alef null) from other "infinite numbers," $\aleph_1, \aleph_2, \cdots$ that Cantor defined.] The letter $\aleph_0$ we will use to represent the smallest of the infinite numbers. Theorem 2 can now be read, "There are $\aleph_0$ positive rationals"; Theorem 3, "$\aleph_0$ times $\aleph_0$ is $\aleph_0$"; and Theorem 7, "There are $\aleph_0$ algebraic numbers."

Any set conumerous with the set of reals, Cantor said, "has $c$ elements" (the "$c$" stands for "continuum"). The Continuum Hypothesis asserted, "Every infinite subset of the reals has either $\aleph_0$ elements or $c$ elements."

Furthermore, Cantor defined a multiplication and addition for these new numbers, using ideas suggested by finite sets.

Cantor's work on infinite sets, carried out at the end of the nineteenth century, not only resolves the riddles that troubled Simplicio, Salviati, and Sagredo in the seventeenth century, but also provides an important tool for twentieth century mathematicians, who, in many branches of their art, must work with infinity.

Infinite sets still offer many challenging problems; and so do finite sets, as E 41–51 illustrate.

### Exercises

1. (a) Show that there are exactly 1000 squares among the first million natural numbers larger than 0.
   (b) Prove that a thousandth part of the first million natural numbers larger than 0 are squares.

2. How many cubes are there among the first million natural numbers larger than 0?

3. Prove that the set of all squares is conumerous with the set of all cubes. ("Squares" and "cubes" here refer to numbers, not geometric figures.)

4. List the 15 proper subsets of $\{1, 2, 3, 4\}$ (remember the empty set).

5. List the 32 subsets of $\{1, 2, 3, 4, 5\}$ (remember the empty set).

6. Fill in the blank and explain: The set $\{1, 2, 3, \ldots, n\}$ has ____ subsets.

7. Prove that the set $\{2, 3, 4, \ldots\}$ and the set $\{1, 2, 3, 4, \ldots\}$ are conumerous.

8. Prove geometrically that the border of a square and the border of a circle are conumerous.

9. Is the set of positive reals conumerous with the set of reals between 0 and 1 (but excluding 0 and 1)?

10. Is the union of two denumerable sets always denumerable? Explain.

11. A very introspective man requires a year to record in his diary the events of a single day of his life. Each day that he is alive he gets further behind. At the end of 100 years he would have written up just the first 100 days of his eventful life. Show that if he lives forever, each day of his life will eventually be recorded.

12. The Fundamental Theorem of Arithmetic can be used to prove that the set of pairs $(a, b)$ is denumerable, where $a$ and $b$ are elements of $\{1, 2, 3, 4, \ldots\}$. The following is an outline of the proof, similar to the proof of Theorem 7.
    (a) Using Theorem 1, prove that the set of natural numbers of the form $2^a \times 3^b$ is denumerable, where $a$ and $b$ are elements of $\{1, 2, 3, 4, \ldots\}$.
    (b) Using the Fundamental Theorem of Arithmetic, prove that the set of natural numbers of the form $2^a \times 3^b$ is conumerous with the set of all pairs $(a, b)$, where $a$ and $b$ are elements of $\{1, 2, 3, 4, \ldots\}$.

13. Using the technique of E 12, prove that the set of all triplets $(a, b, c)$, where $a$, $b$, and $c$ are elements of $\{1, 2, 3, 4, \ldots\}$, is denumerable.

14. Using the technique of E 12, prove that the set of quadruplets $(a, b, c, d)$ that can be formed from the numbers $\{1, 2, 3, 4, \ldots\}$ is denumerable.

15. In (3) and (4), p. 321, what natural number is paired with $\frac{2}{3}$? With $\frac{3}{4}$?

16. In (3) and (4), what rational is paired with 8? With 10? With 20?

17. Prove that the set of integers $\{\ldots, -3, -2, -1, 0, 1, 2, 3, \ldots\}$ is denumerable by pairing the $n$ with the rational $2^n$ and applying Theorems 1 and 2.

18. Prove in the following three ways that the set of regularly spaced dots

extending throughout the plane is denumerable:
(a) by a direct geometric argument like that for Theorem 2;

(b) by the use of Corollary 1 and Theorem 3;

(c) by the technique used in E 12.

19. An infinite motel has single rooms numbered 1, 2, 3, 4, . . . . It is full, and the "no vacancy" sign is lighted. A weary traveler drives in and the sympathetic manager says, "For you we will find a room, though every room is occupied." He goes to the intercom microphone connected directly to all the rooms and orders, "_____," whereupon the new arrival and each of those already registered all have rooms of their own. What was the order?

20. Is the set of positive integers conumerous with the set of all positive even integers?

21. In E 84 of Chapter 16 the set of positive integers is expressed as the union of a denumerable set of endless arithmetic progressions. Use this fact to prove Theorem 3.

22. Cantor's method (2) gives one way of pairing the natural numbers with regularly spaced dots in the plane. The following steps outline another way.
    (a) Prove that every natural number 1, 2, 3, 4, . . . can be written in the form $2^{a-1}(2b - 1)$ for some pair $(a, b)$ of natural numbers larger than 0.
    (b) Prove that for any natural number $n$, larger than 0, there is precisely one pair $(a, b)$ of natural numbers larger than 0 such that $n = 2^{a-1}(2b - 1)$; that is, prove that the representation of part (a) is unique.
    (c) Now pair $n$ with the pair $(a, b)$ given in part (b). We may think of the pair $(a, b)$ as describing a dot in this manner:

23. Why is it that the same argument used in the proof of Theorem 5 does not also prove that the rationals are not denumerable? Point out exactly where the argument breaks down.

24. In a list of all terminating or repeating decimals can the diagonal itself be terminating or repeating?

25. Let $S$ be the set of real numbers between 0 and 1 whose decimal expansion has only 4's and 7's. Is $S$ denumerable? Explain.

26. What are the various ways by which we have proved that there are irrational numbers?

27. Let $S$ be the set of subsets of $N$ that have precisely two elements. Thus $\{3, 4\}$, $\{2, 7\}$, and $\{2, 8\}$ are some elements of $S$. $S$ is infinite. Is it denumerable?

28. Read the proof of Theorem 7.
    (a) Which rational number corresponds to the polynomial $X - 5X^2 + X^7$?
    (b) Which polynomial corresponds to the rational number 400?
    (c) Which polynomial corresponds to the rational number $\frac{5}{34}$?

●

29. By pairing each real number $x$ with $2^x$ what two sets do we show to be conumerous?

30. By pairing each real number $x$ that is positive but less than one with the number $1/x$, what two sets do we show to be conumerous?

31. (a) Argue for the plausibility of this assertion: Every infinite set contains a denumerable subset.
    (b) Let $S$ be an infinite set and let $T$ be a denumerable set disjoint from $S$. Show that the union of $S$ and $T$ is conumerous with $S$.
    (c) Prove that the set of irrational numbers is conumerous with the set of real numbers.
    (d) Which are more plentiful, rational or irrational numbers?

32. (a) When you remove a denumerable subset from a denumerable set, what can you assert about the set remaining?
    (b) When you remove a denumerable subset from the set of real numbers, what can you assert about the set remaining?

33. Let $S$ be the set of all subsets of $N$. Let $T$ be the set of all endless sequences made of 0's and 1's. For instance, one such sequence may begin 00101110...
    (a) Show that $S$ and $T$ are conumerous.
    (b) Using Cantor's diagonal technique, prove that $T$ is not denumerable.

34. See E 33.
    (a) Is the set of all finite subsets of $N$ denumerable?
    (b) Is the set of all subsets of $N$ denumerable?

35. Which sets listed below are conumerous?
    (a) the set of real numbers,
    (b) the set of rational numbers,
    (c) the set of real numbers between 1.3 and 1.4,
    (d) the set of odd integers,
    (e) the set of integers,
    (f) the set of algebraic numbers,
    (g) the set of irrational numbers,
    (h) the set of transcendental numbers.

• •

36. See E 33. An endless sequence of 0's and 1's that is formed by an initial block and then repeats a block endlessly we will call *repeating* (because of the resemblance to a repeating decimal).
    (a) Prove that the set of repeating elements in $T$ is denumerable.
    (b) Prove that the set of elements in $T$ that are not repeating is conumerous with the set of irrational real numbers between 0 and 1.
    (c) Prove that $T$ is conumerous with the set of real numbers.
    (d) Prove that the set of subsets of $N$ is conumerous with the set of real numbers.
    (e) In view of (c) or (d), why may we assert that $2^{\aleph_0} = c$?

37. Just as there are invariably two types of people, let us say that there are two types of sets. Most sets, we expect, are *not* elements of themselves. For instance, the set of all finite sets is not an element of itself, since it is not a finite set. Such a set, which is not an element of itself, we call *ordinary*. But the set of infinite sets *is* an element of itself, since it is an infinite set. A set that is an element of itself we call *extraordinary*. Is the set $S$ of all ordinary sets ordinary, or is it extraordinary?

38. See E 36. We sketch a proof that there are just as many points in the plane as there are in the line.
    (a) Show that there are just as many sequences of 0's and 1's as there are pairs of such sequences. (*Suggestion:* Any such sequence can be broken into two sequences, one formed from the entries at the "even" places, and the other from the "odd" places; that is, the sequence $abcdef\cdots$ may be split into two sequences $ace\cdots$ and $bdf\cdots$.)
    (b) Show that there are just as many pairs of real numbers as there are real numbers.
    (c) From (b) deduce that the plane is conumerous with the line.

39. See E 38. Is the set of points in three-dimensional space conumerous with the set of points in the plane?

40. It is customary to denote the union of the sets $A$ and $B$, $A \cup B$.
    (a) Show that $A \cup A = A$, $A \cup B = B \cup A$, $A \cup (B \cup C) = (A \cup B) \cup C$.
    (b) If $A \cup B = A \cup C$, does it follow that $B = C$?

The following exercises present a problem concerning finite sets that is more than a century old and is far from being completely solved. For another such problem, see Chapter 13.

41. Check that each of the 21 subsets of $\{1, 2, 3, 4, 5, 6, 7\}$ having exactly 2 elements is a subset of exactly one of these 7 sets:

    $\{1, 2, 3\}, \quad \{1, 4, 5\}, \quad \{1, 6, 7\}, \quad \{2, 4, 6\}, \quad \{2, 5, 7\}, \quad \{3, 4, 7\}, \quad \{3, 5, 6\}.$

    For example, the set $\{1, 7\}$ is a subset of $\{1, 6, 7\}$, but of none of the others.

42. Check that each of the 36 subsets of $\{1, 2, 3, 4, 5, 6, 7, 8, 9\}$ having exactly two elements is a subset of exactly one of these 12 sets:

    $\{1, 2, 3\}, \quad \{7, 8, 9\}, \quad \{1, 5, 9\}, \quad \{2, 4, 9\}, \quad \{2, 6, 7\}, \quad \{3, 5, 7\},$
    $\{4, 5, 6\}, \quad \{1, 4, 7\}, \quad \{1, 6, 8\}, \quad \{2, 5, 8\}, \quad \{3, 4, 8\}, \quad \{3, 6, 9\}.$

43. Each of the 78 subsets of $\{a, b, c, d, e, f, g, h, i, j, k, l, m\}$ having 2 elements is a subset of precisely one of these 13 sets:

    $\{a, b, c, j\}, \quad \{a, d, g, k\}, \quad \{a, f, h, l\}, \quad \{a, e, i, m\}, \quad \{d, e, f, j\},$
    $\{b, e, h, k\}, \quad \{b, d, i, l\}, \quad \{b, f, g, m\}, \quad \{j, k, l, m\}, \quad \{g, h, i, j\},$
    $\{c, f, i, k\}, \quad \{c, e, g, l\}, \quad \{c, d, h, m\}.$

    Check the above statement only for the sets $\{a, b\}$, $\{b, f\}$, $\{h, j\}$, and $\{c, m\}$. (We use the letters $a$ through $m$ instead of the integers 1 through 13 for visual simplicity.)

44. Each of the 210 subsets of $\{a, b, c, d, e, f, g, h, i, j, k, l, m, n, o, p, q, r, s, t, u\}$ having two elements is a subset of exactly one of these 21 sets:

$\{a, b, c, d, q\}$,   $\{a, e, i, m, r\}$,   $\{a, f, k, p, s\}$,   $\{a, g, l, n, t\}$,   $\{a, h, j, o, u\}$,
$\{e, f, g, h, q\}$,   $\{b, f, j, n, r\}$,   $\{b, e, l, o, s\}$,   $\{b, h, k, m, t\}$,   $\{b, g, i, p, u\}$,
$\{i, j, k, l, q\}$,   $\{c, g, k, o, r\}$,   $\{c, h, i, n, s\}$,   $\{c, e, j, p, t\}$,   $\{c, f, l, m, u\}$,
$\{m, n, o, p, q\}$,   $\{d, h, l, p, r\}$,   $\{d, g, i, m, s\}$,   $\{d, f, i, o, t\}$,   $\{d, e, k, n, u\}$,
$\{q, r, s, t, u\}$.

Check the above statement for only 5 of the 210 sets in question.

45. If $k$ and $n$ are natural numbers (both at least 1), then by a $(k, n)$-collection we mean a collection of subsets of $A = \{1, 2, 3, \ldots, n\}$ such that
    (i) each member of the collection has exactly $k$ elements, and
    (ii) each subset of $A$ that consists of two elements is a subset of exactly one member of the collection.
    For example, E 41 presents a $(3, 7)$-collection; E 42 presents a $(3, 9)$-collection; E 43 presents a $(4, 13)$-collection; and E 44 presents a $(5, 21)$-collection.
    (a) Construct a $(2, 4)$-collection.
    (b) Show that for each $n$ larger than 1 there is precisely one $(2, n)$-collection.
    (c) Construct a $(4, 4)$-collection.
    (d) Why is there no $(5, 4)$-collection?

46. (a) Prove that a set with $n$ elements has $n \times (n - 1)/2$ subsets having two elements each.
    (b) Prove that if a $(3, n)$-collection has $p$ sets, then $3 \times p = n \times (n - 1)/2$. [*Hint:* Use (a).]
    (c) Using (b), prove that if there is a $(3, 8)$-collection, then 3 must divide 28.

47. Using the ideas of E 46, prove that if there is a $(3, n)$-collection, then 3 divides $n \times (n - 1)/2$.

48. Let $m$ be the number of triplets that belong to a $(3, n)$-collection and that have 1 as one of their elements.
    (a) Check that for the $(3, 7)$-collection of E 41, $m$ is 3.
    (b) Prove that $n - 1 = 2m$ for any $(3, n)$-collection.
    (c) Prove that if $n$ is even, then there is no $(3, n)$-collection. [*Hint:* See (b).]

49. Combining E 47 and E 48(c), we see that if $n$ is a natural number for which there exists a $(3, n)$-collection, then $n$ is odd and 3 divides $n \times (n - 1)/2$.
    (a) Prove that if 3 divides $n \times (n - 1)/2$, then 6 divides $n \times (n - 1)$.
    (b) Prove that if $n$ is odd and 6 divides $n \times (n - 1)$, then $n$ must have the remainder 1 or 3 when divided by 6 [in the language of congruence, $n \equiv 1$ or 3 (mod 6)].
    (c) Prove that there is no $(3, 17)$-collection.
    (d) For which $n$, less than 20, are there definitely no $(3, n)$-collections?
    In 1859 M. Reiss proved that whenever $n$ has the remainder 1 or 3 when divided by 6, there exists a $(3, n)$-collection.

50. Using the same kind of reasoning as in E 47 and E 48, prove that if $n$ is a natural number such that a $(4, n)$-collection exists, then
    (a) 6 divides $n \times (n - 1)/2$,
    (b) 3 divides $(n - 1)$.

51. (a) Using E 50, prove that there is no (4, 10)-collection.
    (b) For which *n*, less than 20, are there definitely no (4, *n*)-collections?

    In 1960 H. Hanani proved that if *n* is any natural number satisfying E 50(a) and (b), then there exists a (4, *n*)-collection. He also determined for which *n* there exists a (5, *n*)-collection. In recent years, (*k*, *n*)-collections have assumed an important role in the design of experiments. For this reason, and to satisfy their curiosity, mathematicians would like to know for which *k* and *n* there exists a (*k*, *n*)-collection. It is not known, for example, whether there is a (6, 46)-collection.

52. In the proof of Cantor's theorem we said, "*B* appears nowhere in list (6), since it differs from the *n*th real number in the list in at least the *n*th digit." But $0.7000\cdots$ differs from $0.6999\cdots$ in the first digit, yet $0.7000\cdots = 0.6999\cdots$. Is Cantor's proof still valid? Explain.

## References

1. Galileo Galilei, *Dialogues Concerning the Two New Sciences*, Dover, New York. (A highly readable account of the beginnings of the sciences: strength of materials and motion.)

2. T. Katigawa and M. Mitome, *Tables for the design of factorial experiments*, Dover, New York, 1955. (Table VII contains many other examples of (*k*, *n*)-collections.)

3. *Briefwechsel Cantor-Dedekind*, edited by E. Noether and J. Cavaillès, Hermann and Co., Paris, 1937. (The correspondence between Cantor and Dedekind, in German.)

4. J. B. Roberts, *The Real Number System in an Algebraic Setting*, W. H. Freeman and Company, San Francisco, 1962. (Infinite sets are discussed on pp. 55–61, 113–125.)

5. T. Dantzig, *Number, the Language of Science*, Macmillan, New York, 1954. (Chapter 6 discusses the algebraic and transcendental numbers.)

6. J. R. Newman, *The World of Mathematics*, Simon and Schuster, New York, 1956, vol. 3, pp. 1576–1611. (Essays by Bertrand Russell and Hans Hahn on infinity.)

7. G. Gamow, *One, Two, Three ... Infinity*, Viking, New York, 1948, pp. 14–23. (Infinite sets.)

8. A. Fraenkel, *Abstract Set Theory*, North-Holland Publishing Company, Amsterdam, 1953. (A good deal of set theory is presented in a leisurely and concrete style on pp. 1–84.)

9. *G. Birkhoff and S. MacLane, *A Survey of Modern Algebra*, Macmillan, New York, 1953. (Pages 356–369, set theory.)

10. M. Kline, *Mathematics in Western Culture*, Oxford University Press, New York, 1953. (Chapter 25 is devoted to infinity.)

11. L. Zippin, *Uses of Infinity*, Random House (New Mathematics Library), 1962. (The use of the pigeonhole principle in infinite sets is discussed on pp. 112–120.)

12. E. T. Bell, *Men of Mathematics*, Simon and Schuster, New York, 1937. (Chapter 29 is devoted to Cantor.)

13. B. Russell, *Introduction to Mathematical Philosophy*, George Allen and Unwin, Ltd., London, 1950. (Chapters 2 and 8 define "number," both finite and infinite.)

14. E. Kamke, *Theory of Sets*, Dover, New York, 1950. (A clear presentation of the arithmetic of infinity is given on pp. 1–51.)

## Chapter 19

# A GENERAL VIEW

The universe of mathematics grows out of the world about us like dreams out of the events of the day. The world serves as a cue, a point of departure, suggesting questions that lure the mathematician. Out of these questions, or their answers, grow more questions. As long as there are questions that demand an answer, mathematics expands.

The universe of mathematics is now so vast that in order to have some idea of its nature and extent, it is necessary to divide it into smaller constellations, even though these constellations are not separated by clearcut barriers. We will describe some of the major mathematical structures, and see where the various topics treated in this book fit into the overall pattern. We will conclude with a brief description of the mathematics that examines the foundations of all mathematical structures; namely, symbolic logic.

We first describe various kinds of geometry. Consider a rigid iron wire bent into a circle. As we move it rigidly through space it remains a circle of the same radius. Notions such as circle, radius, distance, straight line, angle, which are preserved by such rigid motions, are the subject of **congruence geometry.** In particular, the Pythagorean Theorem (Chapter 6) belongs to congruence geometry, since it involves distance, angle, and straight line.

Next, consider a rubber band in the form of a circle. As we move it through space and stretch parts of it, it could become a triangle, a square, an ellipse, or some quite irregular figure. Distance, angle, and straightness are not preserved. But some properties survive all these distortions of the rubber band. For example, the rubber band remains a loop; it never changes into

Moreover, if we have two rubber bands, interlocked like two links of a chain, then, as we move them both about, bending, stretching, or shrinking them, they still remain linked. *The study of properties preserved under such operations is called* **topology** (from the Greek *topo*, space; and *logia*, study). Topology studies such concepts as "loop," "being linked," "able to hold air," "having a hole." Chapters 2, 9, and 14 belong to topology.

Now we come to what might be called the most general possible geometry. Starting again with a rubber band in the form of a circle, this time we permit ourselves to rip the rubber into its individual points and scatter them however we choose. Though we lose distance, angle, straightness, and now, even the property of being a loop, still something is preserved. The number of points remains the same; that is, the set of points occupied by the rubber is always conumerous (in the sense of Chapter 18) with the set of points originally occupied. *The study of properties preserved under all one-to-one correspondences is called* **set theory.** Such properties as "having two elements" or "being finite" or "being infinite" are in the domain of set theory.

So far we have presented three mathematical disciplines: (1) congruence geometry, (2) topology, and (3) set theory, each more general than its predecessors. Set theory, the most fundamental, has the smallest collection of properties to work with, but in return its results have the widest range of application. Both topology and set theory can be considered as examples of geometries.

We will now go in a different direction from the iron wire that gave us congruence geometry. This time, let us mount the wire on a glass slide. If we place a screen parallel to the slide, and shine parallel rays of light through the slide onto the screen, then the shadow of the circle on the screen will be a circle of the same size:

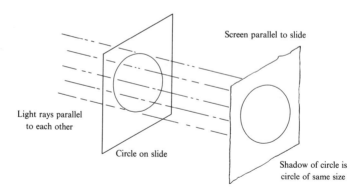

Screen parallel to slide

Light rays parallel
to each other

Circle on slide

Shadow of circle is
circle of same size

Properties preserved under this kind of projection are simply those preserved under rigid motions. Thus we could describe "congruence geometry" as the geometry that studies properties preserved under projection by parallel rays of light on a screen that is parallel to the slide.

But now we follow a different path from the one that led us to topology and set theory. Keep the screen parallel to the slide, but this time use a small bulb as the source of light, in order that the rays of light, instead of being parallel, will all come from one point. The shadow of the circle is still a circle, but perhaps much larger.

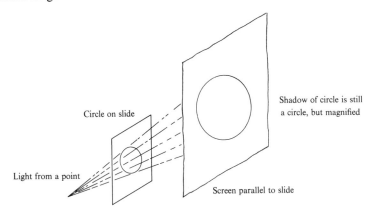

Circle on slide

Shadow of circle is still
a circle, but magnified

Light from a point

Screen parallel to slide

This type of projection, which is just magnification, preserves angle and straightness, but not distance. The shadow of a geometric object under this kind of projection is said to be "similar" to the original figure. *The study of properties preserved under magnifications or under congruences is called* **Euclidean geometry,** part of which we meet in high school.

Before we go on to two more general types of geometry, we will have to define the conic sections: ellipse, parabola, and hyperbola. To do this, take a double, endless (ice cream) cone and place it in a vertical position, as in this diagram:

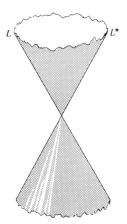

A horizontal or slightly tilted plane cuts the cone in an oval curve, called an *ellipse*. (Note that a circle is a special case of an ellipse.) As an illustration of this figure we have

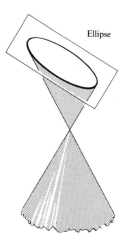

When the plane is tilted further, until it is parallel to one of the lines $L$ or $L^*$, it meets the cone in a *parabola* (shaped like the profile of an unbounded headlight), as is shown by this diagram.

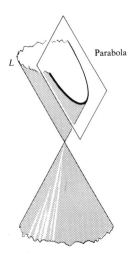

When tilted still more steeply, the plane meets *both* halves of the hourglass in a *hyperbola:*

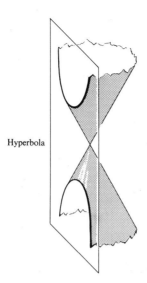

Hyperbola

These three types of curves are called *"conic sections."*

Now that we have defined conic sections, let us consider shadows cast by parallel rays of light, but with the screen no longer parallel to the slide:

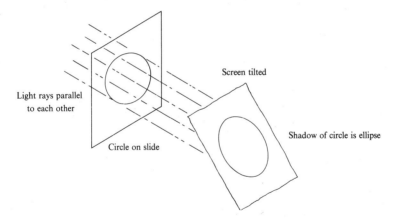

Screen tilted

Light rays parallel to each other

Circle on slide

Shadow of circle is ellipse

The shadow of a circle will now be an ellipse, and the shadow of a square will be a parallelogram. In fact, the shadow of any parallelogram will be a parallelogram; and the shadow of any ellipse, an ellipse. *The study of properties or concepts preserved under the first three types of projections (pp. 340–343) is called* **affine geometry.** Not all of these three types preserve distance and angle, but the notions of "straight line" and of "two lines being parallel" are preserved. For example, in affine geometry the ellipse (but not the circle) and the parallelogram (but not the square or rectangle) are studied.

Finally, let us have the light come from a point and let the screen be tilted:

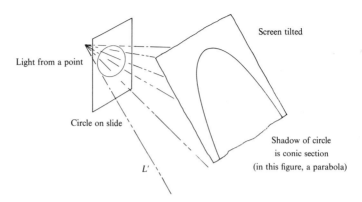

The shadow of a circle need no longer be a circle or an ellipse. However, the shadow will still be a conic section; in the above figure, for example, where line $L'$ and the screen are parallel, the shadow of the circle is a parabola. Moreover, the shadow of any conic section is, again, a conic section. *The study of properties preserved under all four of these types of projections (pp. 340–344) is called* **projective geometry.** Straight lines and their intersections, and conic sections and their tangents are studied in projective geometry, but concepts involving distance, angle, or even parallel lines are not. As an example of an important result in projective geometry, we have

PASCAL'S THEOREM. *If six points labeled 1, 2, 3, 4, 5, and 6 are on a conic section, and if the straight lines 1-2, 2-3, 3-4, 4-5, 5-6, and 6-1 are drawn, then the three points defined by the intersections of*

1-2 and 4-5,          2-3 and 5-6,          3-4 and 6-1

*are all on one straight line.*

(The reader may check Pascal's theorem by selecting six points, located, for convenience, on a circle.)

In this book we have met several of the six branches of mathematics so far described. The Pythagorean Theorem and the tiling of rectangles by squares belong to Euclidean geometry. Complete triangles (Chapter 2), highway systems (Chapter 9), and map coloring (Chapter 14) belong, for the most part, to topology, since the various results never really depend on the lines being straight or the sphere perfectly round. If the sphere were made of rubber and then stretched, regions would remain regions, adjacent regions would remain adjacent, and the degree of each vertex would not change. Cantor's theorem, showing that there are two infinite sets that are not conumerous, belongs to set theory. In fact, Chapter 18 is an introduction to set theory.

*Within set theory we find* **combinatorics,** *which is devoted primarily to finding out how many elements there are in sets that are known to be finite.* As examples of problems belonging to combinatorics, we have these: How long a list of orthogonal tables of order $N$ can be found (Chapter 13); How many memory wheels of $N$-tuplets of 0 and 1 are there (Chapter 10)?

**Algebra,** *a large branch of mathematics, is concerned with operations that assign to each pair of elements of a set an element of that set subject to specified rules.* For example, the operation $+$ assigns to each pair of natural numbers (integers, rationals, reals, and complex numbers) their sum, which is in turn a natural number (integer, rational, real, or complex number, respectively). Moreover, these five examples satisfy such rules as commutativity and associativity. Chapter 12 exhibits operations $\circ$ on finite sets that satisfy not only the usual rules of arithmetic but, in one case, this unusual rule: $X \circ (X \circ Y) = Y \circ X$. Appendix $C$ belongs to algebra.

*Within algebra lies* **number theory,** *the study of* $+$ *and* $\times$, *usually on the integers.* The notions of divisor, prime and composite, even and odd, and congruence modulo $M$ belong to number theory. Thus Chapters 4, 5, and 11 are primarily number theory.

Statements such as "$P_N$ is approximately $N \times (1/1 + 1/2 + \cdots + 1/N)$" (Chapter 4) or "The sequence of numbers $\frac{3}{10}, \frac{3}{10} + \frac{3}{100}, \frac{3}{10} + \frac{3}{100} + \frac{3}{1000}, \cdots$ approaches $\frac{1}{3}$" (Chapter 15) or "$1/1 + 1/2 + \cdots + 1/N$ grows as large as we please" (Appendix E) belong to **analysis.** Analysis, which begins with calculus, studies such topics as the behavior of endless sequences of numbers, the steepness of a tangent to a curve, and the total area under a curve. It is the branch of mathematics most frequently applied to physics and engineering. Indeed, Newton developed the calculus as a tool for studying the motion of the heavenly bodies.

Within analysis is to be found the *theory of* **probability,** *which examines the long-run behavior of events determined in part by chance.* (An example of this theory is to be found in "The average winner's score," treated in Appendix E.)

It would be wrong to take these divisions too seriously, as though they were watertight compartments. Recall how the proof of the topological Five-color Theorem (Chapter 14) depends in part on the addition, subtraction, and multiplication of integers; that is, on number theory. The proof of the two-color theorem exploits properties of the even and odd numbers; hence, number theory. The proof of the geometric theorem that some rectangles cannot be cut into squares (Chapter 8) uses a good deal of algebra, as well as the number theoretic result that there are irrationals. In Chapter 17 we used algebra to answer questions concerning constructions with straightedge and compass.

As Hilbert warned the twentieth century in an address he delivered in 1900:

> ... The question is urged upon us whether mathematics is doomed to the fate of those other sciences that have split up into separate branches, whose representatives scarcely understand one another and whose connection becomes ever more loose. I do not believe this nor wish it. Mathematical science is in my opinion an organism whose vitality is conditioned upon the connection of its parts.

We may take as an expression of Hilbert's scope the fact that he himself helped to develop the mathematics with the broadest vista, the logic of the foundations of all mathematical systems. This part of mathematics, called **mathematical logic,** or **symbolic logic,** *is concerned with such questions as the following: "What do we mean by a mathematical system?" "What is a proof?" "What is a theorem?" "Are the assumptions free of contradiction?" "Is each true statement also a theorem?"*

Let us be a little more specific. Each mathematical structure has its axioms, which describe what we propose to study. For instance, rules I and II of Chapter 8 are axioms for a certain kind of structure (and as examples of such structures we have electrical networks and tilings of rectangles by rectangles). Chapter 12 treats operations satisfying such axioms as $X \circ X = X$, $X \circ Y = Y \circ X$, $X \circ (X \circ Y) = Y \circ X$, or $X \circ (Y \circ Z) = (X \circ Y) \circ Z$. The axioms for a structure record our assumptions and thus help us keep our thinking straight. As mathematicians, we try to discover what conclusions follow from the axioms. As one simple example, we discovered that any table satisfying the axiom $X \circ (X \circ Y) = Y \circ X$ is idempotent. Appendix C derives algebra from eleven simple axioms.

*The reasoning by which we justify our conclusions drawn from the axioms is called a* **proof.** There are various kinds of proofs. For example, we have met on several occasions *proof by contradiction.* In this style of proof we show that a certain statement is true by showing that its denial leads to a contradiction—an inconsistency. For example, in Chapter 1 we proved certain Fifteen puzzles unsolvable by showing that if they were solvable, then a certain natural number would be both odd and even. In Chapter 4 we proved that there is no end to the primes by showing that if there were an end, then there would be a finite complete list of primes, thus contradicting the existence of the Prime-manufacturing Machine. In Chapter 6 we proved various square roots irrational by showing that if they were rational, then a certain natural number would be both odd and even. In Chapter 7 we proved that a box could not be filled with a finite number of cubes of unequal size, for if it could be, then there would be within the finite set of boxes an infinite set of boxes. In Chapter 7 we also proved that a 1 by $\sqrt{2}$ rectangle could not be tiled with squares, for if it could, then $\sqrt{2}$, which we already knew to be irrational, would be rational. In Chapter 12 we proved that there are no idempotent commutative tables of even order, for if there were, then the order would also have to be odd. In Chapter 13 we proved that there is no list of $N$ orthogonal tables of order $N$, for if there were, then two of the tables would not be orthogonal.

We have also met **proof by induction.** *A proof by induction can be described in terms of a fairy princess who lives in a castle with an infinite set of rooms numbered 1, 2, 3, 4,* $\cdots$ . She has in her hand the key to room 1. Moreover, each room contains a key that unlocks the door to the next room. Thus, if the princess can get into room 1, she can also enter room 2. In room 2 she finds the key to room 3; and soon we conclude that she, being immortal, can enter *any* room in the castle.

In a proof by induction we first prove a statement for 1, and then show that if the statement is true for a natural number $N$, it is also true for the next natural number $N + 1$. Recall, for example, our proof in Chapter 16 that every polynomial of degree $N$ has at most $N$ roots. We first proved it for degree 1 (this corresponds to the princess having the key to room 1). Then we used the truth of the statement for degree 1 to prove that it is also true for degree 2 (this corresponds to the princess finding in room 1 the key to room 2). Then we proved, essentially, that room 2 has the key to room 3, and stated that "the same argument then applies, step by step, to equations of degree 4, then 5, then 6, and so on." We were actually outlining a proof by *induction on the degree,* asserting that for each $N$, room $N$ contains the key to room $N + 1$. Just as we concluded that the princess can enter *any* room, we can now conclude that the number of roots of *any* polynomial is less than or equal to its degree.

In Chapter 14, by using a slight variation in proof by induction, we proved LEMMA 6: *Any regular map on a sphere can be colored in five (or fewer) colors.* We first observed that the lemma is true for any map with at most five countries. This corresponds to the princess having the keys to rooms 1, 2, 3, 4, and 5. Then we said, "Let $C$ be a natural number, at least 5, and let us assume that any map with at most $C$ countries can be colored with five colors. Now we will show that any map with $C + 1$ countries can be colored with five colors." Notice the phrase "at most" appearing in the assumption. If the reader looks over the proof he will see that Case 2 uses the fact that any map with $C - 1$ countries can be colored with five colors. Case 1 and Case 2 together tell the princess that she will find the key to room $N + 1$ either in room $N$ or in room $N - 1$. Even so, she would still be able to enter all the rooms of the castle. Our proof of the lemma was an *induction on the number of countries.*

Another technique for proving certain theorems is *exhaustion,* a device of little appeal to a mathematician. In Chapter 13 we mentioned that the proof that there are no orthogonal tables of order 6 was done by listing all possible tables of order 6 and then checking that no two of them were orthogonal. In Chapter 3 we mentioned that every odd natural number through 5775 can be written as the sum of a prime and twice a square. This can be proved by testing, one by one, each odd natural number up through 5775. But this is proof without revelation. Why the theorem should suddenly be false at 5777 is not explained. Nor, for tables of order 6, has any peculiarity of the number 6 been exploited (Euler had thought that the fact that 6 is twice an odd number would be important).

Another defect of proof by exhaustion is that even the fastest electronic computer can check only a finite number of cases. A computer has shown that Goldbach's conjecture is true up to 2,000,000, but no machine will ever prove that Goldbach's conjecture is true by testing *all* even numbers larger than 2.

In a somewhat different vein, we have *disproof by counterexample*. For instance, the conjecture that every odd natural number is the sum of a prime and twice a square is shown to be false by the counterexample, 5777. Euler's conjecture that there are no orthogonal tables of order twice an odd number was disproved by Bose, Parker, and Shrikhande, who constructed counterexamples of all the orders 10, 14, 18, $\cdots$ .

It was by the use of a counterexample that Cohen in 1963 shed new light on the famous continuum problem, first posed by Cantor, namely, "Is every infinite set of real numbers conumerous either with the set of integers or with the set of all real numbers?" (See p. 330) What Cohen showed (completing earlier work of Gödel) was that the continuum problem could not be settled on the basis of a specific set of axioms, the so-called Zermelo-Fraenkel axioms. This result, though austerely technical, is striking because of the power of these axioms. All of ordinary mathematics (including, for example, all of the mathematics in this book) can be deduced from the Zermelo-Fraenkel axioms! So one may conclude from Cohen's work that ordinary mathematics just does not suffice to settle Cantor's problem. Many workers in this field believe that the continuum problem will ultimately be solved by the discovery of some basic new principle that can be added to the Zermelo-Fraenkel axioms. Some others, taking a different philosophical position, feel that Cohen's work has shown that Cantor's question cannot be answered at all.

Cohen's counterexample consisted of a specific family of sets *C*. Any statement about sets may be relativized to *C* (or to any other family of sets) by construing or interpreting every reference to "set" in *that statement* as really meaning "set that belongs to *C*." *C* then provides a *model* of the Zermelo-Fraenkel axioms in the sense that each of these axioms becomes a true statement when relativized to *C*. But the statement *Every infinite set of real numbers is conumerous either with the set of integers or with the set of all real numbers* becomes a *false statement* when relativized to *C*. Earlier (1939) Gödel had found a family *G* of sets that also formed a model for the Zermelo-Fraenkel axioms, but such that the italicized statement becomes *true* when relativized to *G*. The existence of these two models can then be used to show that the italicized statement can be neither proved nor disproved on the basis of the Zermelo-Fraenkel axioms. Moreover, the powerful techniques developed by Cohen for constructing the family of sets *C* have been successfully applied to many other fundamental questions in set theory.

Consider the fate of a conjecture that Polya made in 1919. In order to state it we must define two terms. From the Fundamental Theorem of Arithmetic we know that each natural number larger than 1 is either a prime or is uniquely expressible as a product of primes. Call a natural number *of odd type* if, when expressed in terms of primes, an odd number of primes appear. Thus any prime is of odd type, and so are, for example, $8 = 2 \cdot 2 \cdot 2$ and $12 = 2 \cdot 2 \cdot 3$. Similarly, we shall say that a natural number is *of even type* if it is the product of an even number of primes. Thus $4 = 2 \cdot 2$, $6 = 2 \cdot 3$, $9 = 3 \cdot 3$, and $10 = 2 \cdot 5$ are of even type. Up through 12 there are seven numbers of odd type, 2, 3, 5, 7, 8, 11, and 12, and four numbers of even type, 4, 6, 9, and 10. Note that up through 12 there are more numbers of odd type than of even. We might say that the numbers of odd type are in the lead. Polya checked that up through 1500 the numbers of even type never catch up with the numbers of odd type, and then made this conjecture: the numbers of even type never catch up with the numbers of odd type. Computers later showed that the conjecture is true through 600,000. But, in 1958, Haselgrove, combining computers with theory, proved the conjecture false, but he did not find a specific case where it fails. Finally, in 1960, Lehman, also combining computers with theory, proved that up through 906,180,359 there are exactly as many numbers of even type as of odd type. But whether Polya's conjecture is false for some smaller number was not settled, though there is some numerical evidence for the belief that numbers of odd type hold their lead up to at least 900,000,000.

The fate of Polya's conjecture raises new worries. If we make a conjecture about the natural numbers, we should fear two possibilities. First it might be false, but false only for such large numbers that not even the fastest computer could reach them in a million years. Or it might turn out that the conjecture is true, but true only "by accident," just as the statement "every odd natural number up through 5775 is the sum of a prime and twice a square" is true seemingly without rhyme or reason. Could it perhaps be that a conjecture about the natural numbers is true for each natural number "by accident" in the sense that it "happens" to be true, but for which a proof is logically impossible and hence no one will ever know for sure that it is true?

Consider the three centuries of work put in on a problem posed by Fermat in 1637. He conjectured that there do not exist two cubes whose sum is a cube, or generally, two *Nth* powers whose sum is again an *Nth* power if $N$ is larger than 2. That is, he conjectured that the equation

$$X^N + Y^N = Z^N$$

has no solution in natural numbers $X$, $Y$, $Z$, and $N$ if $N$ is at least 3 and if $X$, $Y$, and $Z$ are at least 1.

Fermat himself proved that his conjecture is true for $N = 4$. Euler, before 1770, proved it for $N = 3$. Dirichlet and Legendre, in 1825, proved it independently for $N = 5$. The case $N = 6$ is implied by the case $N = 3$, since $X^6 + Y^6 = Z^6$ can be rewritten as $(X^2)^3 + (Y^2)^3 = (Z^2)^3$. The case $N = 7$ was settled in 1839 by Lamé. Kummer, in 1857, disposed of all $N$ less than 100. And now, more than 330 years after Fermat made his conjecture, it has been settled for all $N$ up through 25,000. Could his conjecture be true for all $N$ and yet no proof possibly exist for this fact?

Contrast this with our experience in Chapter 5 with proving that every prime is special. We showed that the prime 2 is special by an argument that depended on "even and odd." Another argument, outlined in E 28 of Chapter 4, shows that 5 is special. Exercise 45 in Chapter 5 shows that 3 is special, and the argument used in that exercise may be tried on other primes, one at a time, with no advance assurance that it will work. None of these arguments that show 2 or 3 or 5 to be special refer to the "primeness" of the number in question. There is, however, a proof that *every* prime is special—a single, uniform argument, which in a finite number of words disposes of all primes; it was the substance of Chapter 5. Surely we demand of any proof that it consist of only a finite number of words, for otherwise we would never be able to write it down and decide whether it is correct.

Could it turn out that in Fermat's problem, however, for *each* exponent there exists a proof that shows his conjecture to be correct for *that* exponent, yet there exists no uniform proof that is valid for all exponents from 3 on? Perhaps we cannot insist on a single proof for all exponents. Instead, we might try to classify the exponents into, say, one hundred types, and find for each of these types a separate proof for Fermat's conjecture. But even this may be impossible. No one knows.

Or consider the question of whether there is an end to the twin primes. There is little hope of devising a "twin-prime-manufacturing machine" that would play the role that the Prime-manufacturing Machine did in Chapter 4 in proving that there is no end to the primes. If there is a proof that the set of twin primes is infinite, as most mathematicians believe it is, that proof will probably not resemble the proof we gave for the infinity of primes. Could the set of pairs of twin primes be infinite and yet a proof of this fact be a logical impossibility—forever inaccessible, not because of the limits of man's insight, but because of the limits of the mathematical structure itself that man has created?

What will be the fate of the Four-color problem? Could it, like Polya's conjecture, be true only up to a certain number; say, for all maps on the sphere with at most a billion countries? If so, even the fastest electronic computer might fail to reach a map that cannot be colored with four colors. On the other hand, perhaps the Four-color conjecture is true for all maps on the sphere. But would this imply that a proof could be found that would show *why* the conjecture is true?

In that same address of 1900, Hilbert expressed his views about such problems:

> . . . [Our experience] gives rise to the conviction (which every mathematician shares but which no one has yet supported by a proof) that every definite mathematical problem must necessarily be susceptible of an exact settlement, either in the form of an actual answer to the question asked, or by the proof of the impossibility of its solution and therewith the necessary failure of all attempts. Take any definite unsolved problem, such as the question of . . . the existence of an infinite number of prime numbers of the form $2^n + 1$. However unapproachable these problems may seem to us and however helpless we stand before them, we have, nevertheless, the firm conviction that their solution must follow by a finite number of purely logical processes.
>
> Is this axiom of the solvability of every problem a peculiarity characteristic of mathematical thought alone, or is it possibly a general law inherent in the nature of the mind, that all questions which it asks must be answerable? For in other sciences also one meets old problems which have been settled in a manner most satisfactory and most useful to science by the proof of their impossibility. I instance the problem of perpetual motion. After a vain search for the construction of a perpetual motion machine, the relations were investigated which must subsist between the forces of nature if such a machine is to be impossible; and this inverted question led to the discovery of the law of the conservation of energy, which, again, explained the impossibility of perpetual motion in the sense originally intended.
>
> This conviction of the solvability of every mathematical problem is a powerful incentive to the worker. We hear within us the perpetual call: There is the problem. Seek its solution. You can find it by pure reason, for in mathematics there is no "we will not know."

But this mystical belief was shattered in 1931 in a remarkable work by Gödel, who proved that, in any mathematical system rich enough to include number theory, there is an assertion expressible within the system that is true, yet is not provable within the system. Gödel's reasoning itself was not expressible in terms of the axioms and methods of proof of the system.

The implications of this "Incompleteness Theorem" are vast. First, no set of axioms and rules of proof will be sufficient to make each true statement in a mathematical system a theorem in that system. Second, it suggests that there may be questions whose answers we will never know, no matter what new techniques of proof we may someday discover, or how far we go outside of the structure in which the question was originally phrased. That Fermat's conjecture or the Four-color problem might be essentially unsolvable now looms as a possibility. Though every provable statement is true, it may not be that every true statement is provable. The true and the provable may not coincide. Even mathematics may have its limits as a means for determining what is true and what is false.

In spite of this uncertainty looming in the background, the mathematician, driven by his curiosity and love of order and beauty, will continue to extend his universe. Gauss described this compulsion in a letter to Bolyai, September 2, 1808.

> Does the pursuit of truth give you as much pleasure as before? Surely it is not the knowing but the learning, not the possessing but the acquiring, not the being-there but the getting-there, that afford the greatest satisfaction. If I have clarified and exhausted something, I leave it in order to go again into the dark. Thus is that insatiable man so strange: when he has completed a structure it is not in order to dwell in it comfortably, but to start another. That is the way I imagine a world conqueror must feel; hardly has one empire been subdued, than he immediately stretches his arms towards others.

### Exercises

1. The reader can easily demonstrate the four types of conic sections with a flashlight (or by using a lamp with a dark shade). The light from the flashlight fills out a cone. A wall can serve as the plane cutting the cone; note that it is easier to tilt the flashlight than the wall. With a flashlight in a darkened room, show:
   (a) That when the flashlight is aimed directly at the wall, the bright spot is a circle.
   (b) That when the flashlight is tilted a little, the bright spot becomes an ellipse.
   (c) That when the flashlight is tilted such that it is in this position,

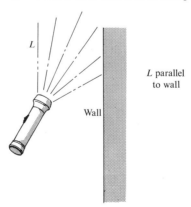

L parallel
to wall

   the bright spot is a parabola.
   (d) That when the flashlight is tilted a little more, we obtain (part of) a hyperbola.

2. Using a flashlight in a darkened room, show that the shadow of a parallelogram is not always a parallelogram. For the parallelogram, the reader could use either a thin book, an index card, or a parallelogram cut out of stiff cardboard. (It is necessary to tilt the book such that it is not parallel to the wall; the book corresponds to the slide, and the wall corresponds to the screen.)

3. (a) Draw a circle, select six points on it, and label them 1, 2, 3, 4, 5, and 6 in some order.
   (b) Check Pascal's Theorem for the six points chosen in (a).

4. Draw an ellipse with the aid of two tacks and a piece of string. Stick the two tacks into a piece of paper in such a way that they are closer together than the length of the string. Tie the two ends of the string to the two tacks. Keeping the string taut with a pencil, as in this diagram,

trace out a curve (to make a complete loop, lift the string over one of the tacks when necessary). It can be proved that this curve is an ellipse. Draw three such curves. If the two tacks coincide, what curve results? Kepler and Newton showed that the orbit of the earth is an ellipse with the sun at one tack.

5. (a) By the method of E 4, use a piece of string about six inches long to draw an ellipse on a piece of paper.
   (b) Shine a flashlight on the paper in such a way that the illuminated region coincides with the ellipse drawn in (a).

6. (a) Using the method of E 4, draw an ellipse.
   (b) Select six points on this ellipse, and check Pascal's Theorem for these points.

7. Another important result from projective geometry is

   DESARGUES' THEOREM. *Let ABC and A'B'C' be two triangles situated in the plane in such a way that the lines AA', BB', and CC' all meet at a point:*

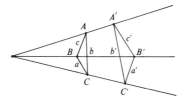

*Then the three points aa', bb', and cc' are all on one line.* (The symbol *aa'* denotes the point where line *a* meets line *a'*.) Check Desargues' Theorem in two examples.

8. Show that the proof of Lemma 2 of Chapter 14 can be phrased as an induction on the number of countries.

9. Using induction, prove that every natural number larger than 1 is either a prime or a product of primes. (*Hint:* The idea behind the proof of Theorem 3 of Chapter 4 can be incorporated into an inductive proof.)

10. (a) Show that the first catless proof of Theorem 1 of Chapter 2 was by induction on the number of dots.
    (b) Did we succeed in finding for Theorem 2 of Chapter 2 a proof by induction on the number of dots?
    (c) Show that the proof of Theorem 2 of Chapter 2 and the proof of E 27 of Chapter 2 suggest the beginning of an induction on the dimension of the figure on which the dots are drawn.

11. *A natural number that is equal to the sum of its divisors other than itself is called* **perfect.**
    (a) Show that the four numbers 6, 28, 496, and 8128 are perfect.
    (b) Show that each of the numbers in (a) is of the form $2^m \cdot$ prime, where the prime is of the form $2^{m+1} - 1$. [For example, $28 = 2^2 \cdot (2^3 - 1)$.]
    There are no odd perfect numbers less than $1,000,000,000,000,000,000$ ($= 10^{18}$). Whether there are any odd perfect numbers is not known. It has been proved that any even perfect number must be of the form described in (b). (Whether there is an infinite set of primes that are one less than a power of 2 is not known.) (See R 1 and R 5. The problem goes back to Euclid.)

12. Prove that Fermat's conjecture is true for $X = 1$. That is, if $N$ is a natural number larger than 2, prove that there are no natural numbers $Y$ and $Z$, both different from 0, such that
$$1^N + Y^N = Z^N.$$

13. In E 41 of Chapter 18 we had the (3, 7)-collection

$$\{1, 2, 3\}, \quad \{1, 4, 5\}, \quad \{1, 6, 7\}, \quad \{2, 4, 6\}, \quad \{2, 5, 7\}, \quad \{3, 4, 7\}, \quad \{3, 5, 6\}.$$

From this collection, which arose from set theory, or combinatorics, obtain a table of order 7 in the following manner:
    (a) Label the guide row and guide column with the numbers 1, 2, 3, 4, 5, 6, and 7.
    (b) For each $N$ from 1 through 7, define $N \circ N$ to be $N$, thus filling in the boxes on the main diagonal.
    (c) If $M$ and $N$ are unequal, define $M \circ N$ as the third number in the member of the collection that contains $M$ and $N$. (For example, $3 \circ 5 = 6$, since 3 and 5 appear in $\{3, 5, 6\}$.)
    (d) Fill in the 49 boxes of the table.

14. Prove that the table built in E 13 satisfies the rules

    (a) $X \circ X = X,$,  (b) $X \circ Y = Y \circ X,$  (c) $X \circ (X \circ Y) = Y.$

The moral of E 13 and E 14 is: there is no barrier between one branch of mathematics and another.

15. We now show that a miniature geometry can be obtained from the (3, 7)-collection of E 13. *A* **point** *will be any one of the natural numbers 1, 2, 3, 4, 5, 6, and 7. Thus our geometry has a total of seven points. By a* **line** *we shall mean one of the sets*

*listed in E 13.* Thus there are seven lines. A line *passes through* a point if that point is an element of the line.

(a) Show that any two distinct points determine a unique line passing through them.

(b) Show that any two distinct lines meet in a unique point. The moral of this exercise is the same as that for E 13 and E 14. (See pp. 94–103 of R 10 for more material in this direction.)

16. Show that in the proof of Theorem 5 of Chapter 11 an induction is used.

17. Show that the proof sketched for E 15 of Chapter 9 can be the basis of a proof by induction on the number of edges.

18. (a) Let $N$ be a natural number. Prove that if no prime less than or equal to $\sqrt{N}$ divides $N$, then $N$ is prime. (*Hint:* Assume that $N$ is composite, and obtain a contradiction.)

(b) Use (a) to show that 211 is prime. (Note that since $15^2 = 225$, we know that $\sqrt{211}$ is less than 15. Thus 211 will be prime if no prime less than 15 divides it.)

(c) Use (a) to show that 307 is prime.

19. Tie the two ends of a piece of string or rubber together such that it forms a self-knotted loop like this:

(a) Show (experimentally) that when you move it through space it remains knotted.

(b) Why does the theory of knots belong to topology but not to set theory?

20. Solve E 20(b) of Chapter 9 by an induction on the number of towns.

21. Using induction, give a complete proof of Theorem 2 of Chapter 5.

22. (a) Punch a small hole in the center of a jar lid or circular piece of stiff paper.

(b) Hold the lid fixed and perpendicular to a flat surface, such as a table. Shine a flashlight on it in such a way that the shadow of the lid is a circle.

(c) When the shadow is a circle, as in (b), does the light that passes through the hole illuminate the center of the shadow?

(d) It can be shown that a bounded shadow of an ellipse is always an ellipse. Is the "shadow" of the center of an ellipse always the "center of the shadow"? Explain.

(e) Would a projective geometer study ellipses? The centers of ellipses?

23. A parallelogram of height $h$ and base $b$

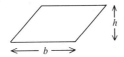

has the same area as a rectangle of height $h$ and base $b$.

(a) Prove this with the aid of pictures in the plane.

(b) Find or make a paper cylindrical tube that has a single seam spiralling from one end to the other. Pull it apart gently along the seam. When you lay the paper flat, it is a parallelogram. How can you use this fact to obtain the result in (a)?

(c) In what way is the proof in (a) more attractive? In what way is the proof in (b) more attractive?

24. Draw some points on the border of a circle, and join each pair of them by a straight line. (Select the points in such a way that no three of these lines meet inside the circle.) For example, with 4 points we may have

In this case we have cut the circle into 8 regions.

(a) Into how many regions do you cut the circle when you draw 2 points, 3 points, 5 points?

(b) Make a conjecture on the basis of the cases 2, 3, 4, 5.

(c) Is your conjecture correct?

25. What role does the distinction between odd and even play in
    (a) the analysis of the Fifteen puzzle (Chapter 1)?
    (b) the labelling of dots in a triangle (Chapter 2)?
    (c) the proof that $\sqrt{2}$ is irrational (Chapter 6)?
    (d) the highway inspector's problem (Chapter 9)?
    (e) the study of idempotent, commutative tables (Chapter 12)?
    (f) coloring a map with two colors (Chapter 14)?
    (g) discussing the real roots of polynomials with real coefficients (Chapter 16)?

26. Obtain a copy of J. D. Watson's *The Double Helix* (Atheneum, New York, 1968), and read at least Chapters 25, 26, and 27. On the basis of this, and other reading perhaps, discuss, "The basis of biology is geometry."

### References

1. E. Nagel and J. R. Newman, *Gödel's Proof*, New York University Press, 1958. (An introduction to the ideas involved in Gödel's proof.)

2. F. DeSua, Consistency and completeness—a resumé, *American Mathematical Monthly*, vol. 63, 1956, pp. 295–305. (Discussion of Gödel's theorem and its implications.)

3. David Hilbert, Mathematical Problems, lecture delivered in 1900, *Bulletin of the American Mathematical Society*, vol. 8, 1902, pp. 437–479. (This is a discussion of the past and future of mathematics; it contains 23 problems, several of which have since been solved.)

4. L. J. Mordell, *Reflections of a Mathematician*, École Polytechnique, Montreal, Canada, 1959. (The use of electronic computers in solving problems is discussed on p. 23.)

5. C. B. Haselgrove, Applications of digital computers in mathematics, *Mathematical Gazette*, vol. 42, 1958, pp. 259–260.

6. E. T. Bell, *Men of Mathematics*, Simon and Schuster, New York, 1937. (Brief biographies of many mathematicians, including Fermat, Pascal, Newton, and Hermite.)

7. M. Kline, *Mathematics, a Cultural Approach*, Addison Wesley, Reading, Mass., 1962. (The chapter "Mathematics and Painting in the Renaissance" discusses the origins of projective geometry.)

8. T. Dantzig, *Number, the Language of Science*, Macmillan, New York, 1954. (Chapter 4 discusses induction.)

9. J. R. Newman, *The World of Mathematics*, vol. 1, Simon and Schuster, New York, 1956. (On pp. 622–641 is reprinted from the Scientific American, 1955, "Projective Geometry," by M. Kline.)

10. D. Hilbert and S. Cohn-Vossen, *Geometry and the Imagination*, Chelsea, New York, 1952. (The conic sections are discussed on pp. 1–11.)

11. W. M. Ivins, Jr., *Art and Geometry*, Harvard University Press, Cambridge, Mass., 1946.

12. M. Kline, *Mathematics in Western Culture*, Oxford University Press, New York, 1953. (Chapters 4, 10, and 11 discuss geometry.)

13. N. Bourbaki, The architecture of mathematics, *American Mathematical Monthly*, vol. 57, 1950, pp. 221–232.

14. A. Weil, The future of mathematics, *American Mathematical Monthly*, vol. 57, 1950, pp. 295–306.

15. D. R. Weidman, Emotional perils of mathematics, *Science*, vol. 49, 1965, p. 1048.

16. E. P. Wigner, The unreasonable effectiveness of mathematics in the natural sciences, *Comm. on Pure and Applied Math.*, vol. 13, 1960, pp. 1–14.

17. Two key mathematics questions answered after quarter century. *New York Times*, November 14, 1963, p. 37.

18. P. J. Cohen and R. Hersh, Non-Cantorian set theory, *Scientific American*, vol. 217, December 1967, pp. 104–116. (This lucid exposition of Cohen's work contrasts the development of modern set theory with the history of non-Euclidean geometry.)

19. O. Ore, *Graphs and Their Uses*, Random House, New York, 1963. (An interesting induction is to be found on pp. 43–46.)

# REVIEW OF ARITHMETIC

For the convenience of the reader whose arithmetic may be rusty, we review the basic ideas and skills of computing.

## THE NATURAL NUMBERS, 0, 1, 2, 3, ...

### *Addition.*

There are two basic operations on the natural numbers: addition and multiplication. We illustrate the definition of *addition* by computing $2 + 3$.

A set with two elements

A set with three elements

Counting the total set of elements, one by one, we obtain "five," and write $2 + 3 = 5$. Clearly $3 + 2 = 5$ also, since it doesn't matter in which order we mention the numbers 2 and 3. More generally, we have for any natural numbers $a$ and $b$,

$$a + b = b + a.$$

This is commonly called the *commutative law of addition*.

Once we *memorize* the sums of two natural numbers in the list 0, 1, ..., 9, we may compute the sums of numbers of any size. (Exercises 1-3)

Note that

$$0 + a = a$$

for any natural number $a$. Since "zero preserves the 'identity' of each number," 0 is called the *identity element for addition*.

## Multiplication.

We illustrate the definition of *multiplication* by computing $2 \times 3$. In this case we take *two* sets, each with *three elements*,

and count, one by one, how many elements are present. In this case, since there are six, we write $2 \times 3 = 6$.

It is illuminating to draw the two sets as neatly arranged rows

In this form, the six dots

may be viewed in two ways: as two rows of three elements, as above, or as three columns of two elements

Thus

$$3 \times 2 = 2 \times 3 = 6.$$

The same argument shows that for any natural number $a$ and $b$

$$a \times b = b \times a.$$

This is commonly called the *commutative law of multiplication*.

Once we *discover* and *memorize* the products of the numbers $0, 1, 2, \ldots, 9$ we may compute the products of larger numbers.

Note that

$$1 \times a = a$$

for any number $a$. Just as 0 is called the identity element for addition, 1 is called the *identity element for multiplication*. (Exercises 4–10)

*Uses of Multiplication.* 1. (Area) The *area* of a rectangle is the product of its width and length. 2. (Levers and balancing) A large boy sitting near the axis of a teeter totter can balance a small boy sitting further away from the axis on the opposite side:

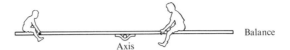

With the aid of multiplication we can compute precisely where two boys must sit in order to balance. A weight of $w$ pounds situated at a distance $d$ from the axis produces a "torque" or "tendency to turn the teeter-totter"

$$w \times d.$$

For the teeter totter to balance, this quantity must be the same on both sides of the axis. Thus, for example, a 90-pound boy 4 feet from the axis balances a 40-pound boy 9 feet from the axis, for the products

$$90 \times 4 \qquad \text{and} \qquad 40 \times 9$$

are equal. (Exercises 11–13)

*The Link Between Addition and Multiplication of Natural Numbers.* If we compute $2 \times (3 + 4)$ and also $(2 \times 3) + (2 \times 4)$, we obtain 14 in both cases. This is not a coincidence. The definition of $2 \times (3 + 4)$ involves this:

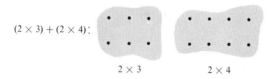

The definition of $(2 \times 3) + (2 \times 4)$ involves this:

So we have another example of two different ways of looking at the same dots. Without any computations, we see that $(2 \times 3) + (2 \times 4) = 2 \times (3 + 4)$. For any natural numbers $a$, $b$, $c$ we have the important rule

$$a \times (b + c) = (a \times b) + (a \times c),$$

commonly called the *distributive* law (Exercises 14–15).

*Notations.* The product of two specific numbers, such as 5 and 7, is denoted $5 \times 7$, $5 \cdot 7$, or $(5)(7)$. If we are using letters to represent numbers, as in the rule $a \times b = b \times a$, we frequently denote the product by placing the letters next to each other, *ab*, and *omit the multiplication sign.*

*Example 1.*

$5 \times 7 = 35; 5 \cdot 7 = 35; (5)(7) = 35$. We do *not* write 57 for this product, for it would be interpreted as fifty-seven.

*Example 2.*

We write $a \times (b + c)$ simply as $a(b + c)$. The rule $a \times (b + c) = (a \times b) + (a \times c)$ we abbreviate by $a(b + c) = (ab) + (ac)$.

What does the notation $3 + 5 \times 7$ mean? Do we mean $(3 + 5) \times 7$, which is 56, or do we mean $3 + (5 \times 7)$, which is 38? It is a general agreement that $3 + 5 \times 7$ shall mean $3 + (5 \times 7)$—the multiplication shall be carried out before the addition. When an expression involving addition and multiplication looks unclear, multiply before adding.

*Example 3.*

$2 \cdot 3 + 5 = 11; 2 \cdot 3 + 4 \cdot 5 = 26$.

*Example 4.*

The rule $a \times (b + c) = (a \times b) + (a \times c)$, which we abbreviated by $a(b + c) = (ab) + (ac)$ now shrinks to

$$a(b + c) = ab + ac.$$

We cannot, however, omit the parentheses in $a(b + c)$, for we would then have $ab + c$, which we interpret as $(ab) + c$. (Exercises 16–21)


## THE INTEGERS

Negative numbers—numbers that are less than zero—appear in weather reports on cold days, the daily stock-market quotations, and summaries of football games. They are situated to the left of zero in the number line:

$$\cdots \quad -3 \quad -2 \quad -1 \quad 0 \quad 1 \quad 2 \quad 3 \quad 4 \quad 5 \quad \cdots$$

The numbers $-1, -2, -3, \ldots$ are called *negative integers*. The natural numbers and the negative integers are called *integers*. Thus 4, 0, and $-7$ are integers. Sometimes $1, 2, 3, \ldots$ are called the *positive integers*. We emphasize that every natural number is also an integer:

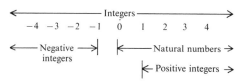

## Addition of Integers.

Either football or the stock exchange suggests how we should add two integers, even if one or both of them are negative. For instance, a gain of 4 yards followed by a loss of 5 yards results in a loss of 1 yard. Thus we will want

$$4 + (-5) = -1.$$

A loss of 2 yards followed by a loss of 3 yards is a total loss of 5 yards. We will want

$$-2 + (-3) = -5.$$

## Subtraction of Integers.

To find $7 - 5$ we fill in the box:

$$5 + \boxed{?} = 7.$$

The answer is "2." To find $5 - 7$ we proceed in the same way:

$$7 + \boxed{?} = 5.$$

Since

$$7 + (-2) = 5,$$

we have

$$5 - 7 = -2.$$

*Example 5.*
$16 - 11 = 5$, $11 - 16 = -5$ (Exercises 22–24).

## Multiplication of Integers.

We already have defined the product of two integers neither of which is negative. But how should we define a product such as $2 \times (-3)$ (a positive integer times a negative interger), or $(-1) \times (-1)$ (a negative integer times a negative integer)?

Football or the stock exchange suggests how $2 \times (-3)$ might be defined. If, on each of 2 successive downs we lost 3 yards, we would *lose*, all told, 6 yards. This indicates that we should define

$$2 \times (-3) = -6$$

("positive times negative is negative").

But neither football nor the stock exchange suggests how we might define $(-1) \times (-1)$. To define the product of two negative integers, we appeal to a general principle that is applicable whenever new definitions are required: if at all possible, we would like to see the same laws continue to be true for the operation to be defined as were true for previously defined operations. In particular, we would like the distributive law to hold also for the multiplication of two negative integers. For example, we would like the following to be true:

$$(-1)(1 + (-1)) = (-1)(1) + (-1)(-1).$$

Now since the left-hand side of this equation is equal to

$$(-1)(0) = 0,$$

we wish the right-hand side to be 0 also; that is,

$$(-1)1 + (-1)(-1) = 0.$$

But $(-1)1 = -1$. Therefore, we wish

$$-1 + (-1)(-1) = 0,$$

which is true only if we define $(-1)(-1)$ to be 1. Consequently we insist: *negative times negative is positive*. For instance, $(-5) \times (-7) = 35$.

We were led to the conclusion "negative times negative is positive" by the desire to preserve a law in a mathematical structure. But even physical considerations will lead us to the same verdict, as we now show.

Imagine now that the two boys have removed the teeter-totter board from its axis and placed it in such a way that it rotates on a vertical pole. Viewed from above, it looks like this

Axis is now vertical

Pole

Now, a boy may push in either direction (North or South) and from either side (East or West). To describe *where the boy is pushing* we mark the board with the integers, placing 0 at the pole

$$-4 \quad -3 \quad -2 \quad -1 \quad 0 \quad 1 \quad 2 \quad 3 \quad 4$$

The board

Pole

A boy may push at any of these locations. Moreover, he may push toward either the *north* or the *south*. Let us use *positive numbers* to describe a push directed *northward*, and *negative numbers* for a push directed *southward*.

For instance, a "push of 30" at "4" tends to move the board counterclockwise

Push of 30 pounds

This will have the same tendency to turn as a push of 60 at 2, since

$$60 \times 2 = 120 = 30 \times 4.$$

A push of $-30$ at 4 tends to rotate the board clockwise; it would cancel with a northward push at 4. The product $(-30) \times 4 = -120$ is therefore a reasonable measure of its tendency to turn, since $(-120) + 120 = 0$.

Now we come to an important new case: a push of $-30$ at $-4$. It has the same effect as a push of 30 at 4; both have counterclockwise influences of the same intensity:

If we wish multiplication to record effects of pushes, we will want

$$\underset{\substack{\uparrow \\ \text{Northward} \\ \text{push}}}{\overset{\substack{\text{Southward push} \qquad \text{at} \\ \downarrow \qquad\qquad \downarrow}}{(-30) \times (-4)}} \underset{\substack{\uparrow \\ \text{at}}}{= 30 \times 4.}$$

Thus $(-30) \times (-4)$ should be 120. Again we conclude that "negative times negative is positive."

*Example 6.*

$$(-2)(-3) = 6, \qquad (-4)(5 + 7) = -48,$$
$$(-2)(-3)(-7) = -42, \qquad (-7)(5 - 8) = 21.$$

*Example 7.*

The rule $a(b + c) = ab + ac$ holds even when some or all of $a$, $b$, or $c$ are negative. For instance,

$$-3(-7 + 5) = (-3)(-2) = 6,$$

while

$$(-3)(-7) + (-3)5 = 21 - 15 = 6.$$

*Example 8.*

The rule $(ab)c = a(bc)$ holds even when some or all of $a$, $b$, or $c$ are negative. For instance,

$$[(-1)(-5)]7 = 5 \cdot 7 = 35,$$

while

$$(-1)[(-5)\,7] = (-1)(-35) = 35$$

This rule is called the *associative* law, since the "$b$ may be associated first with $a$ or first with $c$."

## Division of Integers.

Division is to multiplication as subtraction is to addition.
Consider the question

$$\boxed{?} \times 3 = 6.$$

There is a unique integer that, when placed in the box, yields a valid equation, namely the integer 2. We write

$$6 \div 3 = 2.$$

Note that the equation

$$\boxed{?} \times 3 = 7$$

has no solution in the set of integers.

*Example 9.*

Since the equation $\boxed{?} \times 2 = -6$ has the unique solution $-3$, we write $(-6) \div 2 = -3$.

*Example 10.*

The equation $\boxed{?} \times 7 = 0$ has the unique solution 0. We may say $0 \div 7 = 0$.

*Example 11.*

The equation $\boxed{?} \times 0 = 6$ has no solution. Thus the symbol $6 \div 0$ is without any meaning whatsoever.

*Example 12.*

The equation $\boxed{?} \times 0 = 0$ is valid no matter what integer is placed in the box (for instance, $7 \times 0 = 0$). Thus the symbol $0 \div 0$ is ambiguous, and should not be used (Exercises 31–33).

## Powers.

After addition and multiplication, next in importance is a third operation, called *exponentiation* or, colloquially, "raising to a power."

The product of seven 2's, $2 \times 2 \times 2 \times 2 \times 2 \times 2 \times 2$, we denote for brevity by $2^7$, which we read as "2 to the 7th power." More generally, for an integer $a$ and a positive integer $n$, the product of $n$ of the $a$'s

$$\overbrace{a \times a \times a \times \cdots \times a}^{n \text{ of the } a\text{'s}}$$

we denote by

$$a^n$$

and call the "$n$th power of $a$."

*Example 13.*

$2^3 = 2 \times 2 \times 2 = 8, 2^2 = 2 \times 2 = 4, 2^1 = 2, (-1)^4 = 1, (-1)^2 = 1, 1^5 = 1,$
$0^4 = 0.$

It is not hard to see that $2^4 \times 2^3 = 2^7$, since

$$2^4 \times 2^3 = \underbrace{\underbrace{(2 \times 2 \times 2 \times 2)}_{\text{four 2's}} \times \underbrace{(2 \times 2 \times 2)}_{\text{three 2's}}}_{\text{seven 2's}}$$

We have, therefore, this law of exponentiation

$$a^{m+n} = a^m a^n \qquad \text{($a$ an integer)}$$
$$\text{($m$ and $n$ positive integer)}$$

At this point the symbol $2^0$ is meaningless and requires a definition, since "the product of zero 2's" doesn't make sense. Let us use the same principle, "preserve the rules," that guided us in defining the product of two negative numbers. This time we want the rule $2^{m+n} = 2^m 2^n$ to hold also for the case $m = 0$. In particular, we would like this equation to hold:

$$2^{0+1} = 2^0 2^1;$$

that is,

$$2^1 = 2^0 \cdot 2^1,$$

or

$$2 = 2^0 \cdot 2.$$

We are forced to have $2^0 = 1$, since the only number we may place in the box of the equation $2 = \boxed{?} \times 2$ and obtain a true equation is 1, the identity element of multiplication. Consequently we define $2^0 = 1$.

Generally, for any integer $a$, the symbol $a^0$ "$a$ to the 0th power" is defined as 1, since otherwise the law of exponentiation fails to be true if an exponent is 0. Therefore,

$$a^0 = 1$$

(Exercises 34–36).


## RATIONAL NUMBERS

Cut the line segments between consecutive integers into, say, five equal parts

The small pieces we call "fifths," and name them as follows

Note that $1 = 5/5, 2 = 10/5, 0 = 0/5$

We may divide each interval between integers also into, say, ten parts of equal length. The individual little sections now have length "one tenth," denoted 1/10. Observe that

$$\frac{2}{10} = \frac{1}{5}$$

*Example 14.*

A few diagrams, which the reader may make, show that

$$\frac{1}{5} = \frac{2}{10} = \frac{3}{15} = \frac{4}{20} = \frac{5}{25} = \frac{6}{30} = \cdots$$

$$\frac{5}{7} = \frac{10}{14} = \frac{15}{21} = \frac{20}{28} = \frac{25}{35} = \frac{30}{42} = \frac{35}{49} = \frac{40}{56} = \frac{45}{63} = \frac{50}{70} = \cdots$$

$$\frac{7}{10} = \frac{14}{20} = \frac{21}{30} = \frac{28}{40} = \frac{35}{50} = \frac{42}{60} = \frac{49}{70} = \cdots$$

*Example 15.*

Which is larger, 5/7 or 7/10? Example 14 shows that $5/7 = 50/70$ and $7/10 = 49/70$. Thus 5/7 is a little larger than 7/10.

Symbols that we write, such as 1/5, 2/10, 3/7, are called *fractions* and denote *rational numbers*. The fractions 1/5 and 2/10 denote the *same* rational number. In practice, this distinction is disregarded. In the fraction

$$\frac{a}{b}$$

$b$ is called the *denominator* (colloquially, the "downstairs") and $a$ is called the *numerator* (colloquially, the "upstairs"). (Exercises 37–39)

## *Addition of Rational Numbers.*

Just as 1 orange + 3 oranges = 4 oranges, we have 1 seventh + 3 sevenths = 4 sevenths,

$$\frac{1}{7} + \frac{3}{7} = \frac{4}{7}.$$

Pictorially,

Thus it is easy to add two rational numbers that are described by fractions with the *same denominator*:

$$\frac{a}{d} + \frac{b}{d} = \frac{a+b}{d}.$$

But how shall we compute

$$\frac{5}{7} + \frac{7}{10},$$

which is like the sum of 5 oranges and 7 apples? By a trick: We replace 5/7 by 50/70 and 7/10 by 49/70. Thus

$$\frac{5}{7} + \frac{7}{10} = \frac{50}{70} + \frac{49}{70},$$

and we now have the simpler problem of adding two rational numbers represented by fractions *whose denominators are equal*. This is easy:

$$\frac{50}{70} + \frac{49}{70} = \frac{99}{70}.$$

Thus

$$\frac{5}{7} + \frac{7}{10} = \frac{99}{70}.$$

*Example 16.*

$$\frac{1}{2} + \frac{1}{3} = \frac{3}{6} + \frac{2}{6} = \frac{5}{6},$$

$$\frac{5}{9} + \frac{1}{6} = \frac{10}{18} + \frac{3}{18} = \frac{13}{18}$$

(Exercise 40–42).


## Multiplication of Rational Numbers.

We begin with the simplest cases. First of all, $2 \times \frac{1}{3}$ is "two of the thirds," which we denote $\frac{2}{3}$. Thus

$$\frac{2}{1} \times \frac{1}{3} = \frac{2}{3}.$$

Pictorially,

Two of the thirds

As another instance, "half of a third is a sixth."
Pictorially,

Half of a third
is a sixth

Thus

$$\frac{1}{2} \times \frac{1}{3} = \frac{1}{6}.$$

Both of these instances suggest this rule: To multiply two rational numbers, "multiply the numerators, and multiply the denominators":

$$\frac{a}{b} \times \frac{c}{d} = \frac{ac}{bd}.$$

It is easier to multiply rational numbers than to add them, for we don't have to demand that denominators be equal in order to multiply.

*Example 17.*

$$\frac{7}{5} \times \frac{3}{2} = \frac{21}{10},$$

$$\frac{11}{3} \times \frac{3}{11} = \frac{33}{33} = 1,$$

$$3 \times \frac{7}{3} = \frac{3}{1} \times \frac{7}{3} = \frac{21}{3} = 7.$$

Be careful: $7/5 + 7/5 = 14/5$ (we add the numerators but leave the denominator alone), but $7/5 \times 7/5 = 49/25$ (note the new denominator). (Exercises 43–48)

## Division of Rational Numbers.

We defined $6 \div 2$ with the aid of the equation $6 = \boxed{?} \times 2$. Division of rational numbers is defined similarly.

*Example 18.*

We compute $(5/3) \div (2/7)$. Consider

$$\frac{5}{3} = \boxed{?} \times \frac{2}{7}.$$

Multiplying both sides by $7/2$, we obtain

$$\frac{5}{3} \times \frac{7}{2} = \boxed{?} \times \frac{2}{7} \times \frac{7}{2}.$$

But

$$\frac{2}{7} \times \frac{7}{2} = \frac{14}{14} = 1,$$

so we have the equation

$$\frac{5}{3} \times \frac{7}{2} = \boxed{?} \times 1.$$

But 1 is the identity element of multiplication. Thus

$$\frac{5}{3} \times \frac{7}{2} = \boxed{?}.$$

We have found that

$$\frac{5}{3} \div \frac{2}{7} = \frac{5}{3} \times \frac{7}{2},$$

which equals 35/6. *To divide by a rational number we switch its numerator with its denominator and multiply.*

Example 19.

$$\frac{7}{3} \div \frac{5}{8} = \frac{7}{3} \times \frac{8}{5} = \frac{56}{15}$$

$$\frac{8}{3} \div 7 = \frac{8}{3} \div \frac{7}{1} = \frac{8}{3} \times \frac{1}{7} = \frac{8}{21}$$

Example 20.

$$7 \div 3 = \frac{7}{1} \div \frac{3}{1} = \frac{7}{1} \times \frac{1}{3} = \frac{7}{3}.$$

Thus

$$7 \div 3 = \frac{7}{3}.$$

For this reason it is customary in algebra to denote division, $a \div b$, as

$$a/b \qquad \text{or} \qquad \frac{a}{b}.$$

The symbol $\div$ has almost disappeared from mathematics and science texts.

Example 21.

$$\left(2\frac{1}{4}\right) \Big/ \left(5\frac{1}{7}\right) = (9/4)/(36/7) = \frac{9}{4} \times \frac{7}{36} = \frac{63}{144} = \frac{7}{16}.$$

With this we conclude our review of the addition, subtraction, multiplication, and division of natural numbers, integers, and rational numbers.

## Exercises

Answers to all exercises other than those numbered 3, 6, 9, . . . begin on p. 373.

1. Evaluate (a) $5 + 7$, (b) $0 + 4$, (c) $9 + 8$, (d) $8 + 7$.

2. Using sets, show that (a) $4 + 5 = 9$, (b) $0 + 3 = 3$.

3. Evaluate (a) $17 + 42$, (b) $628 + 74$, (c) $754 + 188$.

4. Draw a picture that shows why $3 \times 4 = 4 \times 3$.

5. Using sets, show that (a) $2 \times 5 = 10$, (b) $3 \times 0 = 0$.

6. Fill out the multiplication table up to $9 \times 9$.

7. Evaluate (a) $3 \times 8$, (b) $6 \times 7$, (c) $7 \times 7$, (d) $6 \times 9$, (e) $7 \times 8$.

8. Evaluate (a) $18 \times 25$, (b) $17 \times 13$, (c) $254 \times 169$.

9. Which is the easier way to compute $5 \times 20 \times 7$: $(5 \times 20) \times 7$ or $5 \times (20 \times 7)$?

10. Evaluate (a) $387 \times 79$, (b) $301 \times 301$, (c) $908 \times 100$, (d) $7003 \times 8007$.

11. Will an 80-pound boy 7 feet from the axis of a teeter-totter balance a 90-pound boy sitting on the opposite side 6 feet from the axis?

12. (a) Find and list various applications of addition.
    (b) Find and list various applications of multiplication.

13. A 60-pound boy is sitting 8 feet from the axis of a teeter-totter. Where should an 80-pound boy sit to balance him?

14. Check these equations by computing both sides:

    (a) $5 \times (3 + 7) = (5 \times 3) + (5 \times 7)$,
    (b) $17 \times (31 + 18) = (17 \times 31) + (17 \times 18)$,
    (c) $(5 \times 11) + (5 \times 9) = 5 \times (11 + 9)$,
    (d) $(12 \times 3) + (17 \times 3) = (12 + 17) \times 3$.

15. Draw a picture that shows that

$$4 \times (3 + 5) = (4 \times 3) + (4 \times 5).$$

Do not compute both sides.

16. Compute (a) $5 \cdot 7$, (b) $(6)(8)$, (c) $5(2 + 4)$, (d) $5(8)$.

17. Compute (a) $8 \cdot 13 \cdot 15$, (b) $(6)(18)(7)$, (c) $11(9 - 7)$.

18. (a) Is $(7 + 5) + 8$ equal to $7 + (5 + 8)$?
    (b) Is $(7 \times 5) \times 8$ equal to $(7 \times 5) \times 8$?
    (c) Is $7 + (5 \times 8)$ equal to $(7 + 5) \times 8$?

19. Compute (a) $7 + 3 \times 5$, (b) $8 \cdot 11 + 9$, (c) $5 \cdot 7 + 6 \cdot 9$, (d) $(5)(8) + (11)(19)$, (e) $5(7 + 11)12$.

20. Compute (a) $8 \times 5 + 7$, (b) $(11)(12) + 9$, (c) $11(12 + 9)$, (d) $(31 + 12)18 + 14$.

21. Check that (a) $6 \times 8 = 7 \times 7 - 1$, (b) $9 \times 11 = 10 \times 10 - 1$. (c) Using the definition of multiplication, show that $6 \times 8 = 7 \times 7 - 1$ *without* computing either side. (d) Use the idea in (a), (b), and (c) to compute $19 \times 21$ by a shortcut.

22. True or false: (a) 7 is an integer, (b) 0 is an integer, (c) 0 is a positive integer.

23. Compute (a) $15 + (-17)$, (b) $-13 + (-19)$, (c) $7 - (-5)$, (d) $-4 - (7)$.

24. True or false: The sum of two negative integers is always a negative integer.

25. Compute (a) $11 \times (-7)$, (b) $(-3)(5)(6)$, (c) $(-8)(5 + 7)$, (d) $(-8)(5 - 7)$.

26. Compute (a) $3 \times 5$, (b) $(-3)(-5)$, (c) $3(-5)$, (d) $-2(11 - 13)$.

27. Draw a diagram of the boys pushing the board, that shows that we will want

$$(-70)(-5) = (70)(5).$$

28. Check that $a(b + c) = ab + ac$ when (a) $a = 5$, $b = -6$, $c = -2$, (b) $a = -7$, $b = 8$, $c = -8$.

29. Check that $a(b - c) = ab - ac$ when (a) $a = 8$, $b = 7$, $c = 3$, (b) $a = -4$, $b = -5$, $c = -7$.

30. Compute $17 \cdot 15 + 17 \cdot 75 + 17 \cdot 10$ in as simple a way as you may devise.

31. Compute and explain: (a) $0 \div 7$, (b) $6 \div 2$, (c) $57 \div 3$, (d) $119 \div 7$, (e) $56 \div 8$.

32. Compute (a) $(-153) \div 3$, (b) $143 \div (-11)$.

33. (a) Is $0 \div 5$ meaningful? Explain your answer.
    (b) Is $5 \div 0$ meaningful? Explain your answer.

34. Compute (a) $(-2)^4$, (b) $(-2)^5$, (c) $3^5$, (d) $5^3$, (e) $10^2$.

35. Compute (a) $4^2$, (b) $5^2$, (c) $5^1$, (d) $5^0$, (e) $4^0$.

36. Which is larger: (a) $2^{10}$ or $10^3$? (b) $2^5$ or $5^2$?

37. Which is larger: $10/21$ or $11/23$?

38. Find three different fractions that name the rational number (a) $14/10$, (b) $-3/5$.

39. Using the number line, show that $3/7 = 9/21$.

40. Compute and express by a single fraction:

(a) $\dfrac{5}{3} + \dfrac{3}{5}$, (b) $\dfrac{-5}{12} + \dfrac{7}{18}$, (c) $6 + \dfrac{5}{3}$, (d) $\dfrac{11}{5} + \dfrac{17}{13}$.

41. Which is larger:

$$\frac{7}{13} + \frac{13}{11} \quad \text{or} \quad \frac{2}{3} + \frac{29}{21}?$$

42. What is the simplest fraction representing $91/26$?

43. Compute (a) $\frac{7}{3} \times \frac{3}{11}$, (b) $\frac{7}{4} \times \frac{9}{4}$, (c) $\frac{7}{4} + \frac{9}{4}$.

44. Compute (a) $\frac{7}{3}(\frac{5}{9} + \frac{4}{13})$, (b) $\frac{7}{3} \times \frac{6}{7}$, (c) $\frac{11}{19} \times \frac{21}{14} \times \frac{7}{9}$.

45. Which is generally easier: To add rational numbers or to multiply them?

46. Compute and express in as simple a fraction as possible (a) $\frac{3}{4} + \frac{1}{2} + \frac{5}{8}$, (b) $\frac{14}{8} \times \frac{12}{7} \times \frac{13}{2}$, (c) $(2\frac{1}{2}) \times (7\frac{1}{3})$.

47. Compute (a) $\frac{4}{7} + \frac{4}{9}$, (b) $\frac{7}{4} + \frac{9}{4}$, (c) $\frac{4}{7} \times \frac{4}{9}$.

48. Check that $a(b + c) = ab + ac$ when

(a) $a = 3, b = 2/3, c = 4/7$, (b) $a = 1/2, b = 2/9, c = 2/11$.

49. Compute (a) $\frac{5}{3} \div \frac{7}{5}$, (b) $(11/4)/(2/9)$, (c) $(6\frac{1}{2})/(4\frac{1}{3})$. (d) Check by multiplication.

50. Compute (a) $\frac{5}{3} \div 7$, (b) $6 \div \frac{4}{7}$. (c) Check by multiplication.

51. Show that $2 \div 3 = 2/3$.

52. Which is larger: $(1 + 1/3)^3$ or $(1 + 1/4)^4$?

53. Replacing $a$, $b$, and $c$ by various positive integers, find out if there is any relation between $a^n$, $b^n$ and (a) $(a + b)^n$, (b) $(ab)^n$.

**Answers**

1. (a) 12, (b) 4, (c) 17, (d) 15.

2. (a) 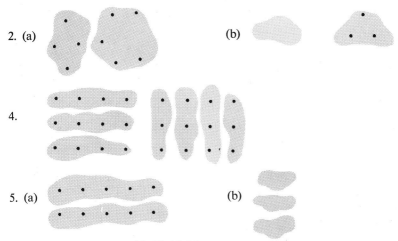 (b)

4.

5. (a) (b)

7. (a) 24, (b) 42, (c) 49, (d) 54, (e) 56.

8. (a) 450, (b) 221, (c) 42,926.

10. (a) 30,573, (b) 90,601, (c) 90,800, (d) 56,073,021.

11. No, $80 \times 7$ is not equal to $90 \times 6$.

13. $60 \times 8 = 80 \times \boxed{?}$, hence $\boxed{?} = 6$ feet.

14. (a) 50, (b) 833, (c) 100, (d) 87.

16. (a) 35, (b) 48, (c) 30, (d) 40.

17. (a) 1560, (b) 756, (c) 22.

19. (a) 22, (b) 97, (c) 89, (d) 249, (e) 1080.

20. (a) 47, (b) 141, (c) 231, (d) 788.

22. (a) True, (b) True, (c) False.

23. (a) $-2$, (b) $-32$, (c) 12, (d) $-11$.

25. (a) $-77$, (b) $-90$, (c) $-96$, (d) 16.

26. (a) 15, (b) 15, (c) $-15$, (d) 4.

28. (a) $-40$, (b) 0.

29. (a) 32, (b) $-8$.

31. (a) $0(7 \times \boxed{?} = 0)$, (b) $3(2 \times \boxed{?} = 6)$, (c) $19(3 \times \boxed{?} = 57)$,
    (d) $17(7 \times \boxed{?} = 119)$, (e) $7(8 \times \boxed{?} = 56.)$

32. (a) $-51$, (b) $-13$.

34. (a) 16, (b) $-32$, (c) 243, (d) 125, (e) 100.

35. (a) 16, (b) 25, (c) 5, (d) 1, (e) 1.

37. Since $10/21 = \dfrac{10 \times 23}{21 \times 23} = \dfrac{230}{483}$ and $\dfrac{11}{23} = \dfrac{11 \times 21}{23 \times 21} = \dfrac{231}{483}$, $11/23$ is larger.

38. (a) For instance, $7/5$, $28/20$, $42/30$. (b) For instance, $-6/10$, $-9/15$, $-12/20$.

40. (a) $34/15$, (b) $-1/36$, (c) $23/3$, (d) $228/65$.

41. The first is $246/143$ (less than 2) while the latter is $129/63$ (larger than 2). So the latter is larger.

43. (a) $7/11$, (b) $63/16$, (c) $16/4$ or 4.

44. (a) $707/351$, (b) 2, (c) $77/114$ (simplest form).

46. (a) $15/8$, (b) $39/2$, (c) $55/3$.

47. (a) $64/63$, (b) 4, (c) $16/63$.

49. (a) $25/21$, (b) $99/8$, (c) $3/2$.

50. (a) $5/21$, (b) $21/2$.

52. $(1 + 1/3)^3 = 64/27$ and $(1 + 1/4)^4 = 625/256$. The latter is larger.

53. (a) $(a + b)^n$ is at least as large as $a^n + b^n$. (b) $(ab)^n = a^n b^n$.

*Appendix* **B**

# WRITING

# MATHEMATICS

We present examples of "bad" and "good" writing, and conclude with some words of advice. Let us discuss five "solutions" to this homework exercise on material in Chapter 5.

Let $A$ and $B$ be natural numbers such that there are integers $M$ and $N$, $MA + NB = 4$. What can be said about $(A, B)$?

Solution 1.　　4;　　　　$12 - 8 = 4$.

Comment: This is unreadable. Where is the scratch work? What is the answer? Perhaps the student took as an example $A = 12$, $B = 8$, $M = 1$, $N = -1$, $1 \cdot 12 + (-1)(8) = 4$, observed that $(8, 12) = 4$, and meant his answer to read: "We may conclude that $(A, B) = 4$."

If so, the solution is wrong; after all, $(2, 3)$ is not 4, yet $5 \cdot 2 + (-2)(3) = 4$. Not even part credit can be given, since the reader doesn't understand what the writer meant.

Solution 2.　　It can be 1, or 4.

Comment: To what does the pronoun "it" refer? What is the reasoning? If "it" refers to "$(A, B)$," does the writer mean "$(A, B)$ can be 1 or 4 and never something else?" If so, the answer is incorrect. For instance, $(4, 6) = 2$ and yet there are integers $M$ and $N$ such that $M \cdot 4 + N \cdot 6 = 2$ [take, for example, $M = (-1)$, $N = 1$]. How did the writer reach his conclusion?

Solution 3.　　$(A, B) = 1, 2, 4$.

Comment: Perhaps this writer meant this as an abbreviation of, "For any natural numbers $A$ and $B$, such that there are integers $M$ and $N$, $MA + NB = 4$, then we have $(A, B) = 1$, 2, or 4. Each of these cases can occur; no other case is possible."

If so, he expects more mind-reading and patience than ordinary mortals possess. Besides, it is unclear how the writer came to his conclusion—did he guess, take a few examples, or did he reason it out?

Solution 4.        $(A, B)$ divides 4, so 1, 2, 4.

Comment: Why does $(A, B)$ divide 4? No explanation is given. How does the phrase "$(A, B)$ divides 4" relate to the following phrase, "so 1, 2, 4"? Argument unclear.

Solution 5. We claim that $(A, B)$ must be 1, 2, or 4 and that each of these cases is possible. First, let us show, by examples, that each case can occur:

> Take $A = 1$,  $B = 1$. Then $3A + 1B = 4$ and $(A, B) = 1$.
> Take $A = 4$,  $B = 2$. Then $1A + 0B = 4$ and $(A, B) = 2$.
> Take $A = 12$, $B = 8$. Then $1A - 1B = 4$ and $(A, B) = 4$.

We now show that $(A, B) = 1, 2, 4$, are the only possibilities. Observe that $(A, B)$ divides $A$ and $B$, hence divides $MA + NB$, which is 4. Thus $(A, B)$, being a divisor of 4, must be 1, 2, or 4.

Comment: This is satisfactory. The reasoning is easy to follow. The scratch work and experiments have not been turned in as part of the final argument.

With these illustrations as background, the reader may appreciate the following admonitions:

## EXPOSITION.

1. Be precise.
2. Be careful of pronouns and their antecedents.
3. Separate ideas into paragraphs.
4. Write legibly.
5. Label problems clearly.
6. Avoid wordiness.
7. Use such words as "if," "then," "thus," "hence," but use them carefully.
8. State what you are doing at each step.
9. Read *The Elements of Style*, by W. Strunk, Jr., and E. B. White, Macmillan paperback, 1959.

## LOGIC.

1. Define your symbols and terms.
2. In a proof by contradiction, state your assumption.
3. Let the final assertion in a proof restate the statement to be proven (if you do this, you won't stop before you are finished; there is less chance you will prove the wrong thing).

4. Check that you haven't assumed what you are trying to prove (even though worded differently) or that you haven't assumed something that depends on what you are trying to prove.

5. A proof by example is not a proof except in the following special cases:
   (a) An example can sometimes be used to illustrate a procedure that is equally applicable to any other example (such as in the proof of lemma 4 of Chapter 5).
   (b) If a statement is to be *dis*proved, a single example for which the statement is false is sufficient.

   For instance, to show that it is not the case that every natural number is expressible as the sum of three squares, it suffices to show that 7 is not the sum of three squares.

6. Show all important steps.

7. Know your definitions. Be able to define "prime number," "special number," "divisor," "irrational," etc.

8. Know the theorems proved previously. Be able to state them precisely, but don't memorize them word for word (it is better to state them in your own words).

*Appendix* **C**

# THE RUDIMENTS
# OF ALGEBRA

Most of this book uses only the arithmetic met in grammar school. However, Chapters 5, 8, 11, 14, and 16 (mainly 8, 11, and 16) use algebra in the manipulations of expressions and equations involving the two fundamental operations, addition and multiplication, and their respective offspring, subtraction and division. This appendix describes and justifies the more common algebraic gymnastics. It is a reference to be read if and when needed. Proofs are included for the convenience of the reader who wants to know the "why."

Just as we can do all of arithmetic by memorizing a few facts (such as the addition and multiplication tables up to nine by nine), so can we reduce all the manipulations of algebra to a few key assertions. Instead of the 162 entries in the addition and multiplication tables, it turns out that for algebra we will have just eleven rules to memorize. With these eleven rules we can do all of elementary algebra: for example, removing parentheses, inserting parentheses, factoring, finding roots, and collecting terms.

We will now state those eleven rules (or axioms) and derive their consequences in the logical style of Euclidean geometry—statement and reason. We choose this formal manner of writing proofs for three reasons: to emphasize how much can lie hidden in a few axioms, to divide the arguments into small enough steps to be justified by a citation of the appropriate axiom, and to show the power of the *axiomatic method* (we will treat an infinite number of algebraic structures simultaneously).

*By a* **field** *we will mean any mathematical structure possessing an addition, denoted* $+$, *and a multiplication, denoted* $\times$, *both defined on some set* $S$ *and satisfying these eleven rules:*\*

---

\* A stands for "addition," M stands for "multiplication," and D stands for "distributivity." A 1 through A 5 refer only to addition; M 1 through M 5 refer only to multiplication (except for M 5, which does mention 0 and hence involves A 4); D links multiplication and addition.

A 1. *If a and b are in S, so is a + b.*

A 2. *For all a and b in S we have*
$$a + b = b + a.$$

A 3. *For all a, b, and c in S we have*
$$a + (b + c) = (a + b) + c.$$

A 4. *There is an element in S, called 0, such that 0 + a = a for all a in S.*

A 5. *For each pair a and b in S there is exactly one x in S such that a + x = b.*

M 1. *If a and b are in S, so is a × b.*

M 2. *For all a and b in S we have*
$$a \times b = b \times a.$$

M 3. *For all a, b, and c in S we have*
$$a \times (b \times c) = (a \times b) \times c.$$

M 4. *There is an element in S, called 1, such that 1 × a = a for all a in S.*

M 5. *For each pair a and b in S (where a is not 0) there is exactly one x in S such that a × x = b.*

D. *For all a, b, and c in S, $a \times (b + c) = (a \times b) + (a \times c)$.*

We meet several *fields* in this book; specifically:

*Example 1.* (Chapter 6)
S is the set of real numbers (usual + and ×);

*Example 2.* (Chapter 6)
S is the set of rationals (usual + and ×);

*Example 3.* (Chapter 16)
S is the set of complex numbers, and the $\oplus$ and $\otimes$ defined in Chapter 16 play the role of + and × respectively;

*Example 4.* (Chapter 11)
S has just five elements {0, 1, 2, 3, 4}, and the $\oplus$ and $\otimes$ defined on p. 163 play the role of + and ×, respectively.

Before checking that Example 4 is truly a field, the reader should first meet some structures that *fail* to qualify as fields. The following four exercises present some structures that satisfy some, but not all, of the eleven rules.

E 1. Let *S* be the set of natural numbers {0, 1, 2, 3, $\cdots$}, and let + and × be the usual addition and multiplication. This structure satisfies all of the eleven rules for a field except A 5 and M 5. Check this assertion by a few examples.

E 2. Let *S* be the set of integers {$\cdots$, −3, −2, −1, 0, 1, 2, 3, $\cdots$}, and let + and × be the usual addition and multiplication. This structure satisfies all of the eleven rules for a field except M 5. Check this assertion by a few examples.

E 3. Let *S* be the set of positive rationals (including 0) and let + and × be the usual addition and multiplication. This structure satisfies all of the eleven rules for a field except A 5. Check this assertion by a few examples.

E 4. Let *S* be the set of irrationals (defined in Chapter 6), and let + and × be the usual addition and multiplication. This structure satisfies only rules

A 2, M 2, A 3, M 3, and D of a field. (*Hint:* Show that the sum and product of irrationals need not be irrational.)

In Example 4 we met a field with only five elements. We now examine a field with two elements. Let $S$ be the set $\{0, 1\}$. Define $+$ and $\times$ by these two arrays

(*1*)

| $+$ | 0 | 1 |
|---|---|---|
| 0 | 0 | 1 |
| 1 | 1 | 0 |

| $\times$ | 0 | 1 |
|---|---|---|
| 0 | 0 | 0 |
| 1 | 0 | 1 |

Clearly $+$ satisfies A 1 and $\times$ satisfies M 1. Since $0 + 1 = 1$ and $1 + 0 = 1$ we see that $+$ satisfies A 2. As the following six exercises show, (*1*) satisfies all eleven rules for a field.

**E 5.** Show that $\times$ of (*1*) satisfies M 2.

**E 6.** (a) Make the (eight) checks necessary to show that (*1*) satisfies A 3.
      (b) Make the (eight) checks necessary to show that (*1*) satisfies M 3.

**E 7.** (a) Make the (two) checks necessary to show that (*1*) satisfies A 4.
      (b) Make the (two) checks necessary to show that (*1*) satisfies M 4.

**E 8.** (a) Make the (four) checks necessary to show that (*1*) satisfies A 5.
      (b) Make the (two) checks necessary to show that (*1*) satisfies M 5.

**E 9.** Make the (eight) checks necessary to show that (*1*) satisfies D.

**E 10.** Make a few sample checks for the fact that Example 4 is a field.

The structures of Example 4 and (*1*) are called Galois fields. Any field for which $S$ is a finite set is called a Galois field. It is known that the number of elements in a Galois field must be either a prime or a power of a prime. For one application of Galois fields see Chapter 13, where they are used in the proof of Theorem 5.

The eleven rules are the *axioms* for *fields*. Any theorems we deduce from them apply to any field, in particular to the fields of Examples 1, 2, 3, and 4.

Axiom A 2 asserts that addition is commutative; axiom M 2 asserts that multiplication is commutative. Several examples of commutative operations are to be found in Chapter 12, where the virtue of commutativity is discussed. Axioms A 3 and M 3 say that both the operations $+$ and $\times$ are associative. Axiom A 4 says that there is an element "with no effect when added to an element." Axiom M 4 says that there is an element "with no effect when multiplying an element."

But when we come to axioms A 5 and M 5 we meet a crucial distinction between $+$ and $\times$. Though we can solve the equation $0 + x = 1$, for example, we cannot solve the equation $0 \times x = 1$. In fact, we have

THEOREM 1. *For any element a in a field, $0 \times a = 0$, and $a \times 0 = 0$.*

PROOF. We will prove only the second assertion, $a \times 0 = 0$.

| STATEMENT | REASON |
|---|---|
| (1) $a \times 0 = a \times (0 + 0)$ | A 4 |
| (2) $a \times (0 + 0) = a \times 0 + a \times 0$ | D |
| (3) $a \times 0 = a \times 0 + a \times 0$ | (1) and (2) |
| (4) $a \times 0 = 0 + a \times 0$ | A 4 |
| (5) $0 + a \times 0 = a \times 0 + 0$ | A 2 |
| (6) $a \times 0 = a \times 0 + 0$ | (4) and (5) |
| (7) $a \times 0 = 0$ | A 5 (If $a \times 0$ were not equal to 0, then (3) and (6) would tell us that the equation $a \times 0 = a \times 0 + x$ has at least two solutions.) |

The reader may easily deduce that $0 \times a = 0$. Theorem 1 is proved.

Note the key role that $D$ plays in the proof of Theorem 1. Rule $D$, the main link between $+$ and $\times$, asserts that "multiplication *distributes* over addition." (The reverse, however, does not hold.) The proof of Theorem 1 shows "why" $a \times 0 = 0$, a fact that might not surprise the reader. But, with no more difficulty, we will be able, in Theorem 11, to show why $(-a) \times (-b) = a \times b$, a fact which many find mysterious.

THEOREM 2. *For all a, b, and c in a field we have $b + (c + a) = a + (b + c)$.*

PROOF.

| STATEMENT | REASON |
|---|---|
| (1) $b + (c + a) = (b + c) + a$ | A 3 |
| (2) $(b + c) + a = a + (b + c)$ | A 2 |
| (3) $b + (c + a) = a + (b + c)$ | (1) and (2) |

This proves Theorem 2.

E 11. Using axioms A 2 and A 3 prove that $a + (b + (c + d)) = c + (b + (a + d))$. (*Hint:* A similar problem is worked out in Chapter 12, p. 184.)

Theorem 2 suggests

THEOREM 3. *The sum of several elements in the field does not depend on the order in which we write them or on the way we insert parentheses.*

The proof of Theorem 3 depends only on A 2 and A 3 but is quite tedious and will be omitted. We also omit the proof of the similar

THEOREM 4. *The product of several elements in a field does not depend on the order in which we write them or on the way we insert parentheses.*

**E 12.** Using rules M 2 and M 3 prove that

$$(a \times b) \times (c \times d) = (a \times c) \times (b \times d).$$

This is a special case of Theorem 4; do not use Theorem 4 in your proof.

THEOREM 5. *For all a, b, and c in S we have*

$$(b + c) \times a = (b \times a) + (c \times a).$$

PROOF.

| STATEMENT | REASON |
|---|---|
| (1) $(b + c) \times a = a \times (b + c)$ | M 2 |
| (2) $a \times (b + c) = (a \times b) + (a \times c)$ | D |
| (3) $a \times b = b \times a$; $a \times c = c \times a$ | M 2 |
| (4) $a \times (b + c) = (b \times a) + (c \times a)$ | (2) and (3) |
| (5) $(b + c) \times a = (b \times a) + (c \times a)$ | (1) and (4) |

This proves Theorem 5.

Theorem 3 says that as long as we work only with addition we can omit parentheses. Theorem 4 says that as long as we work only with multiplication we can omit parentheses. But Theorem 5 says that we cannot omit parentheses in $(b + c) \times a$, since the two possible interpretations of $b + c \times a$, namely $(b + c) \times a$ and $b + (c \times a)$, will usually give different results.

**Agreement.** *It is the custom to omit the multiplication sign $\times$ when such an omission cannot lead to confusion. Thus we frequently write ab instead of $a \times b$. Frequently $a \times a$ is written $a^2$ instead of aa.*

*Example 5.*

Instead of $(M \times A) + (N \times B)$ we might write $(MA) + (NB)$, or simply $MA + NB$ (the gain in brevity more than balances the slight risk that we might try to form the sum of $A$ and $N$ first).

Though in general (as in Example 5) we cannot change the sum of two products very much, we can do a good deal to the product of two sums, as is shown by

THEOREM 6. $(a + b) \times (c + d) = ac + ad + bc + bd.$

PROOF.

| STATEMENT | REASON |
|---|---|
| (1) $(a + b) \times (c + d) = (a + b)c + (a + b)d$ | D |
| (2) $(a + b)c = ac + bc, (a + b)d = ad + bd$ | Theorem 5 |
| (3) $(a + b) \times (c + d) = (ac + bc) + (ad + bd)$ | (1) and (2) |
| (4) $(ac + bc) + (ad + bd) = ac + ad + bc + bd$ | Theorem 3 |
| (5) $(a + b) \times (c + d) = ac + ad + bc + bd$ | (3) and (4) |

This proves Theorem 6.

*Example 6.*

By Theorem 6, $(2a + 1) \times (2b + 1) = (2a)(2b) + (2a)(1) + 1(2b) + 1(1)$. By Theorem 4, $(2a)(2b) = 4ab$. By axioms M 4 and M 2 we see that $(2a)(1) = 2a$. By rule M 4 we have $1(2b) = 2b$ and $1(1) = 1$. Thus $(2a + 1)(2b + 1) = 4ab + 2a + 2b + 1$. This is used in Chapter 5 to prove that the product of two odd numbers is odd.

*Example 7.*

In a similar manner we could show that $(a + 3) \times (a + 3) = a^2 + 6a + 9$. This is used in Chapter 16.

**E 13.** Prove that $(5a + 1) \times (5a + 3) = 25a^2 + 20a + 3$. (Number each step as in our proofs, and give the reason for it.)

**E 14.** $A$, $P$, $Q$, and $M$, are four elements in a field that happen to satisfy the relation $1 = PA + QM$. Prove that for any element $B$ in the field it is then true that $B = BPA + BQM$. (Number the steps, and give reasons.) This result is used in the proof of Theorems 6 and 7 of Chapter 11.

**E 15.** In Chapter 5 we used the fact that if a natural number $D$ is a divisor of a natural number $A$, then it is also the divisor of $AB$ for any natural number $B$. Prove this fact. (*Hint:* Use the definition of divisor given in Chapter 3 or Chapter 4, together with the appropriate rules on multiplication.)

**E 16.** See E 15. Prove that, if $D$ is a divisor of $A$ and also of $B$, then it is a divisor of $A + B$. (*Hint:* This will involve the distributive axiom, $D$.) This is used in Chapter 5.

On a few occasions in this book we manipulated expressions involving "−" (the minus sign). *The trouble with this symbol is that it has three meanings,* just as some words have three meanings. (For example, "bat" means one thing on the baseball diamond, another in a dark cave, and something else to a person working with clay.) Unhappily, the three meanings for "minus" all occur in the same context, arithmetic and algebra.

**First Meaning of "Minus."** The symbol "$-$" is used to name negative numbers in the familiar field of real numbers. This could be done in kindergarten, *before* addition of numbers is mentioned. Rather than using red ink, as on some thermometers, we denote points to the left of zero on the number line by putting a "$-$" in front of a numeral. Only in the field of real numbers, ordinary arithmetic, does the symbol "$-$" have this extra use: part of a name.

**Second Meaning of "Minus."** If $a$ is any element of a field, then according to rule A 5, *there is exactly one element x in the field such that $a + x = 0$. That unique x we give the name $-a$ and call it* **the additive inverse of a**. Thus, by this definition, $a + (-a) = 0$.

*Example 8.*

Since $0 + 0 = 0$ we have $-0 = 0$.

*Example 9.*

In the Galois field of (*1*), p. 380, we have $1 + 1 = 0$. Thus in that field $-1 = 1$.

*Example 10.*

In the Galois field of Example 4 we have $2 + 3 = 0$. Thus in that particular field we have $-2 = 3$.

Note that the additive inverse of the real number 3 is negative 3. This is fortunate, for it asserts that $-3 = -3$; but be careful, for the symbol $-$ has two different uses in this equation.

**E 17.** See Chapter 16 for the definition of complex numbers and their sum.

    (a) Show that if $X$ and $Y$ are complex numbers located such that the midpoint of the line segment with ends $X$ and $Y$ is 0, then $X + Y = 0$.

    (b) Using (a), show that if $X$ is any complex number, then $-X$ is located in such a way that the line segment from $X$ to $-X$ has 0 as its midpoint.

THEOREM 7. $-(-a) = a$.

PROOF. All that we know about $-a$ is that $a + (-a) = 0$. We wish to prove that $a$ is the additive inverse of $-a$. To do so we must prove $(-a) + a = 0$. But this follows by axiom A 2 from the fact that $a + (-a) = 0$.

**E 18.** (a) Check that $-(-4)$ is 4 in the field of rationals.

    (b) Check that $-(-4)$ is 4 in the field of Example 4.

The next few theorems tell how the additive inverse behaves with respect to multiplication.

THEOREM 8. $(-a)b = -(ab)$ and $a(-b) = -(ab)$.

PROOF. We prove only the first assertion.

Justify this "collecting of the 219's" by proving this little theorem: $6a - 19(b - 4a) = 82a - 19b$. Number each step of your proof, and give the reason.

Having obtained the basic properties of $+$, $\times$, and $-$, we next consider division, which we meet occasionally in this book.

**Definition of Quotient.** *If $a$ and $b$ are elements of a field, where $a$ is not $0$, then there is, by rule M 5, exactly one element $x$ in the field such that $ax = b$. This unique $x$ we give the name $b/a$ and call it the **quotient** of $b$ divided by $a$.* Thus, by this definition, $a(b/a) = b$.

Sometimes $b/a$ is written as $\dfrac{b}{a}$ or $b \div a$.

*Example 14.*
Since $(3)(2) = 6$, we have $6/2 = 3$.

**Definition.** *If $a$ is not $0$, then $1/a$ is called the **"reciprocal of $a$,"** or the "multiplicative inverse of $a$."*

E 38. The multiplicative inverse is defined in terms of division. Show how we could have defined the additive inverse in terms of subtraction. (*Hint:* Prove that $-a = 0 - a$.)

E 39. (a) Prove that the reciprocal of the reciprocal of $a$ is $a$.
 (b) What theorem about the additive inverse does (a) resemble?

THEOREM 16. *If $ab = 0$, then at least one of $a$ and $b$ is $0$.*

PROOF. We will prove that if $a$ is not $0$, then $b$ must be $0$.

| STATEMENT | REASON |
|---|---|
| (1) $ab = 0$ | Given |
| (2) $a$ has a reciprocal, denoted $1/a$ | M 5 |
| (3) $(1/a)ab = (1/a)0$ | (1) |
| (4) $(1/a)ab = [a(1/a)]b$ | Theorem 4 |
| (5) $a(1/a) = 1$ | Definition of $1/a$ |
| (6) $1b = b$ | M 4 |
| (7) $b = (1/a)0$ | (3), (4), (5), and (6) |
| (8) $(1/a)0 = 0$ | Theorem 1 |
| (9) $b = 0$ | (7) and (8) |

End of proof.

Theorem 16 is used in Chapter 16, p. 266.
We will now show how division behaves with respect to multiplication.

THEOREM 17. $(a/b)(c/d) = ac/bd$ (*where b and d are not 0*).

PROOF. Call $a/b$ simply $p$. Call $c/d$ simply $q$. We know that $bp = a$ and that $dq = c$. We wish to prove $pq = ac/bd$, that is, that $bd \cdot pq = ac$. This we now do.

| STATEMENT | REASON |
|---|---|
| 1. $bd \cdot pq = bp \cdot dq$ | Theorem 4 |
| 2. $bp = a$ and $dq = c$ | Definition of $p$ and $q$ |
| 3. $(bd)(pq) = ac$ | (1) and (2) |

End of proof.

*Example 15.*

$(\frac{3}{5})(\frac{7}{9}) = \frac{21}{45}$.

THEOREM 18. $(ab)/c = a(b/c)$ (*where c is not 0*).

PROOF. Let $p = b/c$. By definition of $/$ we know that $cp = b$. We will deduce that $c(ap) = ab$, that is, $ap = (ab)/c$. This would prove the theorem.

| STATEMENT | REASON |
|---|---|
| (1) $c(ap) = a(cp)$ | Theorem 4 |
| (2) $cp = b$ | Definition of $p$ |
| (3) $c(ap) = ab$ | (1) and (2) |

End of proof.

**E 40.** Prove that $(b + c)/a = (b/a) + (c/a)$ (for $a$ not 0).
**E 41.** Prove that $-(a/b) = (-a)/b$ (for $b$ not 0).
**E 42.** Prove that $ac/bc = a/b$ (for $b$ and $c$ not 0).
**E 43.** Using E 42, show that $(-5)/(-7) = 5/7$.

The following theorem says that "to divide by a quotient, first turn it upside down, and then multiply."

THEOREM 19. $a/(b/c) = a(c/b)$ (*for b and c not 0*).

PROOF. Let $p$ be $c/b$ and $q$ be $b/c$. We wish to prove that $a/q = ap$. In other words, from $bp = c$ and $cq = b$ we want to deduce that $q(ap) = a$. To do so we will prove that $pq = 1$.

| STATEMENT | REASON |
|---|---|
| (1) $bp = c; cq = b$ | Given |
| (2) $(bp)q = b$ | (1) |
| (3) $b(pq) = b$ | M 3 |
| (4) $b \cdot 1 = 1$ | M 2 and M 4 |
| (5) $pq = 1$ | (3) and (4) |

End of proof.

*Example 16.*

$(\frac{3}{4})/(\frac{9}{7}) = (\frac{3}{4})(\frac{7}{9})$, which, by Theorem 17, is equal to $\frac{21}{36}$.

We conclude with a detailed justification of a typical manipulation made in Chapter 8. We will list twelve steps; in practice, one does several at a time.

*Example 17.* (From p. 113 of Chapter 8.)

We will show that the equation $V_1 - V_2 = 4V_2 - 4V_3 + V_2$ implies the equation $0 = -V_1 + 6V_2 - 4V_3$. In slow motion, we have:

| STATEMENT | REASON |
|---|---|
| (1) $V_1 - V_2 = 4V_2 - 4V_3 + V_2$ | Given |
| (2) $V_1 - V_2 = 4V_2 + (-4V_3) + V_2$ | Theorem 12 |
| (3) $V_1 - V_2 = 4V_2 + V_2 + (-4V_3)$ | Theorem 3 |
| (4) $V_1 - V_2 = 5V_2 + (-4V_3)$ | Theorem 5 |
| (5) $(V_1 - V_2) + V_2 = 5V_2 + (-4V_3) + V_2$ | (4) |
| (6) $V_1 = 5V_2 + (-4V_3) + V_2$ | E 32 |
| (7) $V_1 + (-V_1) = 5V_2 + (-4V_3) + V_2 + (-V_1)$ | (6) |
| (8) $0 = 5V_2 + (-4V_3) + V_2 + (-V_1)$ | First meaning of "minus" |
| (9) $0 = (-V_1) + (5V_2 + V_2) + (-4V_3)$ | Theorem 3 |
| (10) $0 = (-V_1) + 6V_2 + (-4V_3)$ | Theorem 5 |
| (11) $0 = (-V_1) + 6V_2 - 4V_3$ | Theorem 12 |
| (12) $0 = -V_1 + 6V_2 - 4V_3$ | Parentheses dropped in (11) |

End of justification.

**E 44.** (a) Prove that $(a + b) - c = a + (b - c)$.

(b) Use (a) to show why we can be careless and omit parentheses in going from (10) to (11) in the preceding justification.

**References**

1. W. W. Sawyer, *A Concrete Approach to Abstract Algebra*, W. H. Freeman and Company, San Francisco, 1959. (Pages 26–31 discuss fields and give a different proof of Theorem 1; pp. 71–77 discuss finite fields.)

2. G. Birkhoff and S. MacLane, *A Survey of Modern Algebra*, Macmillan, New York, 1953. (A structure weaker than a field is discussed on pp. 1–6; the demand that the equation $ax = b$ have a solution is omitted.)

*Appendix* **D**

# TEACHING
# MATHEMATICS

Next to English, mathematics is the most important subject in the elementary and secondary curriculum. In recent years, the amount of mathematics required for earning a teaching credential has been increased. This appendix is addressed to prospective teachers among those who use this book as a text in a required mathematics course.

I emphasize that this book is not a 'methods' guide; it does not advise a teacher what to do in the classroom. But I did keep the teacher in mind when I wrote it.

For several reasons a teacher should have a greater knowledge of the subject than he will have to teach. First, it will give him more confidence, hence more willingness to encourage an "open" classroom, where questions are welcome and new paths taken. Second, he will be aware precisely how and where the material he teaches is used in more advanced courses and consequently be in a better position to judge what topics he teaches are of particular importance. Third, it will give him a wider perspective, hence a better judgment in evaluating textbooks and reforms, or deciding, from day to day, what to emphasize and what to omit. Fourth, it will make him more resilient; if a pupil does not understand his first explanation he will have the resources from which to draw other approaches. Fifth, he will not convey erroneous impressions, such as "every number is a fraction" or "nothing new is being done in mathematics." And, he will not use "more arithmetic drill" as a punishment for noisy pupils. (This is still done.)

But I also intend this book to have a more direct impact on teaching. For instance, here is how a teacher might apply Chapter 6, Rationals and Irrationals, in the classroom:

*First day or two.* Each pupil cuts eight congruent right triangles out of tagboard. Then he cuts out three squares of different sizes, the side of each corresponding to a side of the triangle. The teacher says, "Can you make a square out of four triangles and the two smaller squares?" The pupils, working alone or in small groups, probably do this in a few minutes. "Can you make a square out of four triangles and the one big square you cut out?" Again they manage. Next the teacher asks, "How do the 'puzzle' squares you assembled compare in size?" After they all agree that they are the same size, then the teacher asks, "What does that tell us about the three squares you cut out?" This opens a discussion, which should be permitted to go on until all the class has obtained the Pythagorean Theorem.

*Next day.* The teacher asks the class to draw right triangles, measure their sides, and check the Pythagorean Theorem, as well as their arithmetic skills permit.

*Next day or two, or more.* The teacher asks, "If the two short sides of a right triangle are 5 inches and 12 inches, how long is the longest side?" If the pupils just measure it, let them. If they don't think of using the Pythagorean Theorem, then ask the same question for a triangle whose short sides are 4 inches and 7 inches. This ought to raise a debate as to the "precise answer" or "who is right," and a pupil or the teacher may bring in the Pythagorean Theorem.

This may be taken further. If the class has *not* studied the multiplication of rational numbers, then it could be asked to find more right triangles whose sides are all "whole numbers." (The teacher may refer to Exercise 70 in Chapter 6 for a method that provides more examples.) Moreover, it may be a good occasion for *introducing* the multiplication of rational numbers (which is easier than their addition).

If the class knows how to multiply rational numbers, then the field is wide open.

*Next few days.* Ask the class to find the long side of a right triangle whose two short sides are 1. They may measure (and then argue.) They may use the Pythagorean Theorem and search for a rational number whose square is 2 (and argue). The teacher may wish, after the argument goes on awhile, to ask a question, "Can the denominator be 5? 6? 7?," and have different groups in the class work on each case, maybe going to some very large denominators. There is no need to *prove* to the class that $\sqrt{2}$ is irrational (but the teacher should know how to prove it, to have confidence in what he is doing).

*Next few days.* If the class has had decimals, then here is a good occasion for estimating $\sqrt{2}$ to several decimal places, first by "trapping" it between 1.4 and 1.5, and so on, then by the method of Exercise 67 of Chapter 6. A meter stick would be handy for introducing decimals to students who want to find $\sqrt{2}$ "exactly."

The resourceful teacher may exploit almost any chapter in this book in the classroom. He may wonder, "But what about drill? What about the curriculum? How can I grade this work? There is no time for such a luxury!"

The answer is simple: There can be lots of "drill" in such an approach. Moreover, the computations are in a context; hence the student has more chance of evaluating his own work. As for the curriculum, every teacher should have a one-page outline of the mathematics curriculum from kindergarten to grade 12. This will assure him that there *is* time and, indeed, that work in the regular class text may be replaced by such projects as the one described.

Is there time for such digressions? I believe that the classroom encounter should consist primarily of such "digressions," and that texts, workbooks, and homework should only supplement this activity.

As for grading the work, I feel that that evaluation should in no way determine content. Grades have two legitimate roles: to inform the student and his parents of his progress, and to indicate for later reference how the student did. In fact, grades frequently serve as a tool of discipline and a substitute for motivation. I urge that a teacher distinguish between scores that are for the student's information (and do not appear in the teacher's roll book) and the less-frequent grade-determining scores.

# THE GEOMETRIC
# AND HARMONIC
# SERIES

In Chapter 15 we used the fact that $3/10 + 3/10^2 + \cdots + 3/10^N$ gets closer and closer to $1/3$ as $N$ grows. This we prove in Theorem 3. In Chapter 4 we used the expression $1/1 + 1/2 + 1/3 + \cdots + 1/N$ as an estimate of the average gap between the first $N$ primes. We shall examine the behavior of this sum in Theorem 1, which is quite a contrast to Theorem 3. Theorem 2 is the basis for the fact that 99,999 is one less than 100,000, which illustrates an important property of the decimal system. In Chapter 17 we used Theorem 2 to show that

$$1 + F + F^2 + F^3 + F^4 = \frac{F^5 - 1}{F - 1}.$$

Consider first the endless sequence of numbers $1, \frac{1}{2}, \frac{1}{4}, \ldots$, each obtained from its predecessor through multiplying by $\frac{1}{2}$. Let $S_N$ be the sum of the first $N$ numbers of that sequence. Thus

$$
\begin{aligned}
S_1 &= 1 & &= 1.000, \\
S_2 &= 1 + \tfrac{1}{2} & &= 1.500, \\
S_3 &= 1 + \tfrac{1}{2} + \tfrac{1}{4} & &= 1.750, \\
S_4 &= 1 + \tfrac{1}{2} + \tfrac{1}{4} + \tfrac{1}{8} &&= 1.875.
\end{aligned}
$$

As $N$ grows, so does $S_N$. Do the $S_N$'s grow without bound, passing 10, even 100, eventually 1000, and so on? Or is their growth restricted by some real number that they are approaching? If so, what is that number?

Two forces influence the growth of $S_N$. First, since we add more and more numbers together, we might suspect that there is no limit to the growth of $S_N$. Second, since the amounts we add are getting smaller and smaller, we might suspect that the $S_N$'s cannot grow very large, and hence must be approaching some number.

Before we jump to any conclusion concerning which force is the more powerful for the sequence $1, \frac{1}{2}, \frac{1}{4}, \frac{1}{8}, \ldots$, it would be wise to examine another sequence, the so-called *harmonic series:*

$$\frac{1}{1}, \frac{1}{2}, \frac{1}{3}, \frac{1}{4}, \frac{1}{5}, \cdots .$$

Let $s_N$ stand for the sum of the first $N$ terms of the harmonic series. Thus

$$
\begin{aligned}
s_1 &= 1 & &= 1.000, \\
s_2 &= 1 + \tfrac{1}{2} & &= 1.500, \\
s_3 &= 1 + \tfrac{1}{2} + \tfrac{1}{3} & &= 1.833\ldots, \\
s_4 &= 1 + \tfrac{1}{2} + \tfrac{1}{3} + \tfrac{1}{4} &&= 2.083\ldots
\end{aligned}
$$

(and the reader may compute a few more such sums).

What happens to $s_N$ as $N$ gets larger and larger? The same two forces are acting as in the first example. Which will be the more powerful? We answer this question with

THEOREM 1. *If $s_N$ is $1/1 + 1/2 + 1/3 + \cdots + 1/N$, then $s_N$ grows without bound as $N$ gets larger and larger.*

PROOF. We shall examine closely $s_2, s_4, s_8, s_{16}$, and so on; that is, $s_N$ when $N$ is a power of 2.

First of all,

$$s_2 = 1 + \tfrac{1}{2} = \tfrac{3}{2}.$$

Next, $s_4 = 1 + \tfrac{1}{2} + \tfrac{1}{3} + \tfrac{1}{4} = s_2 + \tfrac{1}{3} + \tfrac{1}{4}$, which is larger than $s_2 + \tfrac{1}{4} + \tfrac{1}{4}$. Thus,

$$s_4 \text{ is larger than } s_2 + \tfrac{1}{2},$$

and so

$$s_4 \text{ is larger than } \tfrac{4}{2}.$$

Next, $s_8 = s_4 + \tfrac{1}{5} + \tfrac{1}{6} + \tfrac{1}{7} + \tfrac{1}{8}$ is larger than $s_4 + \tfrac{1}{8} + \tfrac{1}{8} + \tfrac{1}{8} + \tfrac{1}{8}$. Thus $s_8$ is larger than $s_4 + \tfrac{1}{2}$, and we conclude that

$$s_8 \text{ is larger than } \tfrac{5}{2}.$$

Similarly, as the reader may show, $s_{16}$ is larger than $s_8 + \tfrac{1}{2}$. Thus,

$$s_{16} \text{ is larger than } \tfrac{6}{2}.$$

Using the same idea, the reader may show that $s_{32}$ is larger than $\tfrac{7}{2}$ and that $s_{64}$ is larger than $\tfrac{8}{2}$.

Since the $s_N$'s for large $N$ can be made larger than the sum of any given number of $(\tfrac{1}{2})$'s, there is no bound to the growth of $s_N$. This proves Theorem 1.

In the harmonic series the first force conquers. Now let us see which of the two forces conquers in the case of our first sequence, $1, \frac{1}{2}, \frac{1}{4}, \frac{1}{8}, \ldots$. Let $S_N$ be the sum of the first $N$ terms of this sequence. Theorem 2 will tell us that

(1)
$$S_N = \frac{1 - (\frac{1}{2})^N}{\frac{1}{2}},$$

a formula which is much shorter than the one we began with,

(2)
$$S_N = 1 + \tfrac{1}{2} + \tfrac{1}{4} + \tfrac{1}{8} + \cdots + (\tfrac{1}{2})^{N-1}.$$

(The reader should note that the $N$th term is $(\frac{1}{2})^{N-1}$; for example, check this for $N = 4$.)

Though (2) gets more and more unwieldy as $N$ grows, (1) remains concise. Using (1) instead of (2), we can easily predict the behavior of $S_N$ as $N$ gets larger and larger. At a glance we can see that $S_N < 1/\frac{1}{2}$; thus as $N$ grows larger, $S_N$ remains less than 2. Moreover, since $(\frac{1}{2})^N$ approaches 0 as $N$ increases, we see that $S_N$ approaches

$$\frac{1 - 0}{\frac{1}{2}},$$

which is 2. For this sequence the force that dominates is the second, "the swift decrease in the size of the terms."

We will obtain a result of which formula (1) is just a special case.

If $a$ and $r$ are two real numbers, then the $N$ numbers

$$a, \quad ar, \quad ar^2, \quad ar^3, \quad \ldots, \quad ar^{N-1}$$

are called a *geometric series* or *geometric progression*. For example, if $a$ is 1 and $r$ is $\frac{1}{2}$, we obtain the geometric progression $1, \frac{1}{2}, \frac{1}{4}, \ldots$. If $a$ is $\frac{3}{10}$ and $r$ is $\frac{1}{10}$, we obtain the geometric series $3/10, 3/10^2, 3/10^3, \ldots$. For any geometric series we have

THEOREM 2. SUM OF A FINITE GEOMETRIC SERIES. *Let $a$ and $r$ be real numbers with $r$ not equal to 1. Let $S_N = a + ar + ar^2 + \cdots + ar^{N-1}$. Then $S_N$ is given by this short formula:*

$$S_N = \frac{a(1 - r^N)}{1 - r}.$$

PROOF. We have

(3)
$$S_N = a + ar + ar^2 + \cdots + ar^{N-1},$$

and

(4)
$$rS_N = ar + ar^2 + \cdots + ar^{N-1} + ar^N.$$

Subtracting (4) from (3) and performing many cancellations, we obtain

$$S_N - rS_N = a - ar^{N-1},$$

which we can simplify to

(5)
$$(1 - r)S_N = a(1 - r^N).$$

Dividing both sides of (5) by $1 - r$ (which fortunately is not 0), we obtain

$$S_N = \frac{a(1 - r^N)}{1 - r},$$

and Theorem 2 is proved.

With the aid of Theorem 2 we will easily obtain

THEOREM 3. SUM OF AN INFINITE GEOMETRIC SERIES. *Let $S_N$ be the sum of the first N terms of the geometric series*

$$a, \quad ar, \quad ar^2, \quad ar^3, \quad \dots,$$

*where r is less than 1 and larger than $-1$. Then as N gets larger and larger, $S_N$ approaches*

$$\frac{a}{1 - r}.$$

PROOF. By Theorem 2 we know that $S_N$ is equal to

$$\frac{a(1 - r^N)}{1 - r}.$$

As $N$ grows, $r^N$ approaches 0, since $r$ is a number between 1 and $-1$. Thus $S_N$ approaches

$$\frac{a(1 - 0)}{1 - r},$$

and Theorem 3 is proved.

Theorem 3 is usually stated in this terse but misleading shorthand:

(6) $$a + ar + ar^2 + ar^3 + \cdots = \frac{a}{1 - r}.$$

The reader should be careful in using this shorthand. It does not describe the addition of an infinite set of numbers; rather, it describes the behavior of sums of finite sets of numbers.

*Example 1.* With $a = 1$ and $r = \frac{1}{2}$, formula (6) gives

$$1 + \frac{1}{2} + \frac{1}{4} + \frac{1}{8} + \cdots = \frac{1}{1 - \frac{1}{2}} = 2.$$

*Example 2.* With $a = \frac{3}{10}$ and $r = \frac{1}{10}$, formula (6) gives

$$\frac{3}{10} + \frac{3}{100} + \frac{3}{1000} + \cdots = \frac{\frac{3}{10}}{1 - \frac{1}{10}} = \frac{1}{3}.$$

(In decimal notation, Example 2 reads: $0.3333\ldots = \frac{1}{3}$.)

We now give an *application of geometric series to probability.* Two equally matched volleyball teams representing Hawaii and Texas play each other until one of them has a lead of two points. For example, if Hawaii wins the first two points, then the game is over quickly; the winner's score being 2, and the loser's score being 0. Or it might happen that Hawaii scores the first point and Texas scores the next three points to get a lead of 2. Then the game would be over; the winner's score would be 3, and the loser's 1.

Using an $H$ to record a point for Hawaii, and $T$ a point for Texas, we could tell the story of each game in a string of $H$'s and $T$'s. For example, $H\,T\,T\,H\,H\,H$ tells us that Hawaii scored the first point, Texas scored the second and third, and Hawaii scored the next three points to win the game. The winner's score was 4; and the loser's, 2.

Some games will be short and some will be quite long. This is the

**Problem.** *If Hawaii and Texas play millions of games what will the average winning score tend to be?*

Even though we have no volleyball or equally matched teams we can still attack the problem experimentally. Select a penny. Flip it. If it turns up heads, give Hawaii a point; if it turns up tails, give Texas a point. With a penny the reader may easily play out ten games and compute their average winning score.

In the ten experimental games I performed with a penny, the winner's score was successively

$$2, 5, 3, 2, 3, 6, 2, 7, 3, \text{ and } 2.$$

In these ten experiments the average winner's score is

(7)  $$\frac{2 + 5 + 3 + 2 + 3 + 6 + 2 + 7 + 3 + 2}{10},$$

which reduces to $\frac{35}{10}$, or simply 3.5. The readers's result need not, of course, be the same as mine.

Now we will figure out what the average winning score will be if more and more games are played. Collecting repetitions in the numerator of (7), we can rewrite (7) as

(8)  $$\frac{4 \cdot 2 + 3 \cdot 3 + 0 \cdot 4 + 1 \cdot 5 + 1 \cdot 6 + 1 \cdot 7}{10}.$$

The numerator of (8) records that the winning score 2 occurred four times; the score 3, three times; the score 4, not at all; the score 5, not at all; the score 6, once; and the score 7, once.

Next we rewrite (8) as

(9)  $$\tfrac{4}{10} \cdot 2 + \tfrac{3}{10} \cdot 3 + \tfrac{0}{10} \cdot 4 + \tfrac{1}{10} \cdot 5 + \tfrac{1}{10} \cdot 6 + \tfrac{1}{10} \cdot 7.$$

In (9), $\tfrac{4}{10}$ is the fraction of the ten games for which the winner's score is 2; $\tfrac{3}{10}$ is the fraction of the games for which the winner's score is 3, and so on.

The expression (9) suggests that we estimate what fraction of the games will have a winning score of 2 (or 3, or 4, . . .) if millions of games are played. What number would replace the fraction $\tfrac{4}{10}$, obtained in our experiment?

Let us work out what fraction of the games would, over the long run, have a winning score of 2. In the first two throws of the penny we have four possible outcomes:

(*10*) $\qquad\qquad H\ H \qquad H\ T \qquad T\ T \qquad T\ H.$

The first and third of these four cases produce the winning score of 2. Since the penny is supposed to be unbiased we would expect each of the four cases in (*10*) to occur equally often if many experiments are performed. Therefore we would expect the winner's score to be 2 in two-fourths of the games. For the long-run estimate we would have $\frac{1}{2}$ instead of the experimental result $\frac{4}{10}$. This is no contradiction; the ten-game experiment serves only as a guide to our thinking, not as a prediction of what would happen over millions of experiments.

Next, what fraction of the games will, in the long run, have a winning score of 3? (In our experiment the winning score of 3 occurred in three-tenths of the games.) To have a winning score of 3, the two teams have to be tied at the end of the first two points, and then one team has to score two points in a row. Inspection of (*10*) shows that the two teams are tied at the end of the first two points in two-fourths of the games. Thus in one-half of the games the winner's score will be more than 2. What fraction of these games that have a winning score of 3 or more will have a winning score of 3? The half for which the third and fourth points are either $H\ H$ or $T\ T$. Since $\frac{1}{2}\cdot\frac{1}{2} = \frac{1}{4}$ we see that in one-fourth of all the games, the winning score will, in the long run, be 3. (It is interesting to compare this $\frac{1}{4}$ to $\frac{3}{10}$ of our experiment.)

The reader may show that the winning score of 4 will tend to occur in one-eighth of all the games. Similarly, the winning score will be 5 in one-sixteenth of the games, and so on.

Taking our cue from (*9*), we suspect that the average winning score will tend to be

(*11*) $\qquad\qquad \frac{1}{2}\cdot 2 + \frac{1}{4}\cdot 3 + \frac{1}{8}\cdot 4 + \frac{1}{16}\cdot 5 + \cdots.$

We are faced with the problem of summing an endless series that resembles a geometric progression but which, unfortunately, is not one. We expect the sum of the first $N$ terms of (*11*) to get closer and closer to the average winning score; and we expect that score to be about 3.5. Pausing for a moment to get a feel for the problem, let us at least sum the first four terms. To three decimals we have

$$\frac{1}{2}\cdot 2 + \frac{1}{4}\cdot 3 + \frac{1}{8}\cdot 4 + \frac{1}{16}\cdot 5 = 1 + \frac{3}{4} + \frac{4}{8} + \frac{5}{16} = 2.562.$$

The reader may compute the sum of the first twelve terms and make a guess as to the value of (*11*).

We will assume that the series (*11*) has a sum (that is, that it does not misbehave like the harmonic series). This assumption can be justified, but doing so would require too long a detour. Let us call the sum of (11) W. We will now find W by exploiting the similarity of (11) to a geometric series.

We have

(12) $$W = 2 \cdot \tfrac{1}{2} + 3 \cdot \tfrac{1}{4} + 4 \cdot \tfrac{1}{8} + 5 \cdot \tfrac{1}{16} + \cdots .$$

Though (12) is not a geometric series, let us try the technique we used to find the sum of the geometric series (3). We now multiply (12) by $\tfrac{1}{2}$. This gives

(13) $$\tfrac{1}{2}W = 2 \cdot \tfrac{1}{4} + 3 \cdot \tfrac{1}{8} + 4 \cdot \tfrac{1}{16} + 5 \cdot \tfrac{1}{32} + \cdots .$$

Just as we subtracted (4) from (3), we will subtract from

(14) $$W = 2 \cdot \tfrac{1}{2} + 3 \cdot \tfrac{1}{4} + 4 \cdot \tfrac{1}{8} + 5 \cdot \tfrac{1}{16} + \cdots$$

the relation

(15) $$\tfrac{1}{2}W = \qquad\quad 2 \cdot \tfrac{1}{4} + 3 \cdot \tfrac{1}{8} + 4 \cdot \tfrac{1}{16} + \cdots .$$

Instead of cancellations, this time we obtain

(16) $$W - \tfrac{1}{2}W = 2 \cdot \tfrac{1}{2} + \tfrac{1}{4} + \tfrac{1}{8} + \tfrac{1}{16} + \cdots .$$

Since $\tfrac{1}{4} + \tfrac{1}{8} + \tfrac{1}{16} + \cdots$ is a geometric progression with $a = \tfrac{1}{4}$ and $r = \tfrac{1}{2}$ we see that (16) simplifies to

(17) $$W - \tfrac{1}{2}W = 2 \cdot \tfrac{1}{2} + \tfrac{1}{2}.$$

Simplifying both sides of (17) we obtain

$$\tfrac{1}{2}W = \tfrac{3}{2},$$

and so W is 3. The problem is solved.


### Exercises

1. Find the sum of these endless geometric series.
   (a) $\tfrac{1}{4} + (\tfrac{1}{4})^2 + (\tfrac{1}{4})^3 + \cdots$ (Answer: $\tfrac{1}{3}$)
   (b) $1 + \tfrac{2}{3} + (\tfrac{2}{3})^2 + \cdots$ (Answer: 3)
   (c) $1 - \tfrac{1}{2} + \tfrac{1}{4} - \tfrac{1}{8} + \cdots$ (Answer: $\tfrac{2}{3}$)

2. Let $S_N = \dfrac{1}{1 \cdot 2} + \dfrac{1}{2 \cdot 3} + \dfrac{1}{3 \cdot 4} + \cdots + \dfrac{1}{N(N + 1)}.$

   For example,

   $$S_3 = \frac{1}{1 \cdot 2} + \frac{1}{2 \cdot 3} + \frac{1}{3 \cdot 4}$$

   which, decimally, is 0.750. Compute to three decimals (a) $S_1$, (b) $S_2$, (c) $S_3$, (d) $S_4$, (e) $S_5$, and (f) $S_6$.

3. See E 2. Prove that $S_N = 1 - (1/(N + 1))$. (*Hint:* Note that $1/(1 \cdot 2) = (1/1) - (1/2)$, $1/(2 \cdot 3) = (1/2) - (1/3)$, $1/(3 \cdot 4) = (1/3) - (1/4)$, and so on.)

4. See E 3. Prove that $S_N$ of E 2 approaches 1.

5. Let $S_N$ equal $1/1^2 + 1/2^2 + 1/3^2 + \cdots + 1/N^2$. For example, $S_4 = \frac{1}{1} + \frac{1}{4} + \frac{1}{9} + \frac{1}{16}$, which, to three decimals, is 1.423. Compute to three decimals (a) $S_5$, (b) $S_6$, and (c) $S_7$.

6. See E 5. Using the facts that $1/2^2$ is less than $1/(1\cdot2)$, $1/3^2$ is less than $1/(2\cdot3)$, $1/4^2$ is less than $1/(3\cdot4)$, and so on, prove that each $S_N$ of E 5 is less than 2. (*Hint:* Use E 2, E 3, E 4.) It can be proved that $1/1^2 + 1/2^2 + 1/3^2 + \cdots = \pi^2/6$.

7. Using E 6 prove that if $S_N = 1/1^3 + 1/2^3 + 1/3^3 + \cdots + 1/N^3$, then $S_N$ is always less than 2. It is not known whether

$$1/1^3 + 1/2^3 + 1/3^3 + \cdots = C$$

is related to $\pi^3$ as intimately as $1/1^2 + 1/2^2 + 1/3^2 + \cdots$ is to $\pi^2$. It is not even known whether $C/\pi^3$ is rational.

8. We proved that $S_N$ for the harmonic series gets arbitrarily large as $N$ increases. Show that $S_N$ grows "slowly." Specifically, prove that
   (a) $S_4$ is less than $1 + \frac{1}{2} + \frac{1}{2} + \frac{1}{2} = \frac{5}{2}$;
   (b) $S_8$ is less than $S_4 + \frac{1}{4} + \frac{1}{4} + \frac{1}{4} + \frac{1}{4}$, hence less than $\frac{7}{2}$;
   (c) $S_{16}$ is less than $\frac{9}{2}$;
   (d) $S_{32}$ is less than $\frac{11}{2}$.

9. In this appendix we have considered sums of numbers. In Chapter 4 (E 51, E 52, E 53, E 54) similar questions were considered for products of numbers. Show that in those four exercises the two forces operating are:
   (1) one force influencing the product $A_N$ (or $B_N$) not to approach 0 (the terms are close to 1);
   (2) one force influencing the product $A_N$ (or $B_N$) to approach 0 (the terms are less than 1).
   Note how they balance.

10. Evaluate
    (a) $1 + 1/3 + 1/9 + 1/27 + \cdots + 1/3^{10}$,
    (b) $1 - 1/3 + 1/9 - 1/27 + \cdots + 1/3^{10}$ (signs alternate),
    (c) $1 + 3 + 9 + 27 + \cdots + 3^{100}$.

11. When a superball is dropped it rebounds 9/10ths of the distance it falls. What is the total distance a superball travels if it is dropped from a height of 6 feet?

12. An object falls $16t^2$ feet in $t$ seconds. How long does the superball in E 11 continue to bounce?

13. Compute the average winning score in 20 experimental games, and compare it with the theoretical result, 3.

14. Compute to three decimals the sum of the first 15 terms of (11).

15. Show why the winning score of 4 will tend to occur in one-eighth of the games. (*Easy way:* The two teams share the first two points, share the second two points, and one team gets both of the third pair of points.)

16. (See E 15.) Show why the winning score of 5 will tend to occur in one-sixteenth of the games.

17. What is the average number of points scored in the volleyball games between Hawaii and Texas? (*Hint:* Why is the average loser's score 1?)

18. A bug is taking a walk on the real line. He starts at 0, and at each step goes either 1 to the right or 1 to the left. Thus, after one step he is either at 1 or $-1$. At 2 and at $-2$ there are bug traps. If many bugs take such promenades, what is the average number of steps they take before they fall into a bug trap? (*Hint:* Show that this is E 17 in disguise.)

The problem of the two teams and the problem of the wandering bugs illustrate the part of probability devoted to *the random walk*. Mathematics of the random walk is applied, for example, to the percolation of a fluid through some object, the spread of smoke through air, or the motion of colliding and rebounding neutrons before capture by the lead casing of a nuclear reactor.

### Reference

1. *W. Feller, *An Introduction to Probability Theory and Its Applications*, vol. 1, 2nd ed., 1957, Wiley, New York. (If a team is to win by $k$ points, then the average winning score is $k(k + 1)/2$; this is implicit in a more general result proved on pp. 317–318. See pp. 311–318 (and the index) for more information concerning the random walk.)

*Appendix* **F**

# SPACE OF ANY
# DIMENSION

In Chapter 16 we referred to the impossibility of building algebras on spaces of dimension other than 1, 2, or 4. This appendix tells what is meant by space of any dimension: 1, 2, 3 or more.

We can describe every point on a straight line by using one real number, telling how far that point is from the point that we choose to label 0, and whether it is left or right of 0. This places on the line a scale such as this,

(*1*)    $\cdots$  −4  −3  −2  −1  0  1  2  3  4  $\cdots$

which tags every point with a real number. In (*1*) we have dotted heavily the points whose tags are integers. Note that the line is *one*-dimensional and that each point is described by using only *one* real number.

When we go to dimension two—the plane—it is the custom to tell where a point is by first selecting two perpendicular lines,

and then describing any point in the plane by its relation to these two lines.

To do this, place on each of the two lines a number scale. Then any point *P* in the plane is described by the two numbers given by the rectangle drawn in this manner:

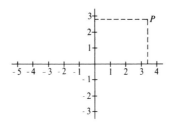

The particular point we chose is described by 3.4 and 2.8. We could call $P$ "the point (3.4, 2.8)," keeping in mind that the first number, 3.4, refers to a point on the horizontal line and 2.8 to a point on the vertical. [The point (2.8, 3.4) is different from the point (3.4, 2.8).] Note that the plane is *two-dimensional* and that *two* real numbers are required to locate a point.

Now let us go to dimension three, ordinary space. To pinpoint a location in space we can proceed in this manner: We select a plane in space and a line perpendicular to that plane.

(2)

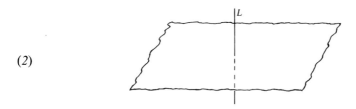

In the plane, we select two perpendicular lines meeting the line $L$ in (2). Then we place on each of the three lines the real number scale. Drawing these lines in perspective, we now have

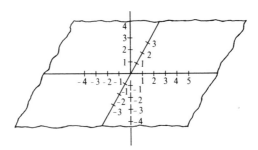

Any point $P$ in space is described by the point $P^*$, which is directly below (or above) $P$, combined with a number on $L$, which tells how high (or low) $P$ is.

For example, consider this $P$:

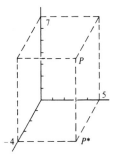

with $P^*$ the point $(5, -4)$ and with height 7. The point $P$ is located by the *three numbers* $(5, -4, 7)$. The order in which the three numbers is written is as important as the order in which we write names: given name, middle name, family name. (John Charles Thomas is a different person from Thomas John Charles or Charles John Thomas.) Thus the $X$ and $Y$ of the plane are written first, in their usual order; the third, or right-hand number, describes the height of the point. Note that space is *three*-dimensional and that *three* real numbers are required to locate a point.

Summarizing, we see that a line corresponds to the set of reals, the plane to the set of pairs of reals, and space to the set of triplets of reals.

It is very easy to conceive of the set of quadruplets of real numbers; for example, $(-7, 6, 0, \sqrt{2})$ is a quadruplet, and so is $(2, \pi, -107, 19)$. The reader is urged to devise a few quadruplets himself. He will find it no harder to write a few quintuplets of real numbers, such as $(10, 15, -\pi, -\sqrt{2}, \frac{1}{2})$, and even toss in a sextuplet, such as $(-1, 2, -\frac{1}{2}, \frac{2}{3}, \frac{3}{4}, 0)$.

It is no harder to think of the set of all quadruplets of real numbers than it is to think of the set of all triplets of real numbers. The set of triplets of real numbers has a convenient geometric model—space. We can "see" the set of reals as the line; the set of couplets of reals as the plane, and the set of triplets of reals as space. But no one, no matter how hard he squints, will "see" the set of quadruplets as some sort of geometric object.

However, if the reader wants to think of four-dimensional space geometrically, visually, here is one way. Consider the set of all arrows that can be drawn in the plane. The feather is described by two real numbers and the point by two real numbers. Thus the set of quadruplets of real numbers can be thought of as representing the set of all arrows in the plane.

It is the custom to call the set of quadruplets of real numbers "Four-dimensional space" or "Four-space," the set of quintuplets "Five-dimensional space" or "Five-space," and so on. With equal ease, we may speak of "Twenty-space"; that is, the set of all 20-tuplets of real numbers.

Any set $S$ in one-to-one correspondence with the set of $k$-tuplets (such that two elements close together in $S$ correspond to $k$-tuplets which are close together, and two $k$-tuplets close together correspond to two elements close together) is also called a $k$-dimensional space. For $k = 1, 2,$ or $3$ we have as examples of such sets, the line, the plane, and the space in which all material things are found. The demand that "closeness be preserved" is essential, as E 38 of Chapter 18 shows.

Even though $S$ can be conumerous with spaces of different dimensions (see E 38 of Chapter 18) it can be proved, with the aid of Sperner's lemma, that $S$ cannot be simultaneously of two different dimensions. The rather technical proof for this fact is to be found in R 1 of Chapter 2.

It is very easy to define an addition, $\oplus$, on $k$-space. For example, with $k = 4$, we can define the "sum" of two 4-tuplets $(a, b, c, d) \oplus (A, B, C, D)$ as $(a + A, b + B, c + C, d + D)$. The reader may check that $\oplus$ is commutative and associative. The same method defines, for $k = 2$, the $\oplus$ that we obtained geometrically in Chapter 16 with the aid of parallelograms. As mentioned in Chapter 16, it is impossible to define a multiplication, $\otimes$, in four-dimensional space such that $\oplus$ and $\otimes$ together satisfy all the rules of algebra.

### Exercises

1. Show that the set of circles in the plane is part of a 3-space. (*Hint:* A circle is described by its center and its radius.)

2. Show that the set of rectangles in the plane whose sides are horizontal and vertical is part of a 4-space by showing that such a rectangle can be described by
   (a) its length, width, and center;
   (b) its lower left corner and its upper right corner; or
   (c) its area, its upper left corner, and the length of its diagonal.
   (Topology guarantees that methods (a), (b), and (c) will all yield the same dimension.)

3. A baseball pitcher determines the speed of the ball, the rate of its spin, the axis of spin, the direction in which the ball is thrown, and the time at which it is thrown. Show that the batter requires seven numbers to describe the pitch.

### Reference

1. D. Hilbert and S. Cohn-Vossen, *Geometry and the Imagination*, Chelsea, New York, 1952. (Further examples of spaces are to be found in pp. 157–164.)

*Appendix* **G**

# SOME TECHNICAL
# TERMS

Readers of the "new" mathematics may meet such terms as "mapping," "function," "one-to-one," and "onto," in the "new" curriculum. These concepts occur in this text, although the particular terms are not used, with the exception of "one-to-one correspondence" (in Chapter 18). In this appendix we define the other terms and illustrate them primarily by examples from the various chapters.

Let $X$ and $Y$ be sets. A rule that assigns to *each* element of $X$ *exactly one* element of $Y$ is called a *function* (or *mapping*) from $X$ to $Y$. Any one-to-one correspondence, defined in Chapter 18, is thus a function.

Functions are quite common. For instance, let $X$ be the set of people with social security. Let $Y$ be the set of natural numbers. Assigning to each person in $X$ his social security number defines a function from $X$ to $Y$.

As a second example, let $X$ be the set of all people and $Y$ be the set of natural numbers. Assigning to each person his weight (rounded to the nearest pound) defines a function from $X$ to $Y$.

A function from $X$ to $Y$ is *one-to-one* if to distinct elements in $X$ it always assigns distinct elements in $Y$. The "social security function" is one-to-one, since different individuals have different social security numbers.

A function from $X$ to $Y$ is *onto* $Y$ if *each* element of $Y$ is assigned to *at least one* element of $X$. Every one-to-one correspondence is onto (by its definition in Chapter 18).

The reader may check that there are eight functions from $X = \{a, b, c\}$ to $Y = \{1, 2\}$; of these, six are onto $Y$, and none are one-to-one. If $X = \{a, b\}$ and $Y = \{1, 2, 3\}$, then there are nine functions from $X$ to $Y$. Six are one-to-one; none are onto. If $X = \{a, b, c\}$ and $Y = \{1, 2, 3\}$ there are twenty-seven functions from $X$ to $Y$; the one-to-one functions (of which there are six) are the same as the onto functions, and are one-to-one correspondences.

For more interesting examples, let us turn to some of the chapters.

## CHAPTER 1

The theory here depends on the function that assigns to each arrangement its backward number. Specifically, let $X$ be the set of all arrangements of the numbers $1, 2, 3, \ldots, n$, without repetitions. Let $Y$ be the set of natural numbers. Assign to each arrangement, $x$, its backward number, which we may denote $B(x)$. The main theorem asserts that if arrangement $x'$ is obtained from arrangement $x$ by a single switch, then $B(x')$ differs from $B(x)$ by an odd number. The function $B$ goes from $X$ to $Y$.

## CHAPTER 4

The Prime-manufacturing Machine describes a function: $X$ is the set of all finite sets of primes; so is $Y$. If $x$ is a finite set of primes (the input), then let $f(x)$ (read "$f$ of $x$") be the output set of primes. The important property of the machine is that $f(x)$ and $x$ do not overlap. It is not one-to-one [for instance $f(5, 7) = (2, 3) = f(11, 13)$]. But it is onto.

Let $X = \{1, 2, 3, \ldots\} = Y$. Assign to the integer $N$ in $X$ the $N$th prime, $P_N$. This function from $X$ to $Y$ is the subject of the Prime Number Theorem. It is one-to-one but not onto.

## CHAPTER 5

Let $X$ consist of pairs of positive integers. Let $Y = \{1, 2, 3, 4, \ldots\}$. The chapter really concerns two functions from $X$ to $Y$. In the first case we assign to each pair, $A$ and $B$, in $X$ their greatest common divisor. In the second case we assign to each pair, $A$ and $B$, the smallest positive number of the form $MA + NB$ (where $M$ and $N$ are integers). It is a consequence of Lemma 4 that these two functions are the same.

## CHAPTER 6

The proof that $\sqrt{2}$ is irrational utilizes a function. Let $X = \{2, 3, 4, \ldots\}$ and $Y = \{1, 2, 3, \ldots\}$. Assign to each $x$ in $X$ the total number of primes in its (unique) factorization into primes. Call this number $f(x)$. Note that $f(xy) = f(x) + f(y)$ and $f(x^2) = 2f(x)$.

## CHAPTER 11

Theorem 7 says that a certain function is one-to-one. Let $X = \{0, 1, \ldots, M-1\} = Y$. Assign to each element $x$ in $X$ the remainder of $Ax$ when divided by *M*. The assertion "$AC \equiv AD$ (mod *M*) implies that $C \equiv D$ (mod *M*)" can be rephrased as "the preceding function is one-to-one."

Exercise 70 (*b*), (*c*) establishes that a certain function is onto by first showing that it is one-to-one. (Here *X* and *Y* are finite and have the same number of elements.)

## CHAPTER 12

Any table describes many functions. For instance, consider Table (*1*). Let *X* consist of the sixteen pairs of letters *A*, *B*, *C*, *D*. Let $Y = \{A, B, C, D\}$. Assign to each pair its "product" according to the table. Thus to (*B*, *C*) we assign *A*.

Each row or column in Table (*1*) describes a one-to-one correspondence between the set $\{A, B, C, D\}$ and itself. For instance, using the *A*-row, we may assign to each *x* in $\{A, B, C, D\}$ the element $A \circ x$. Thus to *A* we assign *A*, to *B* we assign *C*, to *C* we assign *D*, and to *D* we assign *B*. The condition "$A \circ X = A \circ Y$ forces $X = Y$" says that this function is one-to-one.

The proof of Theorem 4 depends on the nonexistence of a one-to-one function from a set of five elements to a set of four elements. The set *X* of five elements consists of the boxes in the first row and second column of each tray. The set *Y* consists of $\{2, 3, 4, 5\}$. We assigned to each element in *X* the number of the muffin in it.

The "pigeonhole principle" asserts that no function from one finite set to another set with fewer elements can be one-to-one.

## CHAPTER 16

Theorem 2 is easily generalized to the following assertion. Let $S = T =$ the set of real numbers. Let *P* be a polynomial of odd degree with real coefficients. To each element *x* in *S* assign the value of *P* when *X* is replaced by *x*. Then the technique that established Theorem 2 shows that the mapping just described is onto *Y*. It need not be one-to-one.

## CHAPTER 18

The concept of one-to-one correspondence is used to define "conumerous" and "denumerable." The proof of Theorem 5 (Cantor's Theorem) shows that any function from a denumerable set to the set *Y* of all real numbers is *not onto*.

# INDEX

413